Getting the Climate Science Facts Right: The Role of the IPCC

RIVER PUBLISHERS SERIES IN CHEMICAL, ENVIRONMENTAL, AND ENERGY ENGINEERING

Indexing: All books published in this series are submitted to the Web of Science Book Citation Index (BkCI), to SCOPUS, to CrossRef and to Google Scholar for evaluation and indexing.

The "River Publishers Series in Chemical, Environmental, and Energy Engineering" is a series of comprehensive academic and professional books which focus on Environmental and Energy Engineering subjects. The series focuses on topics ranging from theory to policy and technology to applications.

Books published in the series include research monographs, edited volumes, handbooks and textbooks. The books provide professionals, researchers, educators, and advanced students in the field with an invaluable insight into the latest research and developments.

Topics covered in the series include, but are by no means restricted to the following:

- Energy and Energy Policy
- Chemical Engineering
- Water Management
- Sustainable Development
- Climate Change Mitigation
- Environmental Engineering
- Environmental System Monitoring and Analysis
- Sustainability: Greening the World Economy

For a list of other books in this series, visit www.riverpublishers.com

Getting the Climate Science Facts Right: The Role of the IPCC

Medani P. Bhandari
Akamai University, USA
Sumy State University, Ukraine
and
Atlantic State Legal Foundation, NY, USA

River Publishers

Published, sold and distributed by:
River Publishers
Alsbjergvej 10
9260 Gistrup
Denmark

www.riverpublishers.com

ISBN: 978-87-7022-186-3 (Hardback)
978-87-7022-185-6 (Ebook)

I dedicate this book to

Prajita Bhandari

To My Friend, Wife, Coauthor in Creating Writings Mrs. Prajita Bhandari, Without Your Help and Support this Book was Impossible to Complete.

Together We Have Miles To Go

Contents

Bishnu Poudel, Professor of International Relation and Diplomacy, VA, USA

Prof. Gopi Upreti, Fairfax, Virginia, USA

Prof. Odeh Al-Jayyousi, Head of Innovation and Technology Management, Arabian Gulf University, Kingdom of Bahrain

Dr. Tulsi Dharel, Professor of Marketing at Centennial College, Toronto, Canada

Dr. Shvindina Hanna, Head of the department of management, Associate Professor at Sumy State University (Ukraine), Fulbright Alumni (Purdue University), USA

Dr. Nadiya Kostyuchenko, Associate Professor at Department of International Economic Relations, Sumy State University, R.-Korsakova st., 2, Sumy, Ukraine

Dr. Denys Smolennikov, Associate Professor at Department of Management, Sumy State University, Ukraine

Foreword

Climate change is a major and disturbing challenge for our world in the new millennium The threats to our human communities, and our physical world are without doubt a vital test to our willingness to sustain and ameliorate the ongoing degradation These matters are effectively addressed in Dr. Medani P. Bhandari's book, Getting the Facts Right: The IPCC and the Role of Science in Managing Climate Change, with Reference to Pollution in Major Cities of Asia.

This essential book provides an historical account of climate science, the role of international organizations, and examination of pollution in Asia's major cities. It thoroughly examines major climate change NGOs, especially the Intergovernmental Panel on Climate Change (IPCC), which is comprised of a great number of climate scientists. IPCC is dedicated to research and validation of climate change studies and essential findings and helps world governments address climate change and its solutions Dr. Bhandari's book clarifies the importance of the mission of IPCC in researching and confirming the climate change investigations and incorporates discussions of counter arguments, and problematic results is the IPCC data interpretation, such as the issues in its 2007 report.

Dr Bhandari's book is valuable, as it helps fill the knowledge gap on climate science, helping to clarify and unravel the facts of climate science, facts which are both complicated and combative.

The book clarifies the challenge of climate change as a vital human priority in the new millennium and how science, politics, and major businesses compete worldwide in the area of resources, policy and decision making Dr. Bhandari discusses the main institution bridging the gaps among these groups, the IPCC. Such international organizations, have set up institutional machinery to provide facts that can serve to build a kind of consensus-development mechanism that the United Nations may use to achieve agreement leading to the formation of effective global policies. Nongovernmental organizations, such as Akamai University have established parallel missions with these global organizations, allowing the integration of knowledge, research and operating structures, permitting higher education in the areas of organizational change, international relations and climate change, as concentrations for study and research.

Douglass Lee Capogrossi, PhD
President
Akamai University, USA

The polar region including the third pole of the world "The Himalayas" are melting, glaciers are retreating at an alarming rate at the higher elevation and similarly the percentage of oxygen in ocean water is declining due to the impacts of rising temperature. This decades has witnessed unanticipated climate related disaster events from Mountains to Oceans and in the ice capped areas of the globe. The human casualty and economic cost is phenomenal which is increasing. Such impacts have been witnessed globally. Virtually we are at the point of climate crises leading to the extinction of many biological species. The root cause of such catastrophic process is related to human greed and infinite desire to unsustainably harvest and use earthly resources to become rich. This process led to an unprecedented release of GHG in the atmosphere. This path of human accelerated emission of pollutants and many different GHGs in the atmosphere is resulting in global temperature rise and eruption of many diseases unknown to the history leading us to the brink of 6th mass extension.

"Getting the Climate Science Facts Right: The Role of the IPCC" book written by Prof. Medani Bhandari covers the history of climate science, role of international organizations in substantiating and bringing the scientific evidences of contemporary 21st century climate change impacts to the policy makers governments and politicians across the globe. The book brings the philosophy of how we *co habit and live in commensal way and cooperation by letting other species also live in harmony* without encroaching their boundary and niche and stay in a state of ecological equilibrium.

IPCC an intergovernmental body, the role of which is "to assess on a comprehensive, objective, open and transparent basis the scientific, technical and socio-economic information relevant to understanding the scientific basis of risk of human-induced climate change, its potential impacts and options for adaptation and mitigation. The author with his 30 years of work experience in environmental protection field has shown the idea of how evidence-based science, policy and politics should work at tandem for the safety of the planet earth. Prof. Bhandari, wandering to attain higher degrees in education, travelled to Europe and north America and other countries, in the book, at the end he seems to have been finally retreated himself back to the original philosophy of *BASHUDAIVAKUTUMBAKKAM*, LIVE AND LET OTHER LIVE" where this used to be the key mantra in eastern Mysticism. A new way of how modern society should change their consumption behavior from individual personal level to the country to the globe must be addressed before it becomes too late. I thank Dr. Bhandari for his affords.

Krishna Prasad Oli, Ph.D.
Member of National Planning Commission
Government of Nepal

There is no question that climate change is one of the international problems that could end human life on the globe. Only nuclear weapons has been a similar threat, but there is a major difference. For nuclear weapons, the actors involved are those States that have the weapons (or are able to get them), while for climate change involvement is universal. What has been called the tipping point, after which the problem cannot really be solved, has been getting more and more likely, as the phenomenon develops in terms of greenhouse gas emissions, ocean temperatures, melting icecaps, and increasingly frequent major weather events.

One problem is that climate change as an issue is relatively recent and this means that the process of creating a response to it has not been complete. As the book notes, this has been a deliberate, but slow process. To place this into context, the history of the climate change response needs to be noted.

States recognized early that weather was something that transcended borders and needed an exchange of information. In 1873, the International Meteorological Organization (IMO) was founded to facilitate the exchange of weather information across national borders. This was somewhat informal and in 1951, the World Meteorological Organization (WMO) became a specialized agency of the United Nations. Its mandate is in the areas of meteorology (weather and climate), operational hydrology and related geophysical sciences.[1] Human effects on environment had been known for some time such as when citizens breathed smog in London or Los Angeles, but it was only in 1972 that the United Nations organized the first World Environment Conference that took place in Stockholm, Sweden. In fact, the main motivation for the conference was phenomena like smog, which were

[1]https://public.wmo.int/en/about-us/who-we-are

increasing and affecting human health. As a result of the Conference, the United Nations Environment Programme (UNEP) was created. UNEP and WMO then combined to create the first international climate program.

One reason that human-caused climate change was not an issue was that there was little scientific evidence about it. For one thing, the thermometer had only been invented in the late 18th century and consistent data on temperatures had only begun to be collected some fifty years later. The relationship between greenhouse gases and temperature had not been carefully studied. Non-governmental international bodies like the International Council of Scientific Unions (ICSU) saw that there was an incentive to undertake an event, the International Geophysical Year. As the United States National Academy of Sciences (NAS) described it[2]

Following a suggestion by NAS member Lloyd Berkner, the International Council of Scientific Unions in 1952 proposed a comprehensive series of global geophysical activities to span the period July 1957-December 1958. The International Geophysical Year (IGY), as it was called, was modeled on the International Polar Years of 1882–1883 and 1932–1933 and was intended to allow scientists from around the world to take part in a series of coordinated observations of various geophysical phenomena. Although representatives of 46 countries originally agreed to participate in the IGY, by the close of the activity, 67 countries had become involved.

The World Climate Programme was set up by WMO to "to determine the physical basis of the climate system that would allow increasingly skillful climate predictions and projections, develop operational structures to provide climate services and to develop and maintain an essential global observing system fully capable of meeting the climate information needs."[3] The WCP ran meetings and encouraged research and, in 1985, organized a conference in Villach, Austria on "Assessment of the Role of Carbon Dioxide and Other Greenhouse Gases in Climate Variation and Associated Impacts". That meeting concluded that there was a connection between greenhouse gases and climate change and that climate change was happening.

In order to address an international issue, a regime needs to be created that specifies what the problem is, who is responsible for addressing it, how that should be done and the institutions that need to be created. I have described this process in a previous study.[4] Creating a regime involves

[2]http://www.nas.edu/history/igy/

[3]http://www.wmo.int/pages/prog/wcp/index_en.html

[4]Mathiason, John, *Invisible Governance: International Secretariats in Global Politics*, Kumarian Press, 2007.

several steps. First, there has to be an agreement on the factual basis for the problem (what is called "principles"). Then, there has to be an agreement on what constitutes state responsibility (what are called "norms"). When that is agreed, the next step is to determine what actions states are expected to take (what are called "rules"). Finally, there needs to be an agreement on what international institutions need to be created to assist and monitor the response to the problem.

One consequence of the Villach conference was for the WMO and UNEP to establish a body in 1988, the Intergovernmental Panel on Climate Change (IPCC) to assess evidence.[5] Over the next thirty years, the IPCC has undertaken five comprehensive assessments of the issue. The first assessment led to the negotiation, in 1992, of a United Nations Framework Convention on Climate Change (UNFCCC) that has provided the institutional context for developing global work to address climate change. In regime theory terms, the UNFCCC was only a partial regime. It agreed that climate change was happening and was human-caused and that States had a responsibility to deal with it. However, it did not agree on the rules or institutions.

The next four assessments led to an increasing agreement about the rules so that in 2015, the UNFCCC adopted the Paris Agreement specifying how to address the problem. While this was a voluntary response rather than one bounded by an international treaty, it suggested movement toward an agreement on rules. In fact, in 2018, states agreed on normative rules for implementing the agreement in the light of a special study by the IPCC on the consequences of reaching a global temperature increase of 1.5 degrees Celsius that gave the world only 10 years to address emissions.[6]

The IPCC works through three working groups dealing with different aspects of the problem. One deals with climate science and has been the main source of information for the assessments. The second deals with adaptation and the third with mitigation. While the science working group will continue to be important in documenting the changes in climate that are occurring, over the next years the working groups dealing with adaptation and mitigation will have an increasing role.

This book describes, in considerable detail, how the process will unfold in the light of its history, as well as the current politics of achieving an agreement among scientists about what to do about the problem. This is a new area of

[5]http://www.ipcc.ch/organization/organization_history.shtml

[6]Intergovernmental Panel on Climate Change, "Summary for Policymakers of IPCC Special Report on Global Warming of 1.5°C approved by governments", IPCC Press Release (2018/24/PR), 8 October 2018.

concern and, as the tipping point approaches, may be the most important. How should adaptation take place and, most importantly, how can mitigation be made to work. Both of these involve presenting facts to a general public in such a way as to generate support for action. In the light of the fact that the largest per capita producer of emissions, the United States, has indicated its intention to withdraw from the Paris Agreement, these new facts become even more critical.

This new book on the IPCC updates previous information by bringing the institution's role into greater clarity in the current stage of dealing with the problem and will contribute to solid understanding of its role and potential.

Prof. John Mathiason,
Professor of International Relations, Cornel University, Ithaca, NY, USA

Climate is understood as the total of weather phenomena occurring in a given area over a multiannual period, determined on the basis of long-term observation of factors, including temperature, precipitation, wind, pressure and air humidity. In many parts of the world, especially in western Europe, the United States and some tropical areas of Africa, the type of weather can vary greatly from year to year. But even then, the climate may change periodically (week to week, month to month, decade to decade or century to century).

Currently, the global climate can be divided into three zones: hot, cold and moderate. Within these zones, however, the climate is extremely balanced. Climatic differences are the main reason for the diversity of vegetation, between regions. The climate determines which plants can be grown and which animals can be bred. The climate also dictates the style of living, including development of seeds as a response to different climatic conditions.

The causes of the climatic differences are to be attributed to the following three factors: latitude, location of the site in relation to the oceans and continents, and its altitude in relation to sea level. In addition to the factors affecting the climatic differences mentioned above, human activities are beginning to play a special role. Their influence on climate change is constantly changing just as human ingenuity is changing. The rapid economic and technological development of countries, globalization processes in the world, and still low public awareness of progressive climate threats are among many accelerators influencing climate change.

The current consumer lifestyle indicates that climate change (including global warming) will be seen significantly over the next decades. The most likely explanation the climate warming, seems to be the increase in gas pollution, which is released into the atmosphere. IPCC draws attention to the rising average temperature and the resulting threat. Scientific research has shown that the faster-than-expected increase in carbon dioxide emissions and other factors will increase temperatures by 1.5 degrees Celsius by 2030 and by 2 degrees Celsius by 2045. IPCC points to the need for decisive action by governments to protect the climate and prevent it from the global catastrophe. It has been highlighted, that climate change can only be halted if greenhouse gas emissions are significantly reduced and global environmental policy should serve this purpose. Attention was drawn to the need to take global actions on climate protection. However, it should be noted, even if policies and measures leading to reduce greenhouse gas emissions have an effect, some climate changes are inevitable due to the current situation.

The book "Getting the Climate Science Facts Right: The Role of the IPCC" by Professor Medani P. Bhandari refers to very significant and timeless issues related to environmental protection, environmental justice, sustainable development, and equity. It clearly presents the role of visionary policy led by IPCC, so important for the future generations. It emphasizes the importance and complexity of the process of managing sustainable development and refers to issues related to climate change, moderated by international, nongovernmental and governmental organizations. It also shows the way in which an IPCC face economic, social and environmental challenges. This book can be recommended for those readers who are interested in the subject of sustainable development and/or for those who are sensitive to the future fate of our planet.

Prof. nadzw. dr hab inż Jacek Binda
Rector, Bielsko-Biala School of Finance and Law, Poland

I feel greatly honored when Professor Medani Bhandari asked me to write a foreword for the book "Getting the Climate Science Facts Right". Prof. Bhandari has undertaken a herculean task and has presented ideas in a very useful form. It explains the very basis of climate change and how does it affect human well-being making the book useful to people of various levels.

The book explains how various scholars, agencies, and stakeholders understand climate change; how international, non-governmental and governmental organizations are responding to the climate change issues in the quest for sustainable resource management. It utilizes climate-related examples from many South Asian countries. Using these examples, this book discusses how science and conscience need to marry seamlessly in order to succeed in any resource management goals Prof, Bhandari incorporates a series of his interviews *Bashudaiva Kutimbakkam* in this book.

By incorporating Prof. Bhandari's interviews, this book justifies that without a proper marriage between the science and conscience, sustainable resource management merely becomes political slogans. It suggests following the guidelines of Intergovernmental Panel on Climate Change (IPCC) to address the most important public priority in the 21st Century.

This book suggests that though the use of 21st-century refined technology might help in sustainable resource management to some extent, raising public awareness is no less important than using the 21st century's technology. Both the use of 21st century refined technology and social awareness are important to bring both hard and soft sciences in the same platform. This book will be very useful to develop climate policy guidelines because it presents real-world problem-solving ideas by incorporating theoretical ideas from both hard and soft sciences. An equally important aspect of this book is the analysis of historical facts and connecting these facts with contemporary politics. Planners and policymakers will find this book very useful as it not only presents the structure of the IPCC but also explains how various organization can work with IPCC to formulate policies for various levels This book warns that policies crafted without the support of evidence-based research become rarely helpful in implementing sustainable programs.

Prof. Keshav Bhattarai, Ph.D.
Professor of Geography
(Program Coordinator)
School of Environmental, Physical & Applied Sciences
University of Central Missouri, Humphreys 223C
Warrensburg, MO 64093, USA

An increase on average global temperature by almost one degree Celsius from preindustrial age resulting in changes on ice melting, sea level rise and changes on precipitation pattern and the incidence of natural disasters such as flood, wildfire, and glacial lake outburst have become more frequent, widespread, and global concern. As climate is the average manifestation of weather conditions especially temperature and precipitation, changes on temperatures and precipitations result on changes on climatic conditions. Climate change is requiring a locality introduction of new crop production techniques, and seasonal adjustments and adaptation to changing conditions by human, animals and other life. Inability to adapt to changing climatic conditions results in loss of agricultural production, poor health, loss of infrastructures, and low ecosystem productivity including land and resources degradation, which will eventually threaten humanity.

For climate change adaptation and resiliency, appropriate policy measures based on accurate data and information is necessary. In order to understand science behind climate change and suggest appropriate policy measures for mitigation of climate change impacts, the Intergovernmental Panel on Climate Change (IPCC), which is an intergovernmental body of the United Nations, was established in 1988. The IPCC produces reports on scientific, technical and socio-economic conditions in relation to climate change, impacts, and mitigation. These reports serve important instruments for policy recommendations.

This extraordinary book 'Getting the Facts Right: The IPCC and the Role of Science in Managing Climate Change" from Professor Medani Bhandiri is a very timely publication. Its value for scientific communities, policy makers, researchers and other stakeholders is as clear as it is from its title in terms of its relevance and importance to countries worldwide in getting facts and figures that are substantiated and valid for necessary policy formulation in climate change adaptation measures. This book provides historical account of climate science, the role of international organization, emerging climate science and cases of many developing countries. This book extraordinarily presents data on climate change of different regions.

Professor Bhandari believes on "BashudaivaKutumbakkam" which means the entire world is our home and all living beings are our relatives. His belief of "BashudaivaKutumbakkam" truly resonates to this treatise of climate change with reference to pollution problems in the major cities of India, Nepal, Bangladesh and Pakistan.

Professor Bhandari has exhibited extraordinary talent and comprehension on the issue of climate change management. He has suggested several approaches, framework and techniques for addressing climate change and environmental quality. Professor Bhandari has published four books on science and four volumes of poetry with Prajita Bhandari. He has published about 100 scientific papers in the international journals, many monographs, and book chapters. He established first Environmental Conservation NGO, APEC-Nepal, in Eastern Nepal- Biratnagar, conducted 100s of environmental projects and campaigns (1985–2002- as a chair of APEC). His motive of life is to give back to the society in fullest which is depicted in his recent interviews from Hungary under tiled "VASHUDAIVAKUTUMBAKKAM" and "LIVE AND LET OTHER LIVE". Bhandari's recent books include

"Green Web-II- Standards and Perspectives from the IUCN: Program and Policy Development in Environment Conservation Domain' and 'Inequality-a divisive factor of the society- how United Nation is trying to overcome this through its SDGs' published by Rivers Publishing house, the Netherlands.

Professor Bhandari is a lifelong conservationist, expert of climate change impact, social empowerment and educationalist, who has devoted his entire life for the conservation of nature and social services. Scholarly publications of Professor Bhandari draw interest of many scholars globally and have very high level of conceptual and philosophical merits. Examples of such publications include, The Development of the International Organization and Organizational Sociology Theories and Perspectives, Sustainable Development: Is This Paradigm the Remedy of All Challenges? Does Its Goals Capture the Essence of Real Development and Sustainability? Climate Change, and pollution problems in the Major Cities of India, Nepal, Bangladesh and Pakistan. His scholarly works in relation to Nepal include the areas of Mainstream Religious Domain and Indigenous Communities in Nepal, Theoretical Route of Green Economy, Agricultural Development, Natural Resources Monitoring, Education Transformation, Standards and Perspectives from the IUCN, and Debate between Quantitative and Qualitative method, The Pedagogical Development of the International Organization and Organizational Sociology Theories and Perspectives, Public Opinion Survey, Tourism, Food Security and Biodiversity Conservation. His publications inspire and educate readers for collective works on solving the global environmental crisis

The book "Getting the Facts Right: The IPCC and the Role of Science in Managing Climate Change" is the product of his more than ten years research on climate change science. Book is embedded within the organizational theories, governance of climate change challenges and shows the practical examples of how international societies are willing to minimize the impact of climate change. Prof. Bhandari evaluates the roles of international organization and mostly, adds the new argument that, international organization are doing their best to overcome with the climate change crisis. The foundation of IPCC is one of the best examples of United Nations contribution to validate the climate change science. Prof. Bhandari argues that whatever, climate change skeptics, deniers or other who only cares present and ignore the fact, do not have any basis of denying fact of climate change challenges. He supports the conclusions of hundreds of scholars like him that the one of the major causes of climate change is due to excessive interventions on the earth ecosystem. Human activities are the major cause of climate change. Prof. Bhandari urges to all stakeholders to think, evaluate, and accept the truth that, I, we, we all are responsible, who destroy the natural balance in the name of progress and development, and it is my, ours responsibility to make efforts to minimize the challenges of climate change. This book is wonderful contribution to the humanity.

Prof. Durga D. Poudel
University of Louisiana at Lafayette, Louisiana, USA

Climate Change has become the defining environmental issue of our times. Although some groups continue to remain climatic sceptic, the vast majority of the scientists, professionals and policymakers have fully accepted that climate change is real, is caused by anthropogenic activities, and it has a high probability of causing adverse impact to the environment, society and economy. In fact, several severe influences of climate change have already become routine events each year.

Unfortunately, many politicians, some highly influential, still deny that human actions are causing climate change, and that climate change will have negative consequences to the society. Some populist politicians still seem to believe in conspiracy theories that climate change is a hoax propagated by some scientists and countries to derail economic development in the West.

Established in 1988, the Intergovernmental Panel on Climate Change (IPCC), a UN agency, is perhaps the most important organization to help establish the scientific basis for understanding climate change. The IPCC does it through documented findings published regularly to show the extent of changes in global temperatures as linked to the greenhouse gas concentrations in the atmosphere. The IPCC publishes its findings of research and writings of hundreds of the most well-known scientists across the world, who volunteer and contribute to the IPCC's regular publications.

The IPCC website depicts the essential functions of the organization as "the United Nations body for assessing the science related to climate change." Further, the IPCC "provides regular assessments of the scientific basis of climate change, its impacts and future risks, and options for adaptation and mitigation."

Prof. Medani Bhandari's book "Getting the Facts Right: The IPCC and the Role of Science in Managing Climate Change" is an important contribution towards understanding the functioning of the IPCC, perhaps the most important international repository of the scientific findings related to climate change.

This book delves into the origins of IPCC by conducting a rigorous review of its history, and the events that led to the establishment of this UN-body. The book's listing of environmental history culminating in the current state of knowledge on climate change is well-researched and comprehensive. The book provides useful analysis of the how the IPCC appraisals, assessments, and reports are produced and how they could be very important resources for our understanding of the science of climate change.

As the impacts of climate change become more visible each day causing severe damages to our lives, properties and the environment, it becomes more important to publicize the scientific background to the global warming causing climate change. Such an understanding will help devise appropriate mitigation and adaptation plans it minimize the negative effects of climate change. Deeper knowledge of organizations such as IPCC, and their outputs becomes ever more important towards the goal of containing climate change and its impacts.

The author, Prof. Bhandari is well suited to write on this topic as he has a first-hand experience about the impacts of environmental degradation and climate change and has a thorough knowledge on the topic. He has worked in poor rural areas in Nepal, has been educated in prestigious universities in the US, and has taught environmental subjects in the US, Bahrain and Ukraine. Prof. Bhandari has written important books and articles related to conservation, the environment, and climate change. He has well utilized his extensive knowledge and wide experience in the preparation of this book.

The author also integrates into the book the concept of interrelatedness of all living plants and animals on earth. It is an extension of the classical eastern philosophy embodied in the Sanskrit phrase "Basudaiva Kutumbakam", meaning all human beings in the world are one family. This is a useful notion to explain and popularize the scientific background related to environmental conservation and protection, and especially climate change.

This book is an important reading for those involved in the science, policy and research related to climate change and appropriate remedies to manage its impacts on human well-beings. Especially, as many politicians in important positions and several groups following them remain doubtful to the validity of global warming, climate change and its anthropogenic roots, a more thorough understanding of the works of IPCC and disseminating it becomes even more urgent. This book fill san important gap in this area by providing a robust analysis of the work of IPPC, and by highlighting the need to promote scientific basis of climate change.

Ambika P. Adhikari, DDes, AICP
Principal Planner, City of Tempe, Arizona
Sr. Sustainability Scientist
Julie Ann Wrigley Global Institute of Sustainability
Arizona State University, USA

Climate change is one of the gravest threats mankind has ever faced. Rising temperatures have worsened extreme weather events around the world. Ever larger chunks of Antarctic ice break free every year. Wildfire seasons are months longer and more devastating. Coral reefs have been bleached of their colors. Mosquitoes are expanding their territory, spreading disease. The seas, on which much of humanity depends, is turning more acidic. The consequences of climate change for humanity could not be more dire.

Dr. Medani P. Bhandari is one of the world's foremost scholars on climate change and environmental issues, particularly in the South Asia region. In his new book, **Getting the Climate Science Facts Right: The Role of the IPCC**, he addresses the difficult subject of reconciling science with world politics.

Getting the Climate Science Facts Right: The Role of the IPCC is a concise, well written, and thoroughly researched work on how the Intergovernmental Panel on Climate Change (IPCC), acting as the main source of facts from which climate change policy is developed, attempts to bridge the difficult gap between science and international politics. Examining the structure of the IPCC, its composition and its procedures, the book gives the reader a better understanding of the IPCC's role and future.

Seldom in history has science had such a direct and confrontational relationship with politics. **Getting the Climate Science Facts Right: The Role of the IPCC** is an extremely important work in understanding how the impacts of climate change have been analyzed by scholars, agencies, and other stakeholders, and what roles international, nongovernmental and governmental organizations play in addressing these pressing issues.

Scott Garner, JD, MTax, MBA, CPA,
CEO at Asia Environmental Holdings Group (Asia ENV Group),
Asia Environmental Daily, Beijing/Hong Kong, People's Republic of China
https://www.linkedin.com/pulse/book-review-getting-climate-science-facts-right-role-scott/?trackingId=

Climate Change. What is it about? Why the book by Prof. Medani P. Bhandari titled "Getting the Facts Right: The IPCC and the Role of Science in Managing Climate Chong" is important?

Unprecedented forest wildfires across USA, Australia and Amazon, high rising record temperatures, unseasonal floods across the continents, untimely storms, melting Arctic, sinking islands due to rising sea water and mercurial temperatures do indicate something have changed in our surroundings. Frequent upheavals in weather occurrences at the local level have resulted in changing patterns in rain, snow, clouds, winds, thunderstorms, heat or cold waves, draughts and floods. Such periodic weather events of the recent past over a period refers to Climate Change. Climate Change is long-term global warming that are happening in the Planet Earth whereas weather is about local changes in the climate that we witness around us on shorter time scales.

Most scientists believe that rapid changes in the Earth's surface since the late 19th century has contributed by a rise of about 1.1 degrees Celsius. This rise in temperature is generally attributed to long-term warming of the planet, rising sea levels, shrinking mountain glaciers, melting down of ice in Antarctica and Greenland, and shifts in plant life and flower blooming times due to climate change.

Scientific community consensus is that the climate is changing, in extreme, and these changes are largely produced by human activities contributing to continuing global warming (i.e. surface temperature increase). Rising temperature has increased carbon dioxide (CO2) and other human-made emissions into the atmosphere, and most of it during the last four decades. Credible evidences are assembled by climate scientists from warming oceans, shrinking ice sheets, glacier retreat, decreased snow which are the trends irrefutable now. This fact establishes the climate change reality. Global warming, as most scientist believe, is now largely irreversible unless some drastic emergency measures are implemented and very quickly.

Climate Change is not a fiction. It is about survival of all living beings, and for the better part quality of human life too. It is not only about saving the planet Earth and eco-systems it is primarily about saving humans and many living organisms. To many deniers it sounds a radical thought. But if there are no living beings what is the purpose of saving the Earth?

How did it all happen? Why now it is an emergency?

There was not enough research on what a 1.5-degree Celsius rise in temperature meant at the time of 2015 Paris Climate Agreement. The Intergovernmental Panel on Climate Change (IPCC) had then reported global warming target at limiting temperatures below 2 degree Celsius while pursuing reduction towards 1.5 degree Celsius. Late in 2019, IPCC report has rung a warning bell with the stark difference of a 1.5-degree Celsius warmer world compared with 2-degree Celsius highlighting benefits and consequences of limiting to 1.5 degree Celsius as opposed to letting the temperature to rise to 2 degree Celsius. Report finding includes for example, survival of some coral reefs at 1.5 degree Celsius while none would survive in a 2 degree Celsius, including social and economic impacts.

The IPCC report predictions are scary to say the least because the world, in recent past, has warmed by 1 degree Celsius, must faster than earlier predicted. We now urgently need to limit to 1.5 degree Celsius instead of 2 degree Celsius in shortest possible time. Scientists believe this target is feasible due to development in science and technology and with news ways of managing social and behavioral economy. The important missing part is this compact is ambivalent political will of world leaders – which seems to be faltering along the pathways.

According to the IPCC report there are two choices in limiting the global warming to 1.5 degree Celsius; (i) reduce energy demand in the short-term by around a half over the next 10/15 years; or (ii) develop the capability for disposing of a large amount of carbon dioxide, particularly the excess amount of CO2 which can't be dumped into the atmosphere either underground or into the ocean. World needs to urgently accelerate reduction of the excess CO2 because getting rid of CO2 to meet the 1.5 degree Celsius is currently slow and lacks urgency, while the clock is ticking.

Carbon Dioxide (is a heat-trapping (greenhouse) gas. CO_2 is generally produced by deforestation, burning fossil fuels, and natural events like volcanic eruptions. Due to greenhouse effect heat is trapped close to the surface of the Earth by greenhouse gases and dust contributing to rise in surface temperature. Increased temperatures on Planet Earth's surface is resulting in weather upheavals affecting all around.

Sea water rise is caused by global warming. Melting water from ice sheets and glaciers, and the expanding warm sea water will have created existential threats to low lying islands states and coastal areas. Likewise, ocean warming will cause survival of marine species. Shrinking Ice Sheets will shrink places like Antarctica and Greenland affecting life and ecosystem. Glacier retreat in the Alps, Himalayas, Andes, Rockies, Alaska and Kilimanjaro will have immeasurable socio-economic impacts. Decreased snow cover will dry up rivers and lakes while melting snow will create floods and eventual shortage of fresh water supply. In nutshell, climate change will have widespread consequences because of changes in sea level, plant life, mass extinctions of many species, and affect human life and societies.

Is Climate Change an alarming challenge and a major threat to humanity?

Climate change is a serious threat to people, socio-economy, ecosystem and the Planet Earth for attaining sustainable economic development. It is considered one of the most challenging threats to modern times and for socio-political and economic stability. Weather events like storms, typhoons, heatwaves, cold-waves, cyclones, hurricanes will reduce productivity when people will work less and less and life of millions will adversely affect contributing to increase in poverty level after inundation of communities, and then widening the inequality gaps. Shrinking harvest due to draughts or floods and storms will reduce food production and availability, complicating feeding the growing world population resulting in hunger and poor health.

Adapting to coastal areas due to rising sea level and need for relocating villages, town and cities will increase maintenance and managing relocation costs. There will also be loss of capacity to work due to extreme heat and cold contributing to productivity loss; more conflicts or wars may ensue for demand for fresh water supply in parts of some countries; diseases may spread due to higher temperature. Extreme weather phenomenon could devastate communities and may lead to widespread poverty in vulnerable communities in many countries.

It is alarming to think of impending consequences of rising global temperatures by just half a degree Celsius could results in the loss of millions of people and countless other living species which will be affected by draughts, sea level rise, heat and cold waves, lack of fresh water supply and widespread food shortages. Climate change will have serious consequences on the environment and people, and seemingly it is one of the biggest threats to economic stability. It is the biggest single threat to living beings of modern human history and triggering imminent destruction of the Planet Earth during the 21st Century whereas nuclear arms and weapons of mass destructions were considered the biggest threats to the world of the recent past.

Climate scientists predict to achieve 1.5-degree Celsius target modern world would need to invest around 2.8% of global GDP into renewable energy systems between now and 2050. It is imperative to reduce greenhouse gas emissions down to zero by mid-century. This appeared to have become a big challenge for governments and some political leaders who are skeptical to what the scientists are telling them. With the rise of political populism, increasingly countries around the globe are appearing reluctant to implement the Paris Agreement (signed by 200 countries and ratified by 181countries) failed to reach a consensus on future action plan at the COP26 Madrid Conference in 2019 so as to build viable options needed to achieve the 1.5 Degree Celsius.

Climate change, single most existential human threat, is a global phenomenon and mitigating actions should not have been confined to the limits of national borders while pursuing the climate agenda. Countries in both hemispheres, developed North and developing South, people are witnessing every day the wrath of nature's calamities due to human-induced rising global warming. It is stunning that political leaders are failing to see this truth as facts presented by scientists, and they are still ambivalent.

The United Nations continues to advocate the fact that climate change is real. It is still not too late to take robust mitigating actions forthwith because the world has organizational and technological capacity to manage the damage and save the humanity from this devastating scourge.

Why Inter-governmental Panel on Climate Change (IPCC) is an important organization?

This is an inter-governmental organization of the United Nations created by the world community to drive the climate agenda process. Since its inceptions in 1988 it has played a lead role and have galvanized nations around the world to the facts emanating from the climate change evidences. IPCC is the only internationally recognized and accepted authority that provides reports that have the climate scientists' agreement and consensus from participating governments. Thousands of scientists and experts from around the globe contribute to the work of IPCC to writing and reviewing reports which are further reviewed by governments.

The principle tasks of the IPCC are to assess the risk of human-induced climate change; its possible impacts, and possible options for prevention. It conducts topics assessment on a comprehensive, objective, open and transparent basis the scientific, technical and socio-economic information relevant to understanding the scientific basis of the issue. But it does not have mandate of implementing policy strategy but could assist governments through influencing relevant policy strategies. IPCC responsibility is to provide scientific information on scientific basis, risks and impacts and possible options for mitigations and generally reports include elements for policymakers.

Why "Getting the Facts Right: The IPCC and the Role of Science in Managing Climate Change" is important?

This incisive book surveys the landscape of historical anecdotes on Climate Change, a global phenomenon, and it is the first of its kind which details the historical account of climate science, the role of international organizations, views of climate scientists, contributing to facts about the climate phenomenon. Both academic and researcher would find this as treasure trove for it lists historical information, including on politics of controversy surrounding the issue and the roles played by IPCC and associated international organization in hammering out the climate issue in international arena. Policymakers and other general enthusiasts will particularly find this encapsulated encyclopedia on climate change worthy of a possession.

This book devotes substantial part narrating the roles of international organization in addressing the Climate Change issue and creation of Intergovernmental Panel on Climate Change (IPCC) very informative and useful reference. Organized listing of evidential papers and reports on climate change and impacts, the timelines on IPCC papers useful. I found this handy and a perfect guide to delving details. Also, narrative on the overview of Climate Change and Science and how it evolved are quite interesting.

Professor Medani Bhandari is an authority in climate issue. He is at the forefront of Climate Change campaign and involved in climate related global issues of concern and contributes to the process. Professor Bhandari, beyond being a climate expert, has devoted substantial part of professional life advocating nature's conservation of nature lover and as conservationist. He has acquired deep knowledge of issues emanating from South-Asian in context of sustainable development conundrum of the sub-region. His familiarity with the climate change campaign and the understanding of conservation related challenges makes him a globalist who could focus on his life-time motto, a broad theme in Sanskrit, "*Vashudhaiva-Kutumbakan* (*The entire world is our home and all living beings are our relatives*)", encapsulating worldly environment that involves nature, resources, conservation, climate change and sustainable development which are key the ingredients of *Let Live and Let Others Live*.

I am deeply honored and gratified by Dr. Bhandari permitting me to glanced through the pages of this book and make commentary. I sincerely hope readers would find this book as immense of value and for ease of reference. This is an important contribution of Professor. Bhandari in the knowledge domain of climate change crusade of this Century.

Kedar Neupane
Retired United Nations Official, and
Founding Board Member, Nepal Policy Institute,
Switzerland.
5 January 2020.

Preface

Addressing climate change is the most important public priority of the 21st Century. Unlike many issues, however, it is being driven by science and its interface with politics. The main institution for making the bridge between science and international politics is the Intergovernmental Panel on Climate Change (IPCC). Founded before the adoption of the United Nations Framework Convention on Climate Change, the IPCC has been the main source of consensus on the science and implications of climate change.

Seldom in history has science had such a direct relationship with politics. The negotiation of an international regime requires, at its outset, an agreement on the facts. In this case, the facts are scientific, complex and contentious. Governments have recognized this and have, using international organizations, set up an institutional machinery to provide facts that they can accept.

The way in which the IPCC functions is also unique in that it melds the way in which science achieves consensus with the way governments do at the international level. Starting with a process to examine, review and debate scientific findings leading to a rough consensus about scientific fact, usually expressed as probabilities that the findings will hold over time, the IPCC then concludes by using the kind of consensus-development mechanism that the United Nations typically uses to achieve agreements leading to the formation of regimes.

The structure of the IPCC, its composition and its procedures have made it successful and these factors are analyzed in order to achieve an understanding of the IPCC role and future. Its target audience includes universities concerned with teaching in the areas of organizational change, international relations, public administration and climate change as a specific subject. It also includes policy makers and administrators who will have to implement agreements reached. This book addresses how climate change science was developed; how climate change impacts have been analyzed by various scholars, agencies (including IPCC), and other concern stakeholders, what roles international, nongovernmental and governmental organizations play to address climate change issues. Book specially, unveils the role of IPCC in creating and validating the climate change knowledge, how IPCC creates climate change regime. There is no doubt that climate change has been accelerating due to anthropological disturbances on earth ecosystem; however, there are still few deniers, who are driving anti current. Book also incorporates deniers' arguments. Book also shows few problematic results (immature data interpretation of IPCC particularly in its 2007 report), which gave points to foster denier's argumentation. However, IPCC have already corrected those errors and have already produced *Fifth Assessment Report* in 2014, Special Report on Global Warming of 1.5°C (SR15) in 2018 and in the process of producing Sixth Assessment Report on 2022. There is no question that climate change is occurring, and direct and indirect impacts are visible mostly on the fragile ecosystems.

The book contents nine forwards (the evaluation of the book) from the prominent scholars of various parts of the world. Each forward incorporate Climate Change. What is it about? *How did it all happen? Why now it is an emergency? Is Climate Change an alarming challenge and a major threat to humanity? Why Inter-governmental Panel on Climate Change (IPCC) is an important organization? Why this book "Getting the Facts Right: The IPCC and the Role of Science in Managing Climate Change" is important?* These forwards present the overall scenarios or climate change challenges, appropriately with the imperial evidences. They have developed their arguments firstly, what the available information, sources say about the current climate change crisis, secondly, they provide their own opinions and thirdly, they conclude why this book "Getting the Climate Science Facts Right: The Role of the IPCC" is important, and how book adds the knowledge in the climate science. Forwards also elaborate about the authors quality, reliability and validity by illustrating authors previous contributions to the science and humanity. These forwards agree that, every individual themselves (or their ancestors) are directly or indirectly responsible for the current global environment crisis; therefore, each of us have responsibility to contribute to minimize the impact of climate change challenges.

In addition to nine forwards, book contents seven endorsements, from the prominent scholars of climate science. Each endorsement writers have evaluated the manuscript and supported the fact that anthropogenic disturbances to the natural ecosystem is the main cause of rapid climate change. Both forward writers and endorsers support the authors conclusions that IPCC assessments are reliable sources to understand the problems and consequences climate change challenges.

Book also present a case study, of climate change impact in the major South Asian cities (air, water, soil pollution).

Acknowledgments

It took me about ten years to complete this book, and there are many people to be acknowledged.

First of all, I would like to thank to Prof. John Mathiason (Cornel University), who actually co-initiated this book project, while we both were at Syracuse University during 2005–2012 (John was professor of international relation and I was doctoral student of sociology). We jointly wrote book proposal and I presented the book concept with the tiled "Governance of Climate Change Science: The Intergovernmental Panel on Climate Change (IPCC) and the New Climate Change Management Regime" at the 2nd UNITAR-Yale Conference on Environmental Governance and Democracy: Strengthening Institutions to Address Climate Change and Advance a Green Economy, Yale University, USA in 2011. This book proposal was published in the conference proceeding. Further, in same year, 2011, we slightly modified the concept-and we presented as conference paper titled "Managing Climate Change: The Institutional Architecture" at the annual meeting of the International Studies Association Annual Conference "Global Governance: Political Authority in Transition" in Montreal, Canada. Finally, with inclusion of those two previously presented papers, we wrote another paper titled "Getting the Facts Right: The IPCC and the Role of Science in Managing Climate Change". We presented this paper in few more conferences. The participants of the conferences including (UNITAR-Yale and Montreal), provided very good comments and insisted, encourage us to develop a book project. Fortunately, one of the Publishing House was even ready to assist us for publication. We accept the offer and complete the agreements etc. However, I was super busy for my PhD dissertation and defense and John moved to Cornel University. So, we were not able to finalize manuscript on time. We have to withdraw the book agreement with that publisher. However, I keep on working on selected chapters. Situation changed again, John got very busy with other tasks and could not contribute. He suggested, insisted and encouraged me to complete the book as sole author. I accept; however, I duly acknowledge John's contributions are embedded in this book, one way or another. I would like to thank to Prof. John Mathiason for his encouragements and guidance and his forward for this book.

During 2011, I was also working with another scholar Dr. Krishna P. Oli (then in ICIMOD, now the member of National Planning Commission, Government of Nepal). We wrote paper titles "roles of NGOs and INGOs in Nepal" and also developed few books proposals. I proposed to be a co-author of this book "Getting the Facts Right: The IPCC and the Role of Science in Managing Climate Change", he happily accepted, and we jointly sign the agreement with the Publisher of this Book, River Publisher. And again, Dr. Oli became so busy that, he also could not contribute for the book; as he has new responsibility as a member of National Planning Commission, Government of Nepal. As john did, Dr. Oli mentioned his inability to contribute for the book and insisted me to go alone. However, his help and guidance has been always with me. His willingness to be the part of the book is illustrated in his forward note in this book. I very sincerely thank both Prof. Mathison and Dr. Oli being part of this book project in the initial stages. Here, I would also like to remember the contributions of the students at the Syracuse University, who involve in initial data collection and coding.

I would like to thank to Prof. Douglas Capogrossi, Prof. Bishnu Paudel, Prof. nadzw. dr hab in£. Jacek Binda, Dr. Ambika Adhikari, Prof. Gopi Uprety, Prof. Durga Poudel, Prof. Keshav Bhattarai, Prof. Tulsi Dharel, Prof. Odeh Al-Jayyousi, Mr. Kedar Neupane, JD. Scott Garner, Prof. Shvindina Hanna, Prof. Nadiya Kostyuchenko, Prof. Denys Smolennikov for your valuable forwards/indorsements, inputs, comments and encouragements. Each of your inputs clearly indicate why climate change is a major challenge of contemporary world and why all concern stakeholders need to work together.

I would also like to thank to Honorable Education Minister Mr. Giriraj Mani Pokhrel (federal Government of Nepal), Chief Minister of Pradesh one-Mr. Sher Dhan Rai, Education and Social Development Minister, Mr. Jivan Ghimire (Pradesh-One-Nepal), for their encouragements to me to contribute for the humanity, fill the knowledge gap of climate change science and evaluate the role of international organizations to minimize the challenges of climate change with the examples from Inter-Governmental Panel on Climate Change (IPCC). I highly appreciate their willingness to support for the knowledge building on one of the major challenges of the contemporary world- Climate Change.

I would like to thank to Prof. Steven R. Brechin, Prof. Cecilia Green, Prof. Lekhnath Sharma, Prof. Baiju Thapaliya, Dr. Dinesh Raj Bhuju, Dr. Ukesh Raj Bhuju, Dr. Rishi Shah, Mr. Ananda Bhandari, Mr. Hikmat Basnet,

Prof. Dhan Pandit, Prof. Volodymyr Boronos, Prof. Tetyana Vasilyeva, Prof. Tulashi Joshi, Mr. Aleksander Skpinski, Dr. Nabin Khanal, Dr. Tara Sigdel, Dr. Santosh Sapkota, Dr. Jan Karey, Dr. Choodamani Khanal, Dr. Bijay Kattel, Dr. Taranath Sharma, Mr. Kshitiz Raj Prasai, Mr. Upendra Basnet- Pagal, Mr. Medini Adhikari and many other scholars for your valuable input directly or indirectly. I would also like to remember to Dr. Rajendra Pauchauri (former chair of the IPCC), who often responded to my emails (during 2007–2014) especially related to errors on 2007 assessment report of the IPCC.

Very special thanks to Prof. Dunlap Riley, for his encouragements, motivation and support throughout and for his valuable inputs on this book. I would also like to thank Prof. Robert Brulle and Prof. Timmons Roberts for providing the useful documents, journal papers and information on IPCC as well as for their views on the future of the IPCC (which I have included in this book as a quote).

I would like to remember my father Lok Nath Bhandari, Grandmother Laxmi Devi Bhandari, my maternal Grandparents Abhikeswar and Jalapa Devi Gautam, my in-law parent Dwarikanath and Durgadevi Devkota, whose blessings are always behind me. I would also like to thank to my mother Hema Devi, My uncle Bhim Prasad Bhandari, My aunty Goma Devi Bhandari (late) Man Prasad Bhandari (cousin uncle) and Yamuna Bhandari (cousin aunty) , my maternal uncle Ram Chandra Gautam, my brothers Krishna, Hari, sisters Kali, Bhakti, Radha, Bindu, Sita and their families for encouraging me by providing peaceful environment. I would also thank to my cousin uncles Khagendra, Govinda, Khem Bhandari, cousin brothers Govinda, Bishnu, Rudra, Bishma, Narendra, Puspa, Chiranjibi, Sureash, Indra, Jay Bhandari, Meghmani, Yogendra and Manish Gautham and cousin sister Sabita, Guna Kumari, Geeta, Kamala, Dipa and all loved one for their best wishes and supports throughout.

I would also like to thank to Sneha Sehgal, Bishal Khanna (Guru Jee) and Subhra Mukherjee (Guru Ma) of Saidam Canada, all members of Sai Group of Virginia and my friends Rajan Adhikari, Govinda Luitel, Tirtha Koirala, Prof. Sanjay Mishra, Dhir Prasad Bhandari and all of my facebook and LinkedIn's friends, who have been always encouraging to us to give back to the society through knowledge sharing.

I would also like to thank to my wife Prajita Bhandari for her encouragements and for giving insightful information on climate change global challenges. I honor your inputs and dedicate this book to you.

I would also like thank to my loved one Neena, Prameya, Kelsey, Manaslu Bhandari and Abhimanyu, Uma and Mahesh Iyer for your varieties of support during this book project.

Thank you to Rajeev Prasad, Junko Nakajima, Rivers Publisher, for encouraging and empowering us to complete this book project on time. Thank you to you all, who have given their inputs for this book project directly or indirectly.

Finally, I would like to declare and acknowledge that chapter two, three, four and last chapter Bashudaiva Kutumbakam contain previously published materials (my copyrights).

Thank you,

Medani P. Bhandari

Endorsements

It has been immense pleasure to comment on the book "Getting the Climate Science Facts Right: The Role of the IPCC" by the globally well-known scholar of climate science Prof. Medani P. Bhandari. His contributions in the field of ravages of climate change for our planet are worldwide known and well documented in his previous works.

This book is another pillar of climate science, which adds new web in the science, bridges the knowledge gap. Prof. Bhandari has done extensive research on how best mankind can prevent climate change devastating challenges to make our earth more ecologically livable. I feel worthy to note that he has been working with the underlying principles of "Vashudhaiva Kuttumbakam" meaning the earth we live in is a family, brings in a human essence of survival to live in ingrained global empathy that is in each of our existence. **What a Profound Proposition**! All saints, sages, and philosophers, thinkers all around the globe from prehistoric to present time have tried to make us aware that "my living as many in one and one in many" enriches me to attain divine platform free from all illusions and unhappiness.

"Getting the Climate Science Facts Right: The Role of the IPCC" unveils the constrains, criticism, stand, validity and reliability of IPCC. Additionally, book also incorporate the views of climate change deniers' arguments and concludes that deniers claim is baseless. Prof. Bhandari support the notion that major cause of climate change is due to anthropogenic disturbances to earth ecosystems; therefore, he urges all concern public of the globe to understand the fact that climate change is a major challenge of contemporary world and it is every human's responsibility to take the response and contribute to overcome from this challenges, from the ground where they are situated.

We have no time to wait for the authority to guide us, but we need to guide to the authority. I commend Prof. Bhandari for his profound contribution. His hard work and perseverance to wake us all is very timely. I urge all to read this very scientific work and utilize the wealth of knowledge incorporated in the book. Thank you.

Bishnu Poudel, Ph.D., Professor of International Relation and Diplomacy, VA, USA

Scientists unanimously believe that climate change is the biggest threat facing the planet Earth. Climate change could make extreme weather more severe, increase droughts in some areas, erratic rainfall, glacier receding, rapid ice melting in Himalaya and Antarctica, change in the distribution of animals and diseases across the globe, and cause low-lying areas of the planet to be submerged in the wake of rising sea levels. The cascade of changes could lead to political instability, severe drought, famine, ecosystem collapse and other changes that make Earth an inhospitable place to live.

In this scenario, the book of Professor Medani Bhandari on IPCC is ground breaking in the sense that the book reveals the facts of science, outlines the history of climate science, the role of International organizations to solve the climate change challenges in the world. The book also provides space to the alternative view of the climate deniers for the benefits of readers and scholars. This is a commendable work important to all stakeholders of the world who care the planet Earth. I highly recommend this book to the prospective students, researchers and anyone who pursues serious scholarship in the area of climate change, its impacts and the global intergovernmental policy approaches to deal with the climate change and its impacts.

Prof. Gopi Upreti, PhD, Fairfax, Virginia, USA

The challenge for environmentalism is to re-define environment issues beyond a confined definition of species and habitat to include human well-being and justice. Science and policy discourse in climate debate should transcend technical fixes and policies to healing of the Earth as a living system, not a machine. The work of Prof. Medani P. Bhandari is about this human blind spot where his narrative addresses the hidden connections humans and nature as One System that needs new consciousness. The debate about climate change is about neoliberal ideology where the adoption of the concept of extractivism limits genuine solution to climate change, since at the core it is simply about distribution of wealth, power and justice. I commend the work of Prof. Medani P. Bhandari since he goes beyond the positivist approach to shed light on the human dimension of this global debate.

Prof. Odeh Al-Jayyousi, Head of Innovation and Technology Management, Arabian Gulf University, Kingdom of Bahrain

Too much consumerism is to be blamed for the climate change. Emerging markets are too eager to get commodities but do not look over the sky, whether it is blue or smokes......I am so glad to read this manuscript written by Prof. Medani P. Bhandari. This book adds a new dimension in the climate change science. This is an outstanding scientific work, which covers how climate science emerged, how public concern begin to worry about climate change and how international organizations begin to find the way to work together to combat with the challenges of climate change. Books links unveils the route of IPCC formation, its importance and also future. As social scientist Prof. Bhandari evaluates IPCC's challenges, its trustworthiness, and its direction. There is no doubt the main cause of climate change occurring due to anthropogenic disturbances on the planet's ecosystem and we human are responsible. Book also warns us that we do not have time to wait and see and urges that we all need to work together to minimize the climate change problem. I wish all the concern people will read this book and utilize the knowledge for the healthy planet.

Dr. Tulsi Dharel, Professor of Marketing at Centennial College, Toronto, Canada

The book 'Getting the Facts Right: The IPCC and the Role of Science in Managing Climate Chong" provides new insights and ideas on Climate change and environmental policy all around the world. Temperature changes and oscillations, climate conditions transformation and even catastrophes namely are consequences of human actions. Globally, the impact of human civilization on the entire planet should not be underestimated.

The book written by Professor Medani Bhandari put the light to many phenomena of climate science. The political polarization between the nations should step back and leave the room for the international cooperation and social partnership on solving the crucial problems of mankind. The Intergovernmental Panel on Climate Change is a serious step towards the establishment of coordinated decisions system on mitigation, adaptation and prevention of climate change. In the era of multiple global transformations, this book gives a balanced and historically ordered explanations of the main changes in the global environment and policymaking. I highly recommend this book to the policymakers, NGOs, activists and scholars who are concern about climate change and our common future.

Dr. Shvindina Hanna, Head of the department of management, Associate Professor at Sumy State University (Ukraine), Fulbright Alumni (Purdue University), USA

The book "Getting the Facts Right: The IPCC and the Role of Science in Managing Climate Change – with reference on pollution in major cities of Asia" is devoted to different aspects of climate change science with reference to Intergovernmental Panel on Climate Change.

This book includes chapters that covered historical aspects of climate change, science and the management of climate change, the origin and composition of Intergovernmental Panel on Climate Change, as well as case studies outlining pollution issues in major cities of Asia.

The book can be helpful for policy makers and university teachers who are working in the sphere of international relations, public administration and climate change. Book unveils the role of IPCC in creating and validating the climate change knowledge.

Dr. Nadiya Kostyuchenko, Associate Professor at Department of International Economic Relations, Sumy State University, R.-Korsakova st., 2, Sumy, Ukraine, 40007

For people who think globally, climate change is not perceived as something remote that does not concern them on an everyday basis. On the contrary, climate change is a problem not only of future, but also of current generations, which can and should be solved now. The Intergovernmental Panel on Climate Change plays a special role in addressing climate change. A general awareness of the problem itself is not enough; a roadmap needs to be developed. The actions need to be applied by each community and by the entire society on Earth, both on local and global levels. In this regards The Intergovernmental Panel on Climate Change is a leading institution in the world.

I am extremely pleased that the book by Professor Medani P. Bhandari will be published, as it provides a detailed assessment of the current state of the planet, as well as summarizes key scientific papers on this issue. The major environmental agreements and climate change theories are provided in the book. The strength of the book is the synergistic combination of historical excursus and modern scientific approaches to climate change issues.

Dr. Denys Smolennikov, Associate Professor at Department of Management, Sumy State University, Ukraine dos@management.sumdu.edu.ua

List of Figures

List of Tables

List of Abbreviations

Adaptation	Any adjustment in natural or human systems in response to a changing or changed climate.
Anthropogenic Climate Change	Climate change with the presumption of human influence, usually warming.
Anthropogenic	Created, caused, or strongly influenced by humans or human activities; man-made.
Anthropogenic Global Warming (AGW)	Global warming with the presumption of human influence.
C^2F^6	Hexafluoroethane
Carbon Footprint	The total set of greenhouse gas emissions caused by an organization, event or product.
Carbon Sequestration	Proposals for removing CO2 from the atmosphere, or for preventing CO2 from fossil fuel combustion from reaching the atmosphere.
CBD	Convention on Biological Diversity
CDM	Clean Development Mechanism "The CDM allows emission-reduction projects in developing countries to earn certified emission reduction (CER) credits, each equivalent to one ton of CO2. These CERs can be traded and sold and used by industrialized countries to a meet a part of their emission reduction targets under the Kyoto Protocol. The mechanism stimulates sustainable development and emission reductions, while giving industrialized countries some flexibility in how they meet their emission reduction limitation targets. The CDM is the main source of income for the UNFCCC Adaptation Fund, which was established to finance adaptation projects and programs in developing country Parties to the Kyoto Protocol

	that are particularly vulnerable to the adverse effects of climate change. The Adaptation Fund is financed by a 2% levy on CERs issued by the CDM" (https://cdm.unfccc.int/about/index.html 2020).
CDR	Carbon dioxide removal
CER	Certified emission reductions
CF^4	Tetrafluoromethane
CFC	Chlorofluorocarbon
CH^4	Methane
Climate Change	Includes both global warming and its effects, such as changes to precipitation, rising sea levels, and impacts that differ by region
Climate	The average and variations of weather in a region over long periods of time.
CO_2	Carbon dioxide
COP	Conference of the Parties
CTBTO	Preparatory commission for the comprehensive nuclear test ban treaty organization.
Ecosystem Services	Benefits humans get from a multitude of resources and processes that are supplied by natural ecosystems
FAO	Food and Agriculture Organization
GEO	Global Environment Outlook
GHG	Greenhouse gas
GIS	Geographic Information System
GSP	High-level Panel on Global Sustainability
GWP	Global-warming potential
HIV	Human Immunodeficiency Virus
IAEA	International Atomic Energy Agency
IBRD	International Bank for Reconstruction and Development (World Bank Group)
ICSU	International Council for Science
ICUN	International Union for Conservation of Nature
IMO	International Maritime Organization
IPCC	Intergovernmental Panel on Climate Change
IPSO	International Programme on the State of the Ocean
ISA	Integrated Sustainability Assessment
ISEW	Index of Sustainable Economic Welfare
ISSC	International Social Science Council

ITC	International Trade Centre
ITU	International Telecommunications Union
LDCs	Least Developed Countries
LLDCs	Landlocked Developing Countries
MDG	Millennium Development Goal
MEA	Millennium Ecosystem Assessment
N20	Nitrous oxide
NAPA	National Adaptation Programme of Action.
NCAR	National Center for Atmospheric Research
NF_3	Nitrogen trifluoride
NGO	Non-Governmental Organization
Nitrous Oxide	(N^2O)- A potent greenhouse gas produced primarily in agriculture, particularly by the livestock sector.
NOAA	National Oceanic and Atmospheric Administration
NRC	United States National Research Council
REDD	Reducing Emissions from Deforestation and forest Degradation mechanisms use market/ financial incentives to reduce the emission of greenhouse gases from deforestation and forest degradation.
SBSTA	Subsidiary Body for Scientific and Technological Advice
SBSTTA	Subsidiary Body on Scientific, Technical and Technological Advice
SCBD	Secretariat of the Convention on Biological Diversity
SD21	Sustainable Development in the 21st Century
UN DESA	United Nations Department of Economic and Social Affairs
UN ESCAP	Economic and Social Commission for Asia and the Pacific
UNAIDS	Joint United Nations Programme on HIV/AIDS
UNCCD	United Nations Convention to Combat Desertification
UN	United Nations
UNEP	United Nations Environment Programme
UNESCO	United Nations Educational, Scientific and Cultural Organization
UNFCCC	United Nations Framework Convention on Climate Change
UNFF	United Nations Forum on Forests

UNHCR	Office of the United Nations High Commissioner for Refugees
UNICEF	United Nations Children's Fund
UNITAR	United Nations Institute for Training and Research
UNRISD	United Nations Research Institute for Social Development
Vasudhaiva Kutumbakam	The entire world is my home and all living being are my relatives
WB	World Bank
WBCSD	World Business Council for Sustainable Development
WHO	World Health Organization
WIPO	World Intellectual Property Organization
WMO	World Meteorological Organization
WTO	World Trade Organization
WWF	World Wildlife Fund

1

Introduction – Getting the Facts Right: The IPCC and the Role of Science in Managing Climate Change-Literature and Chapter Outline

Addressing climate change is the most important public priority of the 21st Century. Unlike many issues, however, it is being driven by science and its interface with politics. The main institution for making the bridge between science and international politics is the Intergovernmental Panel on Climate Change (IPCC). Founded before the adoption of the United Nations Framework Convention on Climate Change, the IPCC has been the main source of consensus on the science and implications of climate change.

Seldom in history has science had such a direct relationship with politics. The negotiation of an international regime requires, at its outset, an agreement on the facts. In this case, the facts are scientific, complex and contentious. Governments have recognized this and have, using international organizations, set up an institutional machinery to provide facts that they can accept.

The way in which the IPCC functions is also unique in that it melds the way in which science achieves consensus with the way governments do at the international level. Starting with a process to examine, review and debate scientific findings leading to a rough consensus about scientific fact, usually expressed as probabilities that the findings will hold over time, the IPCC then concludes by using the kind of consensus-development mechanism that the United Nations typically uses to achieve agreements leading to the formation of regimes.

The structure of the IPCC, its composition and its procedures have made it successful and these factors are analyzed in order to achieve an understanding of the IPCC role and future. Its target audience includes universities concerned with teaching in the areas of organizational change, international relations, public administration and climate change as a specific subject. It also includes policy makers and administrators who will have to implement agreements reached. This book addresses how climate change science was developed; how climate change impacts have been analyzed by various scholars, agencies (including IPCC), and other concern stakeholders, what roles international, nongovernmental and governmental organizations play to address climate change issues. Book specially, unveils the role of IPCC in creating and validating the climate change knowledge, how IPCC creates climate change regime. There is no doubt that climate change has been accelerating due to anthropological disturbances on earth ecosystem; however, there are still few deniers, who are driving anti current. Book also incorporates deniers' arguments. Book also shows few problematic results (immature data interpretation of IPCC particularly in its 2007 report), which gave points to foster denier's argumentation. However, IPCC have already corrected those errors and have already produced Fifth Assessment Report in 2014, Special Report on Global Warming of 1.5 °C (SR15) in 2018 and in the process of producing Sixth Assessment Report on 2022. There is no question that climate change is occurring, and direct and indirect impacts are visible mostly on the fragile ecosystems. Book will also present a case study, of climate change impact in the major South Asian cities (air, water, soil pollution).

There are thousands of scientific papers, books, monographs, videos, commentary, movies, documentaries, personal or societal stories, opinion papers, newspaper articles, interviews, social medias depictions of how climate change has been a major challenge of contemporary world; how it has been impacting the societal equilibrium- economic, social, cultural, environmental and even to some extend international political structure (conflicts in international conventions i.e. Peris, Stockholm etc.). In reality there is no doubt on atmospheric environmental change, temperature increase, and rapid climatic pattern change mostly from industrial era. However, there is still no 100% consciousness on causes of climate change. Ninety nine percent of global scientists, as well as general public firmly trust and believe that, the anthropogenic disturbance is the major cause of rapid climate change (Bhandari 2014, 2018, 2019, Brechin and Bhandari 2011; Bhandari and Bhattarai 2014). However, remaining one percent of scientists (so called), politicians, policy makers, business firms still question on firmly accepted notion- human are responsible on climate change. They do not feel ashamed to argue that climate change is an automatic process of nature (see climate change denier chapter). One should not dismiss counter argument of denier; however, there is no basis of deny because every sensational living being in the planet are witnesses of the impact of climate change (ecosystem change, climatic pattern change, seasonal change etc.). The major victims are the farmers of the developing world, who has firsthand experience as well as the residents of major big cities who hold the experiences of polluted environment (details are explained in the concluding chapter). Also, those who are dying every single moment due to the impacts of environmental hazards (mostly due to polluted water, air and even soil).

The impact on biotic structure (flora and fauna) might be dreadful but we have not quantified, or we have even not considered that domain. However, it is the time to think of our surroundings- natural environment which always supports us by maintaining ecosystems. All ecosystems are disturbed whether we have records or not and disturbed in ecosystems have direct impact on us and each living being in our surroundings (Bhandari 2019). There are thousands of scientists, governmental, non-governmental organizations, local to national and international organizations, social medias, social advocates, and persons and groups; who have been showing the evidences and seriousness of climate change impact. They are also documenting the evidences through research and trying to convince the people, group, community and government to take the necessary action to minimize the impact of climate change. Among such agencies (institutions), founded in 1988, Inter-Governmental Panel on Climate Change (IPCC) has been producing fact base documentary evidences since 1990. The details of working modality of IPCC is described in several chapters in this book. IPCC does not conduct own research; however, uses the documentary evidences from published resources over the given time frame by using hundreds of scientists from United Nations member countries. The publications of IPCC or the periodic reports of IPCC have been major reference of climate science in the contemporary climate change evidence-based research, however, the primary references are the scholarly works of scientists, who have been providing knowledge to all stakeholders with the objectives of protecting planet for generations to come.

Such research papers, books, monographs, videos, commentary, movies, documentaries, personal or societal stories, opinion papers, newspaper articles, interviews, websites, blogs etc. have covered broad spectrum of climate science. The emergence of climate science has long history (outlined of such historical account is described in the chapter titled on history of climate science). However, there are ample of resources on this domain grounded on various perspectives, dominantly humanitarian perspectives. The general purposes of such resources are to make aware to the general public who can influence the local, national to international policy makers, who may understand the severity of the most dangerous problem of current world, created by the human themselves on the name of development, development and development (Bhandari 2019). Here are few such references and perspectives which provide the clear picture of climate change impact in the contemporary world. Scholars have their own way of depiction the facts and one can disagree with the presented literatures, however, the underlined purpose of most of research documents are to reveal the fact of severity of climate change impact on human civilization. Scholars have not only written data driven books, papers, or prepare the audios and videos, blogs, cartoons but also written / presented science fiction, imagery, creative heart feelings on how and what would be the future of the planet if we did not act from now. There are thousands and thousands of such references; and is not possible to gather, table and present them, therefore, the listed references are just like few drops from the ocean. Authors have also tried to show the urgency, using science driven alarms like Bashudaivakutumbakkam- all living beings are our relatives; live and let other live (Bhandari 2019); When Will the Planet Be Too Hot for Humans? Much, Much Sooner Than You Imagine (paper by Machine 2017), which presents a dramatical overview of climate change impact on human civilization. The Collapse of Western Civilization: A View from the Future (a fictional presentation by Oreskes and Conway 2014); The Martian (by Weir 2016); War on Normal People (Young 2019); Can We Solve Global Warming? (Davis 2017); “What Happens AFTER Climate Change?” (Stagar 2012); 5 Ways Climate Change Is Affecting Our Oceans (Fujita 2013) and so on. . . . . Authors have also pin pointed the impact of emerging artificial intelligence with link with development and people, which might have negative connotation to the society and human civilization directly or indirectly related to anthropogenic disturbances on natural environment (Bhandari 2014, 2019; Johannessen 2018; Sterling 1996; Aguirre 2002; Gross and Rayner 1985; Proctor 1998).

There is big worrisome about the environmental disasters occurring due to climate change- hurricanes; floods; and intense heat (Davis 1998; Forest et al 2002); if global warming is allowed to continue, it could have severe environmental effects (Hosansky 2001); climate change induced behavioral change (Jennings 2002); Climate change impact are irreversible problem (Ulph and Ulph 1997); global climate changes are affecting people now (Vitousek 1994); Future Impact of Climate Change is worrisome (Flannery 2005); climate crisis (Lynas 2004), Lynas demonstrates personal experience of climate crisis with the photographic evidences of glacier meltdown. The climate change and its impact on social, economic and cultural environment have largely elaborated by many scholars (Pittock 2005; Ross 1999; Bhandari 2011, 2018,2019; Speth 2004; Stanley 2004; Pendergraft 1998; Yazici, Petry, and Pendergraft 2001;) including country specific case studies, as well as regional and global scenarios analysis (Brechin and Bhandari 2011; Young 2002; UNEP 2010; Rockström 2009; Magistro and Roncoli 2001; Macnaghten and Jacobs 1997; Orr 2003). Since climate change has already, significantly hampered the planet ecosystem, still, scholars have been working extensively to give the theoretical and practical implication of climate change (Proctor 1998; Ungar 2000; William 2003). It is important to note that Williams (2003) discusses "the importance of understanding how humans experience the environment in order to facilitate the resolution of environmental problems. He argues against rational humanism – where humans are viewed as endowed with reason through which the human condition can be improved – as a way of understanding this relationship between humans and the environment and proposes instead that a phenomenology of environmental problems must be formulated. Williams uses the example of grassroots environmentalism as evidence that combining democratic social organization with the real-world consequences of tangible environmental problems can lead to solutions for these" (from abstract). Importantly, McCright and Dunlap (2000) provide a content analysis of social scientific literature on global warming and evaluate the US public opinion on this domain. Similarly, Slocum (2004), analyses climate protection campaign by evaluating 403 cities worldwide. Slocum uses 135 in-person interviews data (including telephone interviews) and provides the responses of how people response on climate protection. Likewise, Antilla (2005), conducts the media analysis and shows the scenario of Climate of Skepticism: US Newspaper Coverage of the Science of Climate Change, in which Antilla presents four frames of climate science-'valid science', 'ambiguous cause or effects', 'uncertain science' and 'controversial science' with the special focus on uncertainty, controversy and climate skepticism. Henderson-Sellers (1998), in 'Climate Whispers: Media Communication about Climate Change' provides how media highlight on the climate change and its impact (including miscommunication). McComas and Shanahan (1999) in 'Telling Stories about Global Climate Change: Measuring the Impact of Narratives on Issue Cycles' identified 8 themes in the media coverage of climate change: 'new evidence or research', 'general science background', 'controversy among scientists', 'consequences of warming', economics/ costs of remedy', 'domestic politics', 'international relations' and 'current weather'.

There is no doubt that climate change has been occurring throughout the geographical process and anthropogenic intervention in the natural process is the triggering factor of its acceleration. There are many books, papers, monographs and reports, and personally experienced factors to present the evidence of climate change. Authors have presented how to understand climate change and what are the evidences of such change. The following table shows few evidential papers / reports of climate change and its impact.

Above listed sources primarily show the reality/evidences and ground truth of climate change is occurring and the one of the major causes is anthropogenic disturbances on the planet ecosystem. One of the major victimized domains due to climate change is biodiversity (both flora and fauna's ecosystem). Many scholars have pinpointed how biodiversity is responding to climate change. Scholars have analyzed climate change impact on ecosystems (Agrawal 1998; Folke, and Carpenter, et al. 2004; Saunders, and Easley, et al. 2007; Millar, and Stephenson, et al. 2007; Hobbs, and Cramer 2008; Frelich, and Reich 2009; Galatowitsch, and Frelich, et al. 2009) to species (Parmesan 2006; Heller and Zavaleta 2009; Hannah 2008; Inkley, and Anderson, et al. 2004; Hoegh-Guldberg, and Hughes, et al. 2008). Scholar have also analyzed the climate change and invasive species (Dukes and Mooney 1999; Brain et al. 2008; Bradley, and Blumenthal, et al. 2009; Hellmann, and Byers, et al. 2008; Mainka, and Howard 2010; Botkin 2001; Bentz, and Regniere, et al. 2010), climate change and fire (Flannigan, and Stocks, et al. 2000; Whitlock, and Shafer, et al. 2003; Fulé 2008; Noss 2001; Bachelet, and Lenihan, et al. 2000); climate change and water (Strayer, and Dudgeon 2010).; Heino, and Virkkala, et al. 2009; Wilby, and Orr, et al. 2010; Abell 2007; Pearlstine, and Pearlstine, et al. 2010).

Authors also have shown the rational of global climatic condition how, when and why change became threat to the planet and why we should worry. Scholarly world, as well as all concerned stakeholders (except climate change deniers), including governments, general public accept that we human and our behavior is responsible for this unwanted change and impact in the planet due to climate change. However, the evidence also shows that, there is still hope of improvement. Many governmental, nongovernmental, organizations, academic and non-academic agencies (including government, public, private) have been working to find out the way to slow down the change (Brower and Leon 1999; Dauncey, and Mazza 2001; Heede, 2002; Flannery 2005; Gershon 2006; Kolbert 2006; Linden 2006). Various tools and mechanisms have been introduced and policies have been formulated from local to international levels. United Nations is the key player to formalize

Table 1.1 Evidential papers / reports of climate change and its impact.

Subject/Issues	Author/Authority	Comment/Arguments/Evidences/Outcome
Models Underestimate Warming, Sea Level Rise Warming pushes the upper bound	IPCC (2007); Rahmstorf, S., et al. (2007); Rahmstorf, S. (2007); Meier, M., et al. (2007); NOAA (2007)	Sea level rise so far has also been significantly faster than predicted in the 2007 IPCC reports. The 2007 IPCC report is the first to give a best estimate – 5.4 degrees Fahrenheit – of "climate sensitivity," the global mean temperature rise that is likely to result from a doubling of carbon dioxide levels.
Ice Sheet Melt Concerns and Confounds Scientists	Cazenave (2006); World Bank (2007); Luthcke, et al. (2006); Chen, et al. (2006	The hardest part of predicting sea level rise is figuring out how fast land-based ice sheets will melt. World Bank estimates that a much smaller 1-meter increase in sea level would displace at least 56 million people living in coastal regions, and a 5-meter rise would displace more than 240 million people.
Sea Ice Is Reaching a Breaking Point	Holland et al. (2006); Winton, (2006); IPCC (2007)	Multiple climate models predict that sea ice will disappear as the earth warms, and recent observations provide alarming evidence that these predictions are being realized.
Polar Bears and Other Species Stressed from Global Warming	Parmesan, and Yohe (2003); Stirling, and Parkinson (2006).	Many plants and animals are stressed due to habitat destruction and other human-induced pressures. Global warming presents another threat, and warming is occurring at a rate that may prove too fast for species to adapt-polar bear is an example.
Carbon Sinks Reconsidered Tropical forest sink More bad news about smog Old growth forests continue to accumulate carbon	Gullison, et al. (2007); Stephens, et al (2007); Sitch, et al. (2007); Zhou, et al (2006)	About half of carbon dioxide emitted into the atmosphere from human activities-stays in the atmosphere. As the temperature rises, drought-induced tree mortality, logging, and fire may double these emissions, and loss of land-sink capacity as forest area decreases may further increase atmospheric carbon dioxide levels. Impact on plants, smog causes more carbon dioxide to build up in the atmosphere and accelerates global warming.

(*Continued*)

Table 1.1 *Continued*

Global Warming Affects Weather Extremes	Hansen, et al. (2006)	Super El Niño Increases in hurricanes
Example from -Yellowstone Area Grizzly Bears Suffer as Whitebark Pines Decline	Powell, and Logan (2005); Felicetti, et al. (2003); Interagency Grizzly Bear Study Team. (2004); Mattson, Blanchard and Knight (1992); Mattson, (1998); Schwartz, et al. (2005)	Once whitebark pine forests are damaged, the grizzly bears that depend on this critical food source would face food shortages for many years because it takes 60 to 80 years for the pines to begin producing large cone crops.
Hotter and Hotter	Hansen, Ruedy, Sato and Lo (2006)	GISS Surface Temperature Analysis, Global Temperature Trends- shows accelerating hotter temperature every year
The Hockey Team Keeps Winning	National Research Council (2006); Osborn and Briffa (2006); Thompson, et al (2006)	Data available from instrumental records, reconstructions of past temperature from temperature-sensitive proxies, such as tree rings, ice cores and coral reefs indicate that the globally-averaged surface temperature has been greater in the past few decades than during any comparable period during the preceding 400 years. The graph of these reconstructions resembles a hockey stick with the blade pointing up and to the right – in other words, global temperatures were relatively steady for about 900 years and then turned up sharply during the late 20th century
More Players on the Hockey Team	Wahl and Ammann (xxxx); Mann, Bradley, and Hughes (1998)	New reconstructions of the earth's temperature for the last 1000 years confirm that the last several decades are most likely the warmest such period not just of the last century, but of the entire millennium. Results resemble a hockey stick with the blade pointing up and to the right – in other words, global temperatures were relatively steady for about 900 years and then turned up sharply during the 20th Century. This strongly reinforces the conclusion from other evidence that recent global warming is due to heat-trapping pollution.

(*Continued*)

Table 1.1 *Continued*

Rapid Build Up	Siegenthaler, et al. (2005); Tans, NOAA/ESRL (2006)	Recent data indicate that carbon dioxide is accumulating in the atmosphere at a greater rate than in the past. an increase of more than 100 ppm since the start of the Industrial Revolution, and ice core records show that it is the highest concentration of atmospheric carbon dioxide for at least the last 650,000 years
A Tipping Point at the Poles?	Lawrence and Slater (2006); Otto-Bliesner, et al. (2006); Rignot and Kanagaratnam (2006); Velicogna and Wahr (2006)	Science magazine – "Climate Change – Breaking the Ice." The edition included articles covering important new research on warming at both poles that is leading to changes in the ice system. These changes are occurring faster than previously observed or expected, therefore indicating that both the Arctic and Antarctic may be approaching a "tipping point" after which dangerous transformations will become unavoidable.
A Future without Summer Sea Ice in the Arctic?	National Snow and Ice Data Center, NASA and University of Washington (2005); National Snow and Ice Data Center (2006); Overpeck, et al. (2006); Stroeve, et al (2005)	Arctic could be completely free of summer sea ice well before the end of this century, a state that has not occurred over at least the last million years.
Are We Underestimating Global Warming?	Brinkhuis, et al. (2006); Moran, et al. (2006); Torn and M.S. and Harte (2006); Scheffer, Brovkin and Cox (2006); Sluijs, et al. (2006)	Warming associated with doubling of carbon dioxide due to human activities is amplified from the range of 1.5 - 4.5°C to 1.6 - 6.0°C. Rising temperatures on atmospheric carbon dioxide concentrations will further enhance warming by an extra 15 to 78 percent Polar stratospheric clouds and hurricane-induced ocean mixing could have led to the high-latitude warming and tropical cooling found in the records

(*Continued*)

Table 1.1 *Continued*

It's Official: Satellite and Surface Temperature Records Agree	Christy and Spencer (2005); Karl, et al. (2006); Kerr (2006)	"No Doubt About It, the World is Warming." On the basis of over the 25-year satellite record, the surface and mid-troposphere have both warmed by approximately 0.15°C per decade.
Warmer Seas Mean More Intense Hurricanes	Mann and Emanuel (2006); Sriver and Huber (2006); Trenberth and Shea (2006)	Warming of the tropical Atlantic due to human activity is responsible for the recent increase in tropical cyclone activity... As the background levels of global sea surface temperatures continue to climb, we can expect greater hurricane activity in future.
Mountain Snowpack Declines Mean Trouble for the Future of the World's Water Supply	Ghan and Shippert (2006)	As the world warms, we are likely to see declines in mountain snowpack, a key water source for many areas of the world which could result in severe impacts on the world's water supply because many regions are dependent on spring and summer runoff for irrigation and drinking water.
Further Evidence for Extinctions Due to Global Warming	Pounds et al. (2006); Both et.al. (2006)	Global warming can have a profound impact on population dynamics and ecosystem functioning.
Avoiding Dangerous Climate Change	Meinshausen, (2006); Schellnhuber, et al. (2006); Warren, (2006);	Warren analyzed global warming impacts at levels at or above 2°C and, among other results, found that: 1 to 2.8 billion people will experience an increase in water stress; Up to 26 million people will be displaced by sea level rise and increased storm intensity; Up to 220 million additional people will be at risk of hunger as agricultural yields fall; There will be a total loss of summer Arctic sea ice and we will have likely committed to the complete meltdown of the Greenland Ice Sheet; 97 percent of the globe's coral reefs will be lost; 50 percent of major wetlands in Bangladesh and Australia will be lost; And, ocean acidification may disrupt ocean ecosystem function.

(*Continued*)

Table 1.1 *Continued*

UNFCCC target	UNFCCC (2019)	The United Nations Framework Convention on Climate Change (UNFCCC) has the objective of preventing "dangerous anthropogenic interference with the climate system." While a "non-dangerous" concentration level has not been defined under the UNFCCC, the European Union has set a goal of avoiding an increase of more than 2°C from pre-industrial levels, in order to avoid the most dangerous changes to climate
Wildfires in the West	Running (2006); Westerling, et al. (2006)	Wildfire activity to human-induced global warming, increases in spring and summer temperatures and earlier spring snowmelt are tell-tale signs of global warming.
Pumping Up the Volume	EIA (2005, 2006)	Global warming pollution emitted in the United States reached the highest level ever recorded in 2004, according to an Energy Information Administration (EIA) report.
Arctic Meltdown	Monnett, Gleason, and Rotterman (2005); Overpeck, et al. (2005); National Snow and Ice Data Center (2005); Lawrence and Slater (2005)	"Arctic sea ice is likely on an accelerating, long-term decline." Permafrost on land is also melting at an alarming rate. New simulations that include permafrost dynamics in a global climate model project that with continued rapid growth of global warming pollution, two-thirds of the northern hemisphere's near-surface permafrost (to a depth of 11 feet) would melt by 2050 and more than 90 percent would melt by the end of the century.
It's the Heat and the Humidity	Soden, et al. (2005); EIA (2005)	Warmer air can hold more water, this implies that global warming results in an increase in the amount of heat trapped by water vapor, fueling additional warming.
The Core of the Problem	Siegenthaler et al. (2005)	New results from the European ice coring project in Antarctica extend the record back to 650,000 years before the present and confirm that the current atmospheric carbon dioxide concentration of 380 parts per million (ppm) is much higher than anything on record.

(*Continued*)

Table 1.1 *Continued*

Putting a Fine Point on Climate Extremes	Diffenbaugh et al. (2005)	All parts of the United States would experience at least a doubling in the frequency of extremely hot days (defined as hotter than 95 percent of days in the current climate), with the Southwest being particularly hard hit with up to 100 additional extremely hot days each year.
Satellite and Surface Temperature Records Reconciled	Mears and Wentz (2005); Revkin (2005)	Scientists at Remote Sensing Systems reanalyzed the raw satellite data and found that the lower atmosphere is actually warming slightly faster than the surface, in agreement with theory and models.
Warm Seas Fuel More Destructive Storms	Emanuel (2005); Webster et al. (2005)	Warm Seas Fuel More Destructive Storms
Science Academies Call for Prompt Action	Gleneagles G8 Summit (2005)	"The scientific understanding of climate change is now sufficiently clear to justify nations taking prompt action." The statement on global warming called on world leaders to acknowledge that the threat of climate change is clear and increasing, to recognize that delayed action will increase the risk of adverse effects and likely increase costs, to identify cost-effective steps that can be implemented to substantially reduce global greenhouse gas emissions, to develop and deploy clean energy technologies along with approaches to energy efficiency, and to work with developing nations to enable them to develop innovative solutions for mitigating and adapting to climate change.
Ocean Warming – The Smoking Gun of Global Warming	Hansen et al. (2005); Barnett et al. (2005); NASA (2005)	New precise measurements of the accumulation of heat in the earth's ocean confirm that heat-trapping pollution is the primary cause of global warming. not only are the earth's land and ocean surfaces warming, but that the heating has penetrated more than 1,000 feet into the oceans' depth. "This energy imbalance is the 'smoking gun' that we have been looking for".

(*Continued*)

Table 1.1 *Continued*

Impacts of a Warming Arctic: Arctic Climate Impact Assessment	Arctic Council (2004)	The Arctic is warming almost twice as fast as the rest of the world, with significant impacts apparent now, according to a major new four-year study conducted by an international team of 300 scientists. The Arctic Climate Impact Assessment found that in Alaska, western Canada and eastern Russia average winter temperatures have increased as much as 4 degrees to 7 degrees Fahrenheit in the past 50 years.
Impact of Global Warming on Hurricane Intensity	Knutson and Tuleya (2004)	If the frequency of tropical cyclones remains the same, global warming would result in a significant increase in the most destructive Category 5 storms. Meanwhile, sea level rise due to global warming will push shorelines inland by 400 feet or more in low-lying areas, making storm surges even more damaging.
Emissions Pathways, Climate Change, and Impacts on California	Hayhoe et al. (2004)	Unmitigated global warming would have severe consequences for the Golden State according to a comprehensive study published in the Proceedings of the National Academy of Sciences.
Dissolving Corals	Feely et al. (2004)	Coral reefs are in serious jeopardy due to carbon dioxide emissions and global warming. Severe bleaching has been observed in reefs throughout the tropics due to excessively warm ocean temperatures and other environmental stresses
Heat Advisory: How Global Warming Causes More Bad Air Days	Patz, et al. (2004)	A new analysis by some of the nation's top medical experts' projects that hotter temperatures caused by global warming will speed formation of the lung-damaging pollution commonly known as smog
Satellite Data Confirms Climate Change	Nature (2004)	Measuring temperature trends using satellites suggested that the lower atmosphere is warming upper atmosphere is cooling apparently due to increased heat trapping in the lower atmosphere and stratospheric ozone depletion

(*Continued*)

Table 1.1 *Continued*

Inside the Greenhouse: The Impacts of CO2 and Climate Change on Public Health in the Inner City	Harvard Medical School (2004)	Residents of the inner city are particularly vulnerable to the effects of climate change and global warming
Climatology: Threatened Loss of the Greenland Ice-Sheet	Nature (2004)	Unless heat-trapping emissions are reduced substantially, Greenland is likely to warm by at least 3 degrees Celsius by the year 2100, enough to trigger the complete and irreversible meltdown of the Greenland ice sheet, reported scientists in the April 8 issue of Nature.
Defusing the Global Warming Time Bomb	Scientific American (2004) Hansen	Earth's energy balance has already been altered by pollution
Global Warming: The Imperatives for Action from the Science of Climate Change	Sir David King (2004)	Efforts can make change
The Effects of Climate Change on Water Resources in the WestAs the West Goes Dry	Climate Change (2004); Science (2004)	The American West will have more wintertime floods and summertime droughts if the climate continues to warm Throughout the West, higher temperatures will decrease snowpack and cause spring runoff to start 30 to 40 days earlier than it does today
Hottest year on record	NOAA (2004)	The most recent data show that 2003 tied 2002 as the second hottest year on record, following 1998. The five hottest years have all occurred since 1997 and the 10 hottest since 1990. Extreme heat waves caused more than 20,000 deaths in Europe and more than 1500 deaths in India during 2003.
Extinction Risk from Climate Change	Nature (2004)	Extinction risk from global warming, found that more than 1 million species could be committed to extinction by 2050 if global warming pollution is not curtailed

(*Continued*)

Table 1.1 *Continued*

Modern Global Climate Change	Science (2003)	"In the absence of climate mitigation policies . . . the likely result is more frequent heat waves, droughts, extreme precipitation events and related impacts [such as] wildfires, heat stress, vegetation changes and sea-level rise."
Human Impacts on Climate	American Geophysical Union (2003)	"Scientific evidence strongly indicates that natural influences cannot explain the rapid increase in global near-surface temperatures observed during the second half of the 20th century."
Fingerprints of Global Warming on Wild Animals and PlantsA Globally Coherent Fingerprint of Climate Change Impacts Across Natural Systems	Nature (2003)	The relatively small global warming that has occurred to date has already changed the habits or forced significant shifts in the range of many species of birds, insects, fish and plants, according to the authors of these two studies published in the prominent scientific journal Nature.
Satellite Spies on Doomed Antarctic Ice Shelf	National Snow and Ice Data Center, British Antarctic Survey (2002)	Over the last 50 years, average temperatures in the Antarctica Peninsula have risen by 4.5 degrees Fahrenheit (2.5 degrees Celsius), four times the global average.
We do have global warming, and that most of this is due to human action	National Climate Data Center (2001); World Meteorological Organization (2001)	Temperatures are currently rising three times as fast as in the early 20th century The hottest year on record, according to the organizations, was 1998, when average global temperatures were 58.1 degrees Fahrenheit. Average temperature for 2001 was 57.8 degrees, according to the National Climatic Data Center.
Climate Change Science: An Analysis of Some Key Questions	National Academy of Sciences (2001)	"Greenhouse gases are accumulating in earth's atmosphere as a result of human activities, causing surface air temperatures and subsurface ocean temperatures to rise." "Temperatures are, in fact, rising,"
Pollution (mainly from the burning of fossil fuels) causes climate change	IPCC (2001)	"Emissions of greenhouse gases . . . due to human activities continue to alter the atmosphere in ways that are expected to affect the climate," the study says. Global warming has caused sea levels to rise, ocean heat content to increase, and snow cover and ice extent to decrease, according to the study.

(*Continued*)

Table 1.1 *Continued*

		The report also predicts that earth's average temperature could rise by 3 to 10 degrees Fahrenheit over the next 100 years. That increase would mark the most rapid change in 10 millennia. It would also be as much as 60 percent higher than the IPCC predicted less than six years ago. The study found that warming in the 20th century was most likely the greatest of the last 1,000 years, and that the 1990s was the hottest decade of the last millennium. Rising temperatures caused by the burning of fossil fuels could cause large-scale and irreversible climate changes. Those changes include altered ocean currents, slowed circulation of warm water in the North Atlantic and a vast reduction of mountain glaciers and the Greenland ice sheet. The study also warns of savage floods, disrupted water supplies, droughts, violent storms and the spread of cholera and malaria as temperatures rise over the next century. Poor countries, particularly those in Latin America, Africa and Asia would bear most of the burden of extreme climate changes, which would further widen the gap between poor nations and rich ones, the report concludes
How human-induced global warming will affect	National Assessment Synthesis Team (2000)	Increasingly, there will be significant climate-related changes that will affect each one of us, increase will cause, for example, alpine meadows in the Rocky Mountains to disappear, sugar maple trees to vanish in the Northeast, and greater risk from storm surges in the Southeast. Rising temperatures will also exacerbate water shortages (especially in the West) and cause New York City to steam in the summer like Atlanta does now. Other likely impacts: coastal erosion, destructive storm surges and the disappearance of barrier islands, all due to rising sea levels.
Climate Extremes: Observations Modeling and Impacts	Science (2000); National Climatic Data Center (2000)	Extreme weather events such as droughts, floods, heat waves and heavy rainfall are expected to increase over the next 100 years, according to a team of scientists from the National Climatic Data Center.

(*Continued*)

Table 1.1 *Continued*

		Changes will continue to increase with the rise of "ever greater amounts of GHGs in the atmosphere."
Causes of Climate Change Over the Past 1,000 Years	Science (2000)	Humans are the dominant force behind the sharp global warming trend seen in the 20th century, according to this analysis of the climate over the last 1,000 years.
Adapting to Climate Change	Raworth (2007); Oxfam International (2007)	What's Needed in Poor Countries and Who Should Pay-Tackling climate change will require an unprecedented level of global cooperation. Rich countries, the ones primarily responsible for the problem, must not only cut greenhouse gas emissions, but also pay for their fair share of helping poor countries adapt to climate change. In this new report, Oxfam estimates the cost will be at least $50 billion each year
Africa – Up in Smoke?	Simms et. al., (2005)	The Second Report from The Working Group on Climate Change and Development explains that the world's wealthiest countries have emitted more than their fair share of greenhouse gases. Resultant floods, droughts and other climate change impacts continue to fall disproportionately on the world's poorest people and countries, many of which are in Africa.

Note: It is firmly acknowledged that above listed reference and illustrations are directly taken from abstracts, as well as from the comments made by scholars on annotated bibliographies created whose names were not available or I missed to note.

the global policy to combat with the climate change induced challenges, by encouraging, empowering governments to create policy instruments and implementation plans to its member countries (Bhandari 2012, 2018, 2019). Additionally, United Nations through its agencies - The United Nations Environment Programme (UNEP) and World Meteorological Organization (WMO) establish the Intergovernmental Panel on Climate Change (IPCC) in 1988. The 70th United Nations General Assembly of December 6, 1988 endorses the action of UNEP and the WMO in setting up the IPCC.

> "In 1990, the First IPCC Assessment Report (FAR) underlined the importance of climate change as a challenge with global consequences and requiring international cooperation. It played a

decisive role in the creation of the UNFCCC, the key international treaty to reduce global warming and cope with the consequences of climate change. The Second Assessment Report (SAR) (1995) provided important material for governments to draw from in the run-up to adoption of the Kyoto Protocol in 1997. The Third Assessment Report (TAR) (2001) focused attention on the impacts of climate change and the need for adaptation. The Fourth Assessment Report (AR4) (2007) laid the ground work for a post-Kyoto agreement, focusing on limiting warming to 2°C. The Fifth Assessment Report (AR5) was finalized between 2013 and 2014. It provided the scientific input into the Paris Agreement. The IPCC is currently in its Sixth Assessment cycle where it will prepare three Special Reports, a Methodology Report and the Sixth Assessment Report. The first of these Special Reports, Global Warming of 1.5°C (SR15), was requested by world governments under the Paris Agreement. In May 2019, the IPCC finalized the 2019 Refinement – an update to the 2006 IPCC Guidelines on National Greenhouse Gas Inventories. The Special Report on Climate Change and Land (SRCCL) will be finalized in August 2019 and the Special Report on the Ocean and Cryosphere in a Changing Climate (SROCC) will be finalized in September 2019. The Sixth Assessment Report (AR6) is expected to be finalized in 2022 in time for the first global stock take the following year" (IPCC 2019, Homepage)."

There are hundreds of papers on IPCC itself or related to IPCC which affirms that accelerating climate change is caused by anthropogenic disturbances on planet ecosystems (IPCC AR5 WG3; 2014; IPCC AR5 WG2 A 2014; IPCC AR5 WG1 2013; IPCC AR4 SYR 2007; IPCC AR4 WG1 2007; IPCC AR4 WG2 2007; IPCC AR4 WG3 2007; IPCC TAR SYR 2001; IPCC TAR WG1 2001; IPCC TAR WG2 2001; IPCC TAR WG3 2001; Davidson,; Swart,; and Pan 2001; IPCC SRES 2000; IPCC SAR WG1 1996; IPCC SAR WG3 1996; ITGP 2010; Jones.; Briffa,; Barnett, and Tett, 1998; Lambeck, 2007; Mann, Bradley, Raymond and Hughes 1999; Meinshausen, et al. 2011; NASAC 2007; Osborn, and Briffa, 2006; PAGES 2k Consortium 2013; Parson, et al. 2007; PBL et al. 2009; PBL 2010; Pollack, Huang, and Shen 1998; Rive, Jackson, Rado, and Marsh, 2007; Stern, 2006; US NRC 2001; US NRC 2010; Hines, Heath and Birdsey 2010; Wahl, and Ammann 2007; IPCC 2018). "The IPCC provides regular assessments of the scientific basis of climate change (Agrawal 1998; Cicerone *et.al.* 2001; Oppenheimer *et.al.* 2007; Jones *et.al.* 2009), its impacts and future risks, and options for adaptation and mitigation" (IPCC 2019, home page). The table 2, shows the IPCC timeline.

On the basis scholarly works listed above and other, this book fills the knowledge gap on how science developed, and the climate change issue became the part of scientific research. Book describes how scientific (innovation of technology) from Greek era to the 20th Century helped to understand climate change and why scientists recommended the IPCC formation. Book unveils the scientists 'efforts to give the real picture of climate science, through research outcomes throughout the history. Book further, elaborates how climate change regime has been formed, following by the role of international organization to form IPCC its origin of IPCC. Book begins with the origin of the international organization, how the international organization's role became important, particularly in the 20th century. The origin of international bounding conventions and treaties, the first government's initiation to address climate change (of course they are USA and Switzerland). It also unveils the formation of the League of Nations, United Nations, and its agencies and also the formation of international scientific union, IUCN, its role to establish UNEP and UNEP and WMO role to establish IPCC, including a list of the major international agreements and treaties related to environment and climate change.

On the basis of various arguments and anti-arguments (climate change deniers' opposition) book further analyzes the major five errors of the IPCC 2007 report: Himalayan glaciers, The Netherlands land surface, rain fed agriculture in Africa and the impact of Amazonian forest loss. IPCC made a citation error in most of the cases and the errors are serious. However, the core scientists still do not question IPCC's assessment but want to improve scientific knowledge. The East Angelia University hacked emails and document does not make a difference in the outcome of the IPCC result. I have noted major problematic correspondences anonymously. It is known that it was the ethical violation by the climate change skeptics. Book also shows the seriousness of the United Nations to address the issues. It also recommends that IPCC not be superficial in such serious cases. Further, book explains that all members of WMO and UNEP are the participants, but the scientists are not equally represented from north and south. Book also unequal distributions of lead authors in the IPCC reports mostly until 2014.

The book contents nine forwards (the evaluation of the book) from the prominent scholars of various parts of the world. Each forward incorporate

Table 1.2 IPCC time line

Year	Functions
1988	The United Nations Environment Programme (UNEP) and the World Meteorolog-ical Organization (WMO) establish the Intergovernmental Panel on Climate Change
	The United Nations General Assembly endorses the action of UNEP and the WMO in setting up the IPCC.
1990	The IPCC publishes its First Assessment Report (Working Group I – Climate Change- Scientific Assessment; Working Group II – Climate Change: The IPCC Impacts Assessment; Working Group III – Climate Change: The IPCC Response Strategies).
	The UN General Assembly notes the report findings and decides to initiate negoti-ations for a framework convention on climate change.
1992	The IPCC publishes Supplementary Reports (Working Group I – Climate Change
	The Supplementary Report to the IPCC Scientific Assessment; Working Group II – Climate Change
	The Supplementary Report to the IPCC Impacts Assessment; Climate Change: The IPCC 1990 and 1992 Assessments).
	The United Nations Framework Convention on Climate Change (UNFCCC) opens for signature at the UN Conference on Environment and Development in Rio de Janeiro.
1995	The IPCC publishes its Second Assessment Report (Working Group I – Climate Change The Science of Climate Change; Working Group II – Climate Change 1995: Impacts, Adaptations and Mitigation of Climate Change: Scientific-Technical Analyses; Working Group III – Climate Change 1995: Economic and Social Dimensions of Climate Change; IPCC Second Assessment: Climate Change 1995 (includes Synthesis Report).
1996	The IPCC issues the Revised 1996 IPCC Guidelines for National Greenhouse Gas Inventories.
1997	The UNFCCC's Kyoto Protocol is adopted. It comes into force in 2005.
1998	The IPCC sets up the Task Force on National Greenhouse Gas Inventories (TFI) to oversee the National Greenhouse Gas Inventories Programme. Since 1999 the Task Force has been supported by the Government of Japan.
2000	The IPCC issues the Good Practice Guidance and Uncertainty Management in National Greenhouse Gas Inventories.
2001	The IPCC publishes its Third Assessment Report (Working Group I – Climate Change The Scientific Basis; Working Group II – Climate Change 2001: Impacts, Adap-tation, and Vulnerability; Working Group III – Climate Change 2001: Mitigation; Climate Change 2001: Mitigation; Climate Change 2001: Synthesis Report).

(*Continued*)

Table 1.2 *Continued*

2003	The IPCC issues the Good Practice Guidance for Land Use, Land-Use Change and Forestry
2006	The IPCC issues the 2006 Guidelines for National Greenhouse Gas Inventories.
2007	The IPCC publishes its Fourth Assessment Report (AR4) (Working Group I – Climate Change 2007: The Physical Science Basis; Working Group II – Climate Change 2007: Impacts, Adaptation and Vulnerability; Working Group III – Climate Change 2007: Mitigation of Climate Change; Climate Change 2007: Synthesis Report).
	The IPCC shares the Nobel Peace Prize which is awarded for its "efforts to build up and disseminate greater knowledge of man-made climate change, and to lay the foundations for the measures that are needed to counteract such change".
2009	The IPCC approves the outlines of the three Working Group contributions to the Fifth Assessment Report (AR5), due to be finalized in 2013 and 2014.
2010	The three Working Groups complete the selection of the 831 authors for the AR5 and work on the assessment starts.
	The IPCC starts a review of its processes and procedures, completed in 2012, based on recommendations from the Inter Academy Council.
2011	The IPCC approves the Special Report on Renewable Energy Sources and Climate Change Mitigation (SRREN), prepared by Working Group III.
2012	The IPCC approves the Special Report on Managing the Risks of Extreme Events and Disasters to Advance Climate Change Adaptation (SREX), prepared by Working Group II and Working Group I.
2013	The IPCC approves Climate Change 2013: The Physical Science Basis, the Working Group I contribution to AR5.
	The IPCC approves two Methodology Reports: the 2013 Supplement to 2006 IPCC Guidelines for National Greenhouse Gas Inventories: Wetlands (Wetlands Supplement) and the 2013 Revised Supplementary Methods and Good Practice Guidelines Arising from the Kyoto Protocol (KP Supplement).
2014	The IPCC approves Climate Change 2014: Impacts Adaptation and Vulnerability and Climate Change Mitigation of Climate Change, the Working Group II and Working Group III contributions to AR5. The Fifth Assessment Report was completed in November 2014 with the Synthesis Report.
2018	Special Report on Global Warming of 1.5 °C (SR15) – "Global Warming of 1.5 °C, an IPCC special report on the impacts of global warming of 1.5 °C above pre-industrial levels and related global greenhouse gas emission pathways, in the context of strengthening the global response to the threat of climate change, sustainable development, and efforts to eradicate poverty"

(*Continued*)

Table 1.2 *Continued*

2019	Special Report on climate change and land (SRCCL) – "Special Report on climate change, desertification, land degradation, sustainable land management, food security, and greenhouse gas fluxes in terrestrial ecosystems"
2019	Special Report on the Ocean and Cryosphere in a Changing Climate (SROCC)-report concluded that sea level rises could be up to two feet higher by the year 2100, even if efforts to reduce greenhouse gas emissions and to limit global warming are successful; coastal cities across the world could see so-called "storm[s] of the century" at least once a year.
2022	Will come Sixth Assessment Report

Source: (IPCC 2019, Homepage).

Climate Change. What is it about? *How did it all happen? Why now it is an emergency? Is Climate Change an alarming challenge and a major threat to humanity? Why Inter-governmental Panel on Climate Change (IPCC) is an important organization? Why this book "Getting the Facts Right: The IPCC and the Role of Science in Managing Climate Change" is important?* These forwards present the overall scenarios or climate change challenges, appropriately with the imperial evidences. They have developed their arguments firstly, what the available information, sources say about the current climate change crisis, secondly, they provide their own opinions and thirdly, they conclude why this book "Getting the Climate Science Facts Right: The Role of the IPCC" is important, and how book adds the knowledge in the climate science. Forwards also elaborate about the authors quality, reliability and validity by illustrating authors previous contributions to the science and humanity. These forwards agree that, every individual themselves (or their ancestors) are directly or indirectly responsible for the current global environment crisis; therefore, each of us have responsibility to contribute to minimize the impact of climate change challenges.

In addition to nine forwards, book contents seven endorsements, from the prominent scholars of climate science. Each endorsement writers have evaluated the manuscript and supported the fact that anthropogenic disturbances to the natural ecosystem is the main cause of rapid climate change. Both forward writers and endorsers support the authors conclusions that IPCC assessments are reliable sources to understand the problems and consequences climate change challenges. Book also present a case study, of climate change impact in the major South Asian cities (air, water, soil pollution).

References

Abell, R. (2007), “Unlocking the potential of protected areas for freshwaters.” Biological Conservation 134: 28–63.

Agrawala, S. (1998), “Context and Early Origins of the Intergovernmental Panel on Climate Change”. Climatic Change. 39 (4): 605–620. doi:10.1023/A:1005315532386.

Agrawala, S. (1998), “Structural and Process History of the Intergovernmental Panel on Climate Change”. Climatic Change. 39 (4): 621–642. doi:10.1023/A:1005312331477.

Aguirre, Benigno. (2002), Cultural Dimensions of Global Environmental Change “Sustainable Development” as Collective Surge’. Social Science Quarterly. 83:1: 101–118.

American Geophysical Union (2003), Human Impacts on Climate, American Geophysical Union (December 2003)

Antilla, Liisa. (2005), Media Analysis and Climate Change ‘Climate of Skepticism: US Newspaper Coverage of the Science of Climate Change’. Global Environmental Change. 15:4: 338–352.

Arctic Council (2004), Impacts of a Warming Arctic: Arctic Climate Impact Assessment, (November 2004) For more information: Arctic Climate Impact Assessment website

Bachelet, D., J. Lenihan, et al. (2000), “Interactions between fire, grazing, and climate change at Wind Cave National Park, SD.” Ecological Modelling: 299–244.

Barnett T.P. et al., (2005), Ocean Warming – The Smoking Gun of Global Warming, Science 2005 309:284; (June 2, 2005)

Bentz, B. J., J. Regniere, et al. (2010), “Climate change and bark beetles of the western United States and Canada: direct and indirect effects.” Bioscience 60(8): 602–613.

Bhandari, MP.(2012), The Intergovernmental Panel on Climate Change (IPCC) [The forth assessment reports on climate change: a factual truth of contemporary world], in Ritzer, George (Ed.) Blackwell Encyclopedia of Globalization, Wiley-Blackwell Publication Volume 3, 1063–1067.

Bhandari, MP.(2019), Live and let other live – the harmony with nature /living beings-in reference to sustainable development (SD)- is contemporary world’s economic and social phenomena is favorable for the sustainability of the planet in reference to India, Nepal, Bangladesh, and

Pakistan? Adv Agr Environ Sci. (2019);2(1): 37-57. DOI: 10.30881/aaeoa.00020 http://ologyjournals.com/aaeoa/aaeoa_00020.pdf

Bhandari, MP.(2019), "BashudaivaKutumbakkam"– The entire world is our home and all living beings are our relatives. Why we need to worry about climate change, with reference to pollution problems in the major cities of India, Nepal, Bangladesh and Pakistan. Adv Agr Environ Sci. (2019);2(1): 8-35. DOI: 10.30881/aaeoa.00019 http://ologyjournals.com/aaeoa/aaeoa_00019.pdf

Bhandari, MP. (2018), Green Web-II: Standards and Perspectives from the IUCN, Published, sold and distributed by: River Publishers, Denmark/the Netherlands ISBN: 978-87-70220-12-5 (Hardback) 978-87-70220-11-8 (eBook). http://www.riverpublishers.com/book_details.php?book_ id=568

Bhandari, MP. (2018), "Climate Change Impacts on Food Security, a Brief Comparative Case Study of Bangladesh, India, Nepal and Pakistan". Acta Scientific Agriculture 2.8 (2018): 136–140. https://www.actascientific.com/ASAG/pdf/ASAG-02-0157.pdf

Both, C. et al., (2006), Further Evidence for Extinctions Due to Global Warming, Nature, 441, 81 (May 4, 2006)

Botkin, D. B. (2001), "The naturalness of biological invasions." Western North American Naturalist 61(3): 261–266.

Bradley, B. A., D. Blumenthal, *et al.* (2009), "Predicting plant invasions in an era of global change." Trends in Ecology & Evolution 25(5).

Brian C.; Webster, Mort; Agrawal, Shardul (2008), "The limits of consensus". In Donald Kennedy and the Editors of Science (ed.). Science Magazine's State of the Planet 2008-2009: with a Special Section on Energy and Sustainability. ISBN 9781597264051.

Brinkhuis, H.S. et al., (2006), Are We Underestimating Global Warming? Nature, 441

British Antarctic Survey (2002), Satellite Spies on Doomed Antarctic Ice Shelf, British Antarctic Survey, National Snow and Ice Data Center (March 2002)

Brower, Michael and Warren Leon, (1999), The Consumer's Guide to Effective Environmental Choices: Practical Advice from The Union of Concerned Scientists, New York: Three Rivers Press

Cazenave, A. (2006), Ice Sheet Melt Concerns and Confounds Scientists, Science. Nov. 24, 2006

Chen, J., et al. (2006), Ice Sheet Melt Concerns, Science. Sept. 29, 2006

Christy J.R. and R.W. Spencer, (2005), It's Official: Satellite and Surface Temperature Records Agree, Science, 310, 972 (November 11, 2005)

Cicerone, Ralph J.; Barron, Eric J.; Dickinson, Robert E.; Fung, Inez Y.; Hansen, James E.; Karl, Thomas R.; Lindzen, Richard S.; McWilliams, James C.; Rowland, F. Sherwood; Sarachik, Edward S.; Wallace, John M.; Turekian, Vaughan C. (2001), "Assessing Progress in Climate Science". Climate Change Science: An Analysis of Some Key Questions. National Academies Press. Committee on the Science of Climate Change, Division on Earth and Life Studies, National Research Council. ISBN 978-0-309-07574-9.

Climatic Change (2004), The Effects of Climate Change on Water Resources in the West, Climatic Change 62 (1-3): 1–11(January 2004)

Dauncey, Guy and Patrick Mazza, (2001), Stormy Weather: 101 Solutions to Global Climate Change, Gabriola Island, BC: New Society Publishers.

Davis, Lee. (1998), Environmental Disasters: A Chronicle of Individual, Industrial, and Governmental Carelessness. New York: Facts on File, 126–128.

Davis, Sean (2017), Can We Solve Global Warming? Lessons from How We Protected the Ozone Layer. TED, 2017, www.ted.com/talks/sean_davis_can_we_solve_global_warming_lessons_from_how_we_protected_the_ozone_layer.

Diffenbaugh et al., (2005), Putting a Fine Point on Climate Extremes, Proceedings of the National Academy of Sciences102, 15774, (November 1, 2005)

Dukes, J. S. and H. Mooney (1999), "Does global change increase the success of biological invaders?" Trends in Ecology & Evolution 14(4).

EIA (2005), Emissions of Greenhouse Gases in the United States 2004, (December 2005) and EIA, Annual Energy Outlook 2006 (December 2005)

Emanuel, K. (2005), Warm Seas Fuel More Destructive Storms, Nature (online) (July 31, 2005)

Feely et al., (2004), Dissolving Corals, Science 2004 305: 362-366 (July 16, 2004)

Felicetti, L.A., et al. (2003), Yellowstone Area Grizzly Bears Suffer as Whitebark Pines Decline, Canadian Journal of Zoology, Vol. 81: 763-770.

Flannery, Tim (2005), The Weather Makers: How Man is Changing the Climate and What it Means for Life on Earth, New York: Atlantic Monthly Press.

Flannery, Tim. (2005), Climate Change: Popular Non-Fiction the Weather Makers: The History and Future Impact of Climate Change. Melbourne: Text, 2005.

Flannigan, M. D., B. J. Stocks, et al. (2000), "Climate change and forest fires." Science of the Total Environment 262(3)

Folke, C., S. Carpenter, et al. (2004), "Regime shifts, resilience, and biodiversity in ecosystem management." Annual Review of Ecology Evolution and Systematics

Forest, Chris E., Peter H. Stone, and Henry D. Jacoby. (2002), "Human Influence on Climate." Forum for Applied Research and Public Policy 16.4 47–51.

Frelich, L. E. and P. B. Reich (2009), "Wilderness Conservation in an Era of Global Warming and Invasive Species: a Case Study from Minnesota's Boundary Waters Canoe Area Wilderness." Natural Areas Journal 29(4): 385–393.

Fujita, Rod (2013) "5 Ways Climate Change Is Affecting Our Oceans." Environmental Defense Fund, 2013, www.edf.org/blog/2013/10/08/5-ways-climate-change-affecting-our-oceans.

Fulé, P. (2008), "Does it make sense to restore wildland fire in changing climate?" Restoration Ecology 16(4): 526–531.

Galatowitsch, S., L. Frelich, et al. (2009), "Regional climate change adaptation strategies for biodiversity conservation in a midcontinental region of North America." Biological Conservation 142(10): 2012–2022.

Geophysical Research Letters (2003), Offsetting the Radiative Benefit of Ocean Iron Fertilization by Enhancing N2O Emissions, Geophysical Research Letters, vol. 30, no. 24, 2249 (December 2003)

Gershon, David (2006), Low Carbon Diet: A 30 Day Program to Lose 5,000 Pounds, Woodstock, NY: Empowerment Institute.

Ghan S.J. and T. Shippert, (2006), Mountain Snowpack Declines Mean Trouble for the Future of the World's Water Supply, Journal of Climate, 19, 1589 (2006)

Gross, J.L and S. Rayner. (1985), Measuring Culture: A Paradigm for the Analysis of Social Organization. New York: Columbia University Press

Gullison, Raymond E., et al. (2007), Carbon Sinks Reconsidered, Science, Vol. 316: 18. May 2007

Hannah, L. (2008), Protected areas and climate change. Year in Ecology and Conservation Biology 2008. 1134: 201–212

Hansen J. et al., (2005), Ocean Warming – The Smoking Gun of Global Warming, Science 2005 308:1431; (April 28, 2005)

Hansen, J., R. Ruedy, M. Sato and K. Lo (2006), GISS Surface Temperature Analysis, Global Temperature Trends: 2005 Summation, NASA's Goddard Institute for Space Studies, available at: http://data.giss.nasa.gov/gistemp/2005/

Hansen, James, et al. (2006), Global Warming Affects Weather Extremes, Proceedings of the National Academy of Science.

Harvard Medical School (2004), Inside the Greenhouse: The Impacts of CO2 and Climate Change on Public Health in the Inner City, Harvard Medical School (April 2004)

Hayhoe et al., (2004), Emissions Pathways, Climate Change, and Impacts on California Proceedings of the National Academy of Sciences (August 2004)

Heede, Richard (2002), "Cool Citizens: Everyday Solutions to Climate Change," Rocky Mountain Institute, <http://www.rmi.org/images/other/Climate/C02-12_CoolCitizensBrief.pdf>

Heino, J., R. Virkkala, et al. (2009), "Climate change and freshwater biodiversity: detected patterns, future trends and adaptations in northern regions." Biological Reviews 84(1): 3954.

Heller, N. E. and E. S. Zavaleta (2009), "Biodiversity management in the face of climate change: A review of 22 years of recommendations." Biological Conservation 142(1): 14–32.

Hellmann, J. J., J. E. Byers, et al. (2008), "Five potential consequences of climate change for invasive species." Conservation Biology 22(3): 534–543.

Henderson-Sellers, A. (1998), 'Climate Whispers: Media Communication about Climate Change'. Climate Change. 40:3-4: 421–456.

Hines, Sarah J., Linda S. Heath and Richard A. Birdsey (2010), An Annotated Bibliography of Scientific Literature on Managing Forests for Carbon Benefits, United States Department of Agriculture Forest Service Northern Research Station, General Technical Report NRS-57, Published by: U.S. Forest Service, 11 Campus Blvd Suite 200, Newtown Square PA 19073–3294.

Hobbs, R. J. and V. A. Cramer (2008), "Restoration Ecology: Interventionist Approaches for Restoring and Maintaining Ecosystem Function in the Face of Rapid Environmental Change." Annual Review of Environment and Resources 33: 39–61.

Hoegh-Guldberg, O., L. Hughes, et al. (2008), "Assisted colonization and rapid climate change." Science 321(5887): 345–346

Hosansky, David (2001), The Environment A to Z. Washington: Congressional Quarterly Inc., 2001. 106–10.

Hoyos, C.D., et al. (2006), Science, Global Warming Affects, Vol. 312: 94–97.

Inkley, D., M. G. Anderson, et al. (2004), Global climate change and wildlife in North America. K. E. M. Galley. Bethesda Maryland, The Wildlife Society.

Interagency Grizzly Bear Study Team. (2004), Annual Report. Interagency Grizzly Bear Study Team, USGS Northern Rocky Mountain Science Center, Montana State University, Bozeman.

IPCC (2001), Climate Change 2001: The Scientific Basis, Climate Change 2001: Impacts, Adaptation and Vulnerability, Intergovernmental Panel on Climate Change (January and February 2001)

IPCC (2007), IPCC Fourth Assessment Report (AR4), 2007 (http://www.ipcc.ch/)

IPCC (2018), Summary for Policymakers. In: Global warming of 1.5°C. An IPCC Special Report on the impacts of global warming of 1.5°C above pre-industrial levels and related global greenhouse gas emission pathways, in the context of strengthening the global response to the threat of climate change, sustainable development, and efforts to eradicate poverty [V. Masson-Delmotte, P. Zhai, H. O. Pörtner, D. Roberts, J. Skea, P. R. Shukla, A. Pirani, W. Moufouma-Okia, C. Péan, R. Pidcock, S. Connors, J. B. R. Matthews, Y. Chen, X. Zhou, M. I. Gomis, E. Lonnoy, T. Maycock, M. Tignor, T. Waterfield (eds.)]. World Meteorological Organization, Geneva, Switzerland, 32 pp.

IPCC AR4 SYR (2007). Core Writing Team; Pachauri, R.K.; Reisinger, A. (eds.). Climate Change 2007: Synthesis Report (SYR). Contribution of Working Groups I, II and III to the Fourth Assessment Report (AR4) of the Intergovernmental Panel on Climate Change. Geneva, Switzerland: IPCC. ISBN 978-92-9169-122-7.

IPCC AR4 WG1 (2007). Solomon, S.; Qin, D.; Manning, M.; Chen, Z.; Marquis, M.; Averyt, K.B.; Tignor, M.; Miller, H.L. (eds.). Climate Change 2007: The Physical Science Basis. Contribution of Working Group I to the Fourth Assessment Report of the Intergovernmental Panel on Climate Change. Cambridge University Press. ISBN 978-0-521-88009-1. (pb: 978-0-521-70596-7)

IPCC AR4 WG2 (2007). Parry, M.L.; Canziani, O.F.; Palutikof, J.P.; van der Linden, P.J.; Hanson, C.E. (eds.). Climate Change 2007: Impacts, Adaptation and Vulnerability. Contribution of Working Group II to the Fourth Assessment Report of the Intergovernmental Panel on Climate Change. Cambridge University Press. ISBN 978-0-521-88010-7. (pb: 978-0-521-70597-4)

IPCC AR4 WG3 (2007). Metz, B.; Davidson, O.R.; Bosch, P.R.; Dave, R.; Meyer, L.A. (eds.). Climate Change 2007: Mitigation of Climate Change. Contribution of Working Group III to the Fourth Assessment Report of the Intergovernmental Panel on Climate Change. Cambridge University Press. ISBN 978-0-521-88011-4. (pb: 978-0-521-70598-1).

IPCC AR5 WG1 (2013), Stocker, T.F.; et al. (eds.), Climate Change 2013: The Physical Science Basis. Working Group 1 (WG1) Contribution to the Intergovernmental Panel on Climate Change (IPCC) 5th Assessment Report (AR5), Cambridge University Press. Climate Change 2013 Working Group 1 website.}}

IPCC AR5 WG2 A (2014), Field, C.B.; et al. (eds.), Climate Change 2014: Impacts, Adaptation, and Vulnerability. Part A: Global and Sectoral Aspects. Contribution of Working Group II (WG2) to the Fifth Assessment Report (AR5) of the Intergovernmental Panel on Climate Change (IPCC), Cambridge University Press, Archived from the original on 28 April 2016. Archived

IPCC AR5 WG3 (2014), Edenhofer, O.; et al. (eds.), Climate Change 2014: Mitigation of Climate Change. Contribution of Working Group III (WG3) to the Fifth Assessment Report (AR5) of the Intergovernmental Panel on Climate Change (IPCC), Cambridge University Press, Archived from the original on 29 October 2014. Archived

IPCC SAR WG1 (1996). Houghton, J.T.; Meira Filho, L.G.; Callander, B.A.; Harris, N.; Kattenberg, A.; Maskell, K. (eds.). Climate Change 1995: The Science of Climate Change. Contribution of Working Group I to the Second Assessment Report of the Intergovernmental Panel on Climate Change. Cambridge University Press. ISBN 978-0-521-56433-5. (pb: 0-521-56436-0) pdf.

IPCC SAR WG3 (1996). Bruce, J.P.; Lee, H.; Haites, E.F. (eds.). Climate Change 1995: Economic and Social Dimensions of Climate Change (PDF). Contribution of Working Group III (WG3) to the

Second Assessment Report (SAR) of the Intergovernmental Panel on Climate Change (IPCC). Cambridge University Press. ISBN 978-0-521-56051-1. (pb: 0-521-56854-4)

IPCC SRES (2000), Nakićenović, N.; Swart, R. (eds.), Special Report on Emissions Scenarios: A special report of Working Group III of the Intergovernmental Panel on Climate Change, Cambridge University Press, ISBN 978-0-521-80081-5, archived from the original (book) on 3 February 2017, 978-052180081-5 (pb: 0-521-80493-0, 978-052180493-6). Also published in English (html) (PDF) on the IPCC website. Summary for policymakers available in French, Russian, and Spanish.

IPCC TAR SYR (2001). Watson, R. T.; the Core Writing Team (eds.). Climate Change 2001: Synthesis Report. Climate Change 2001: Synthesis Report. Contribution of Working Groups I, II, and III to the Third Assessment Report of the Intergovernmental Panel on Climate Change. Cambridge University Press. Bibcode:2002ccsr.book.....W. ISBN 978-0-521-80770-8. (pb: 0-521-01507-3)

IPCC TAR WG1 (2001). Houghton, J.T.; Ding, Y.; Griggs, D.J.; Noguer, M.; van der Linden, P.J.; Dai, X.; Maskell, K.; Johnson, C.A. (eds.). Climate Change 2001: The Scientific Basis. Contribution of Working Group I to the Third Assessment Report of the Intergovernmental Panel on Climate Change. Cambridge University Press. ISBN 978-0-521-80767-8. Archived from the original on 30 March 2016. (pb: 0-521-01495-6)

IPCC TAR WG2 (2001). McCarthy, J. J.; Canziani, O. F.; Leary, N. A.; Dokken, D. J.; White, K. S. (eds.). Climate Change 2001: Impacts, Adaptation and Vulnerability. Contribution of Working Group II to the Third Assessment Report of the Intergovernmental Panel on Climate Change. Cambridge University Press. ISBN 978-0-521-80768-5. Archived from the original on 14 May 2016. (pb: 0-521-01500-6)

IPCC TAR WG3; Davidson, Ogunlade; Swart, Rob; Pan, Jiahua (2001), Metz, B.; Davidson, O.; Swart, R.; Pan, J. (eds.), "Climate Change 2001: Mitigation", Climate Change 2001: Mitigation, Contribution of Working Group III to the Third Assessment Report of the Intergovernmental Panel on Climate Change, Cambridge University Press, Bibcode:2001ccm..book.....M, ISBN 978-0-521-80769-2, archived from the original on 27 February 2017 (pb: 0-521-01502-2)

ITGP (2010), THIRTY-SECOND SESSION OF THE IPCC, held in Busan, (IPCC-XXXII/INF. 4 (27.IX.2010)). Review of The IPCC Processes

And Procedures: Notes on the Informal Task Group on Procedures (ITGP) (PDF), IPCC, Archived from the original on 6 July 2013, retrieved 1 February 2014.

Jennings, Lane. (2002), "Climate Change: What we can do." The Futurist 36.1

Jones, P. D.; Briffa, K. R.; Osborn, T. J.; Lough, J. M.; Van Ommen, T. D.; Vinther, B. M.; Luterbacher, J.; Wahl, E. R.; Zwiers, F. W.; Mann, M. E.; Schmidt, G. A.; Ammann, C. M.; Buckley, B. M.; Cobb, K. M.; Esper, J.; Goosse, H.; Graham, N.; Jansen, E.; Kiefer, T.; Kull, C.; Kuttel, M.; Mosley-Thompson, E.; Overpeck, J. T.; Riedwyl, N.; Schulz, M.; Tudhope, A. W.; Villalba, R.; Wanner, H.; Wolff, E.; Xoplaki, E. (2009), "High-resolution palaeoclimatology of the last millennium: a review of current status and future prospects" The Holocene. 19 (1): 3–49. Bibcode:2009Holoc.19....3J. doi:10.1177/0959683608098952.

Johannessen, Jon-Arild (2018), Automation, Innovation and Economic Crisis: Surviving the Fourth Industrial Revolution, Routledge, 2018. ProQuest Ebook Central, https://ebookcentral.proquest.com/lib/nyulibrary-ebooks/detail.action?docID=5394166.

Joint Science Academics (2005), Science Academies Call for Prompt Action, (June 7, 2005) For more information: Joint science academies' statement

Jones, P. D.; Briffa, K. R.; Barnett, T. P.; Tett, S. F. B. (1998), "High-resolution palaeoclimatic records for the last millennium: interpretation, integration and comparison with General Circulation Model control-run temperatures", The Holocene, 8 (4): 455–471, Bibcode:1998 Holoc...8..455J, doi:10.1191/095968398667194956

Karl, T.R. et al., eds. (2006), A Report by the Climate Change Science Program and the Subcommittee on Global Change Research, (2006)

Kerr, J. (2006), It's Official: Satellite and Surface Temperature Records Agree, Science, 312, 825 (May 12, 2006)

King, Sir David (2004), Chief Scientific Adviser to the U.K. Government; Address to the AAAS: Global Warming: The Imperatives for Action from the Science of Climate Change, (February 2004)

Knutson and Tuleya, (2004), Impact of Global Warming on Hurricane Intensity, Journal of Climate Vol. 17, No. 18, (September 15, 2004)

Kolbert, Elizabeth (2006), Field Notes from a Catastrophe: Man, Nature, and Climate Change, New York: Bloomsbury.

Lambeck, K. (2007), Science Policy: On the edge of global calamity, Canberra: Australian Academy of Science, archived from the original on 6 August 2013.

Lawrence D. and A. Slater, (2006), A Tipping Point at the Poles? Stanford EE Computer Systems Colloquium (April 5, 2006)

Lawrence D.M. and A.G. Slater, (2005), Arctic Meltdown, Geophysical Research Letters 32, L24401 (December 17, 2005)

Linden, Eugene (2006), The Winds of Change: Climate, Weather and the Destruction of Civilizations, New York: Simon and Schuster.

Linder, Stephen H. (2006), 'Cashing in on Risk Claims: On the For-Profit Inversions of Signifiers for "Global Warming"'. Social Semiotics. 16:1: 103–133.

Luthcke, S., et al. (2006), Ice Sheet Melt Concerns, Science. Nov. 24, 2006

Lynas, Mark. (2004), High Tide: The Truth About Our Climate Crisis. New York: Picador

M.E. Mann and K.A. Emanuel, (2006), Satellite and Surface Temperature Records, Eos, 87 (24), 233 (June 13, 2006)

Machine, Heartless. (2017), "When Will the Planet Be Too Hot for Humans? Much, Much Sooner Than You Imagine." Intelligencer, 10 July 2017, nymag.com/intelligencer/2017/07/climate-change-earth-too-hot-for-humans.html.

Macnaghten, Phil and Michael Jacobs. (1997), 'Public Identification with Sustainable Development: Investigating Cultural Barriers to Participation'. Global Environmental Change. 7:1: 5–24.

Magistro, J. and C. Roncoli. (2001), 'Anthropological Perspectives and Policy Implications of Climate Change Research'. Climate Research. 19:2: 91–96.

Mainka, S. and G. Howard (2010), "Climate change and invasive species: double jeopardy." Integrative Zoology 5: 102–111.

Mann, Michael E.; Bradley, Raymond S.; Hughes, Malcolm K. (1999). "Northern hemisphere temperatures during the past millennium: Inferences, uncertainties, and limitations". Geophysical Research Letters. 26 (6): 759–762. Bibcode:1999GeoRL.26..759M. doi:10.1029/1999 GL900070.

Mattson, D.J. (1998), Grizzly Bears, Ursus. Vol. 10: 129–138.

Mattson, D.J., B.M. Blanchard and R.R. Knight (1992), Grizzly Bears, Journal of Wildlife Management, Vol. 56: 432–442.

McComas, Katherine and Shanahan, James. (1999), 'Telling Stories about Global Climate Change: Measuring the Impact of Narratives on Issue Cycles'. Communication Research. 26:1: 30–57.

McCright, Aaron and Riley Dunlap. (2000) 'Challenging Global Warming as a Social Problem: An Analysis of the Conservative Movement's Counter-Claims'. Social Problems. 47:4: 499–522.

Mears C.A. and F.J. Wentz, (2005), Satellite and Surface Temperature Records Reconciled,Science (August 11, 2005)

Meier, M., et al. (2007), Climate Change, Science, Vol. 317: 1064. Aug. 24, 2007

Meinshausen, M. (2006), in H. Schellnhuber, et al., (eds.) Avoiding Dangerous Climate Change, Cambridge University Press (2006)

Meinshausen, M.; et al. (2011), "The RCP greenhouse gas concentrations and their extensions from 1765 to 2300 (open access)", Climatic Change, 109 (1–2): 213–241, doi:10.1007/s10584-011-0156-z.

Millar, C. I., N. L. Stephenson, et al. (2007), "Climate change and forests of the future: Managing in the face of uncertainty." Ecological Applications 17(8): 2145–2151.

Monnett C., J.S. Gleason, and L.M. Rotterman. (2005), Arctic Meltdown, Presentation at the Society for Marine Mammalogy 16th Biennial Conference on the Biology of Marine Mammals(December 12–16, 2005)

Moran, K.J. et al., (2006), Are We Underestimating Global Warming? Nature, 441 (June 1, 2006)

NASAC (2007), Joint statement by the Network of African Science Academies (NASAC) to the G8 on sustainability, energy efficiency and climate change (PDF), Nairobi, Kenya: NASAC Secretariat, archived from the original (PDF) on 1 May 2014, retrieved 10 September 2013. Statement website.

National Academy of Sciences (2001), Climate Change Science: An Analysis of Some Key Questions, National Academy of Sciences (June 2001)

National Assessment Synthesis Team (2000), Climate Change Impacts on the United States, National Assessment Synthesis Team, (December 2000)

National Climate Data Center (2001), Climate of 2001 – Annual Review, National Climate Data Center

National Research Council, (2006), The Hockey Team Keeps Winning, The National Academies Press, (2006)

National Snow and Ice Data Center (2005), press release, Arctic Meltdown, (September 28, 2005)

National Snow and Ice Data Center, NASA and University of Washington (2005), Sea Ice Decline Intensifies, A Future without Summer Sea Ice in the Arctic? joint press release (September 28th, 2005)

Nature (2004), Climatology: Threatened Loss of the Greenland Ice-Sheet, Nature 2004 428: 616 (April 2004)

Nature (2004), Extinction Risk from Climate Change, Nature 2004 427:145–148 (January 2004)

Nature (2004), Satellite Data Confirms Climate Change, Nature 2004 429:7 (May 2004)

NOAA (2007), 2006 Annual Climate Review Summary, updated June 21, 2007, NOAA, National Climatic Data Center, available at: http://www.ncdc.noaa.gov/oa/climate/research/2006/perspectives.html

Noss, R. (2001), "Beyond Kyoto: Forest management in a time of rapid climate change." Conservation Biology 15(3): 578–590.

Oppenheimer, Michael; O'Neill, Brian C.; Webster, Mort; Agrawal, Shardul (2007), "Climate Change, The Limits of Consensus". Science. 317 (5844): 1505–06.

Oreskes, Naomi, and Erik Conway (2014), The Collapse of Western Civilization: A View from the Future, Columbia University Press, 2014. ProQuest Ebook Central, https://ebookcentral.proquest.com/lib/nyu library-ebooks/detail.action?docID=1684958.

Orr, Matthew. (2003), 'Environmental Decline and the Rise of Religion'. Zygon. 38:4: 895–910.

Osborn T.J. and K.R. Briffa (2006), Hockey Team Keeps Winning,Science, 311, 841 (February 10, 2006)

Osborn, T. J.; Briffa, K.R. (2006). "The Spatial Extent of 20th-Century Warmth in the Context of the Past 1200 Years". Science. 311 (5762): 841–844. Bibcode:2006Sci... 311..841O. doi:10.1126/science. 1120514. ISSN 0036-8075.

Otto-Bliesner B. L., et al., (2006), A Tipping Point at the Poles? Science, 311, 1751 (March 24, 2006)

Overpeck, J. T. et al., (2005), Arctic Meltdown, EosEos 86, 309 (August 23, 2005)

Overpeck, J.T. et al., (2005), A Future without Summer Sea Ice in the Arctic? EOS, 86, 309

Overpeck, J.T. et al., (2006), A Tipping Point at the Poles? Science, 311, 1747 (March 24, 2006)

PAGES 2k Consortium (2013), "Continental-scale temperature variability during the past two millennia" (PDF), Nature Geoscience, 6 (5): 339–346, Bibcode:2013 NatGe...6..339P, doi:10.1038/ngeo1797 (78 researchers, corresponding author Darrell S. Kaufman).

Parmesan, C. (2006), "Ecological and evolutionary responses to recent climate change." Annual Review of Ecology Evolution and Systematics 37: 637–669.

Parmesan, C., and G. Yohe (2003), Polar Bears and Other Species Stressed from Global Warming, Nature. Jan. 2, 2003

Parson, E.; et al. (2007), Global Change Scenarios: Their Development and Use. Sub-report 2.1B of Synthesis and Assessment Product 2.1 by the U.S. Climate Change Science Program and the Subcommittee on Global Change Research, Washington, DC., USA: Department of Energy, Office of Biological & Environmental Research, archived from the original on 30 June 2013

Patz, et al. (2004), Heat Advisory: How Global Warming Causes More Bad Air Days, Science, (July 2004)

PBL (2010), Assessing an IPCC assessment. An analysis of statements on projected regional impacts in the 2007 report. A report by the Netherlands Environmental Assessment Agency (PBL) (PDF), Bilthoven, Netherlands: PBL. Report website.

PBL; et al. (2009), News in climate science and exploring boundaries: A Policy brief on developments since the IPCC AR4 report in 2007. A report by the Netherlands Environmental Assessment Agency (PBL), Royal Netherlands Meteorological Institute (KNMI), and Wageningen University and Research Centre (WUR) (PDF), Bilthoven, Netherlands: PBL, archived from the original (PDF) on 1 May 2014. Report website.

Pearlstine, L., E. Pearlstine, et al. (2010), "A review of the ecological consequences and management implications of climate change for the Everglades." Journal of the North American Benthological Society 29(4): 1510–1526.

Pendergraft, Curtis. (1998), 'Human Dimensions of Climate Change: Cultural Theory and Collective Action'. Climatic Change. 39: 643–666.

Pittock, A Barrie. (2005), Climate Change: Turning up the Heat. Collingwood, Victoria: CSIRO Publishing

Pollack, H. N.; Huang, S.; Shen, P.Y. (1998), "Climate Change Record in Subsurface Temperatures: A Global Perspective", Science, 282

(5387): 279–281, Bibcode:1998Sci...282..279P, doi:10.1126/science.282.5387. 279

Pounds J.A. et al., (2006), Further Evidence for Extinctions Due to Global Warming, Nature, 439, 161 (January 12, 2006)

Proctor, James. (1998), 'The Meaning of Global Environmental Change – Retheorizing Culture in Human Dimensions Research'. Global Environmental Change. 8:3: 227–248.

Proctor, James. (1998), 'The Meaning of Global Environmental Change – Retheorizing Culture in Human Dimensions Research'. Global Environmental Change. 8:3: 239.

Rahmstorf, S. (2007), Climate Change problem, Science. Jan. 19, 2007 (http://www.pik-potsdam.de/ stefan/)

Rahmstorf, S., et al. (2007), Climate Change impact, Science. May 4, 2007

Raworth, Kate et. al, (2007), "Adapting to Climate Change: What's Needed in Poor Countries and Who Should Pay," Oxfam International <http://www.oxfam.org/en/policy/briefingpapers/bp104_climate_change_0705>

Revkin, A. (2005), "Errors Cited in Assessing Climate Data," New York Times(August 12, 2005)

Rignot E. and P. Kanagaratnam, (2006), A Tipping Point at the Poles? Science, 311, 986 (March 24, 2006)

Rignot et al., (2004), Melting Ice Caps, Geophysical Research Letters, Vol. 31, No. 18, L18401 (September 22, 2004)

Rive, N.; Jackson, B.; Rado, D.; Marsh, R. (2007). "Complaint to Ofcom Regarding "The Great Global Warming Swindle" (final revision)". Ofcom Swindle Complaint website. Also available as a PDF

Rockström J et. al. (2009), Planetary Boundaries: Exploring the Safe Operating Space for Humanity, Ecology and Society 14(2): 32

Rockström J, Karlberg L. The Quadruple Squeeze: Defining the safe operating space for freshwater use to achieve a triply green revolution in the Anthropocene. Ambio 2010; 39: 257–265

Ross, Andrew. (1991), Strange Weather: Culture, Science and Technology in the Age of Limits. New York/London: Verso

Running, S.W. (2006), Wildfires in the West, Science, published in Science Express (July 6, 2006)

Saunders, S., T. Easley, et al. (2007), "Losing ground: western national parks endangered by climate disruption." The George Wright Forum 24(1): 41–81.

Scambos et al., (2004), Melting Ice Caps, Geophysical Research Letters, Vol. 31, No. 18, L18402 (September 22, 2004)

Scheffer, M., V. Brovkin and P.M. Cox, (2006), Are We Underestimating Global Warming? Geophysical Research Letters, 33 L10702 (2006)

Schwartz, C.C., et al. (2005), Grizzly Bears, Wildlife Monographs, Vol. 161.

Science (2000), Causes of Climate Change Over the Past 1,000 Years, Science v. 289:270–277 (July 2000)

Science (2000), Climate Extremes: Observations Modeling and Impacts, Science v. 289: 2068–2074 (September 2000)

Science (2003), Modern Global Climate Change, Science 2003 302: 1719–1723 (December 2003)

Science (2004), As the West Goes Dry, Science 2004 303: 1124–1127 (February 2004)

Science Daily (2007), article summarizing Ecological Society of America presentation. Aug. 7, 2007

Scientific American (2004), Defusing the Global Warming Time Bomb, Scientific American(March 2004)

Siegenthaler U. et al., (2005), The Core of the Problem, Science 310, 1313 (November 25, 2005)

Siegenthaler, U. et al., (2005), Rapid Build Up, Science, 310, 1313 (November 25, 2005)

Simms, Andrew et. al., (2005), "Africa – Up in Smoke?" New Economics Foundation, <http://www.neweconomics.org/gen/uploads/4jgqh545jc4sk055soffcq4519062005184642.pdf>

Sitch, S., et al. (2007), Carbon Sink, Nature. Published online July 25, 2007. doi:10.1038/

Slocum, Rachel. (2004), 'Consumer Citizens and the Cities for Climate Protection Campaign'. Environment and Planning A. 36:5: 763–782.

Sluijs, A. et al., (2006), Are We Underestimating Global Warming? Nature, 441 (June 1, 2006)

Soden, B. et al., (2005), It's the Heat and the Humidity, Science310, 841, (November 4, 2005)

Speth, James Gustave (2004), Red Sky at Morning: America and the Crisis of the Global Environment. New Haven: Yale University Press

Sriver R. and M. Huber, (2006), Satellite and Surface Temperature Records, Geophysical Research Letters, 33 (June 8, 2006)

Stager, Curt (2012), "What Happens AFTER Climate Change?" Nature News, Nature Publishing Group, 2012, www.nature.com/scitable/knowledge/library/what-happens-after-global-warming-25887608.

Stanley Robinson, Kim. (2004), Forty Signs of Rain. New York: Bantam

Stephens, B.B., et al. (2007), Carbon Sinks Reconsidered, Science. June 22, 2007

Sterling, Bruce. (1996), Heavy Weather. New York: Bantam

Stern, N. (2006). "Stern Review Report on the Economics of Climate Change (pre-publication edition)". London, UK: HM Treasury. Archived from the original on 7 April 2010. Retrieved 8 March 2017.

Stirling, I., and C.L. Parkinson (2006), Polar Bears and Other Species Stressed from Global Warming, Arctic. September 2006

Strayer, D. L. and D. Dudgeon (2010), "Freshwater biodiversity conservation: recent progress and future challenges." Journal of the North American Benthological Society 29(1): 344-358.

Stroeve, J.C. et al., (2005), A Future without Summer Sea Ice in the Arctic? Geophysical Research Letters, 32, L04501 (February 25, 2005)

Tans, P. (2006), Rapid Build Up, NOAA/ESRL (2006)

Thomas et al., (2004), Melting Ice Caps, Science 2004 306: 255–258 (October 8, 2004)

Thompson, L.G. et al., (2006), Proceedings of the National Academy of Sciences, 103(28), 10536 (May 12, 2006)

Tilman, David, et al. (2006), Sustainable Biofuels, Science, Vol. 314: 1,598–1,600. doi:10.1126/science.1133306.

Torn M.S. and M.S. and J. Harte, (2006), Are We Underestimating Global Warming? Geophysical Research Letters, 33, L10703 (2006)

Trenberth K.E. and D.J. Shea (2006), Satellite and Surface Temperature Records, Geophysical Research Letters, 33 (June 8, 2006)

UK Royal Society, Climate Change: A Summary of the Science (PDF), London: Royal Society. Report website.

Ulph, Alistair, and David Ulph. (1997), "Global Warming, Irreversibility and Learning." The Economic Journal 107.442: 636–50.

UNEP. (2010), Background paper for XVII Meeting of the Forum of Ministers of Environment of Latin America and the Caribbean, Panamá City, Panamá, 26–30 April 2010, UNEP/LAC-IG.XVII/4, UNEP, Nairobi, Kenya http://www.unep.org/greeneconomy/AboutGEI/WhatisGEI/tabid/29784/Default.aspx.

UNEP. Keeping Track of Our Changing Environment: United Nations Environmental Program From Rio to Rio+20 (1992-2012). Retrieved November 4, 2015, from http://www.unep.org/geo/pdfs/Keeping_Track. pdf.

Ungar, Sheldon. (2000), 'Knowledge, Ignorance and the Popular Culture: Climate Change versus the Ozone Hole'. Public Understanding of Science. 9:3: 297–312.

US NRC (2001), Climate Change Science: An Analysis of Some Key Questions. A report by the Committee on the Science of Climate Change, US National Research Council (NRC), Washington, D.C., USA: National Academy Press, ISBN 978-0-309-07574-9, archived from the original on 5 June 2011

US NRC (2010), America's Climate Choices: Panel on Advancing the Science of Climate Change; A report by the US National Research Council (NRC), Washington, D.C.: The National Academies Press, ISBN 978-0-309-14588-6, archived from the original on 29 May 2014

Velicogna I. and J. Wahr, (2006), A Tipping Point at the Poles? Science, 311, 1754 (March 24, 2006)

Vitousek, Peter (1994), "Beyond Global Warming: Ecology and Global Change." Ecology 75.7 (Oct.1994): 1861-76.

Wahl E. R. and C.M. Ammann, (2007), More Players on the Hockey Team Robustness of the Mann, Bradley, Hughes Reconstruction of Surface Temperatures: Examination of Criticisms Based on the Nature and Processing of Proxy Climate Evidence, Climate Change

Wahl, Eugene R.; Ammann, Caspar M. (2007), "Robustness of the Mann, Bradley, Hughes reconstruction of Northern Hemisphere surface temperatures: Examination of criticisms based on the nature and processing of proxy climate evidence". Climatic Change. 85 (1–2): 33–69. Bibcode:2007ClCh...85...33W. doi:10.1007/s10584-006-9105-7. ISSN 0165-0009.

Warren, R. (2006), in H. Schellnhuber, et al., (eds.) Avoiding Dangerous Climate Change, Cambridge University Press (2006)

Webster P. et al., (2005), Warm Seas Fuel More Destructive Storms, Science 309: 1844; (September 16, 2005)

Weir, Andy (2016), The Martian. Del Rey, 2016.

Westerling, A., et al. (2006), Warming Will Lead to Frequent and More Intense Wildfires in the Western United States, Science, Vol. 313: 940-943. doi:10.1126/science.1128834

Westerling, A.L. et al., (2006), Wildfires in the West,Science, published in Science Express (July 6, 2006)

Whitlock, C., S. L. Shafer, et al. (2003), "The role of climate and vegetation change in shaping past and future fire regimes in the northwestern US and the implications for ecosystem management." Forest Ecology and Management 178: 5-21.

Wilby, R., H. Orr, et al. (2010), "Evidence needed to manage freshwater ecosystems in a changing climate: Turning adaptation principles into practice." Science of the Total Environment 408: 4150–4164

Williams, Jerry. (2003), 'Natural and Epistemological Pragmatism: Democracy and Environmental Problems'. Sociological Inquiry.37:4: 529–544.

WMO (2001), WMO Statement on the Status of the Global Climate in 2001, World Meteorological Organization (December 2001)

World Bank (2007), World Bank Policy Research Working Paper S4136: "The impact of sea level rise on developing countries: a comparative analysis." February 2007

Yang, Andrew (2019). War on Normal People: The Truth about America's Disappearing Jobs and Why, Universal Basic ... Income Is Our Future. Hachette Books.

Yazici, Adnan, Fred Petry, and Curt Pendergraft. (2001), 'Fuzzy Modelling Approach for Integrated Assessments using Cultural Theory'. Turkish Journal of Electrical Engineering. 9:1: 31–42.

Yellowstone Area Grizzly Bears Suffer as Whitebark Pines DeclinePowell, J.A., and J.A. Logan. (2005), "Insect seasonality: circle map analysis of temperature-driven life cycles." Theoretical Population Biology, Vol. 67: 161–179. 2005

Young, O. R. (2002), The institutional dimensions of environmental change: Fit, interplay, and scale. Cambridge, MA: MIT Press.

Zhou, G., et al. (2006), Carbon Sink, Science, Vol. 314: 1,417.

2

Climate Change and Science

This chapter provides a brief overview of the scientific developments regarding climate change, showing how the science evolved slowly, but by the end of the 20th century began to predict problems.

> *"Philosophy of science without history of science is empty; history of science without philosophy of science is blind"* (Imre Lakatos, 1970:91).
>
> *"Knowledge and power meet in one; for where the cause is not known, the effect cannot be produced. Nature to be commanded must be obeyed; and that which in contemplation is as the cause is in operation as the rule"* Francis Bacon (1561–1626)

2.1 The Historical Outline of Climate Change Science

The climate change science has a relatively long history. The word climate is derived from the Greek word Klima, which meant "inclination, the supposed slope of the earth toward the pole." Several editions of The Encyclopedia Britannica make essentially the same statement but add, "or the inclination of the earth's axis" (Longwell, 1954; Fleming, 1998; Bolin, 2007). Vhe concept of science and technology began in the Greek era. Technology, which came through the legacy of the term *"techne,"* was a major product of Greek era. Techne was used to illustrate the rational method to achieve a perceived goal or objective (Dunne, 1997). "For Aristotle, techne was a very particular kind of knowledge...this was a kind of knowledge associated with people who were bound to necessity (Young, 2009: 190). This indicates that during the Greek era, techne was not as we understand it today, but was a research system or knowledge production to attain a certain goal. "Heidegger identifies Plato's articulation of *techne* as the foundation upon which contemporary technology builds" (Tabachnick, 2006). However, Heidegger's notion is concentrated to articulate the anti-technology theses (Zimmerman, 1990), which is largely in academic discussion. Heidegger himself did not accept he was against the technological enhancement[1]. The notion of technology, which developed throughout human history, is backed by the need of human society. In this connection, James Robert Flaming (2004) nicely depicts the situation.

> *Since the dawn of the nineteenth century, the world has experienced at least three major technological revolutions which have transformed the way humans live, work, and play, while providing the technological infrastructure for the development of modern science, and also changing the meaning and experience of the weather and climate. Following an extremely long (but not at all static) period of human experience before 1800 that may be broadly characterized as agricultural or pastoral, the rate of technological change accelerated dramatically with two major technological eras – the industrial and the post-industrial –occurring in the past two centuries. These revolutions have fundamentally altered humanity's interaction with nature, both in the sense of the built environment, which mediates our perceptions of the weather, and in the sense of instruments that transform scientists' ability to observe, analyze, and predict* it (Flaming, 2004: 2)

The climate change detection was only possible through the technological enhancement and use of enhanced tools to detect the change. The Greek began to explore the position of the Earth and atmospheric variation through Geology and Geography (Geology: Greek meaning Earth and its speed and Geography[2]: "ge" for earth and "grapho" for "to write"); from where the exploration of climate variation and change came into the research agenda.

[1] *"I think about what is developing today as biophysics, that in the foreseeable future, we will be in a position to make man in a certain way i.e., to construct him, purely in his organic being, according to the way we need him: skilled and unskilled, intelligent and... stupid. It will come to that!....So, above all, the misunderstanding that I am against technology is to be rejected. I see technology in its essence as a power which challenges man and, in opposition to which, he is not free any longer–that something is being announced here, namely a relationship of Being to man–and that this relationship, which is concealed in the essence of technology, may come to light someday in its undisguised form. I do not know whether it is going to happen! (Conversation 43) (from Tabachnick, David Edward 2006: 104).*

[2] *Tuan, Yi-Fu (1991) A View of Geography, Geographical Review, Vol. 81, No. 1 pp. 99–107, in page 99 state "What is the intellectual character and core of geography? An answer, from a broadly humanist viewpoint, that may satisfy the genuinely curious and literate public*

The concern about the environmental change can be seen from the Greek Era; however, it was only within a certain group of people. The geological and geographical study of the Earth's system paved the ground for research on scenarios of climate variation; these are the oldest disciplines of the academic world. Longwell (1954) examines the root of the geological exploration - the first step in the detection of climate change.

Longwell (1954) states that:

> "*A current textbook of Geology states categorically that the word climate commemorates one of the oldest scientific discoveries, that the earth's axis is inclined to the ecliptic. Which of these two widely different interpretations has the better basis*" (Longwell 1954:355).

He further says that the concept of climate was to identify the Earth's position and originator including Hipparchus, of the second century B.C.; Eratosthenes, of the third century B.C.; Ptolemy of Alexandria (mathematician and astronomer of the second century A.D) and others. The pioneers saw that a change in latitude means a change in atmospheric conditions as well as in length of day....Probably the present meaning of climate developed gradually; but it is significant that divisions into torrid, temperate, and frigid zones still are made formally by parallels of latitude, although climatic belts in the modern sense are highly irregular in form (Longwell, 1954:355).

Change of the atmosphere condition directly relates with what we understand about the global climate today.

> "*Climate change is, by definition, detected when a statistically significant variation in mean climate, or in its variability, persists for an extended period (typically 30 years)*" (UNESCO, 2009: 2); *the climate system is a complex, interactive system consisting of the atmosphere, land surface, snow and ice, oceans and other bodies of water, and living things. The atmospheric component of the climate system most obviously characterizes climate; climate is often defined as 'average weather'. Climate is usually described in terms of the mean and variability of temperature, precipitation*

lies in the definition of the field as the study of the earth as the home of people. Home is the key, unifying word for all the principal subdivisions of geography, because home, in the large sense, is physical, economic, psycho- logical, and moral; it is the whole physical earth and a specific neighborhood; it is constraint and freedom-place, location, and space" http://www.jstor.org/stable/pdfplus/215179.pdf (accessed on 04/02/2010).

> *and wind over a period of time, ranging from months to millions of years (the classical period is 30 years)* (IPCC, 2007: 96).... and *Climate is generally defined as average weather, and as such, climate change and weather are intertwined. Observations can show that there have been changes in weather, and it is the statistics of changes in weather over time that identifies climate change. While weather and climate are closely related, there are important differences* (IPCC, 2007: 104).

To come up with this definition, several scientists have developed various tools and technologies and applied them to compare statistics over centuries to explore the truth that climate change is happening; however, where statistical history is not available, it remains unavoidable. The human civilization developed as a process along with the innovation of science and technology. The following paragraphs provide a brief outline of such a development process in major five geological eras (Flaming 1998).

Paleolithic before 10,000 BCE: human muscle was the energy; stone and animal bone was used for human survival and the artifact was sharp stone. Similarly, in the Eco-techniques era from 10,000 BCE to 1750 AD, people developed the technique to use water and wind as energy. They produced wood, animal leather and cloth to protect the body and developed the Clipper Ship, Water Wheel and Violin. This helped human civilization enter a new way of life such as agriculture, Pastoralists and Artisans.

The Paleo-technic era (1750-1930) holds the fast-forward development scenario in technological enhancement and application. With the use of coal and iron as material and the steam engine as energy, this era accelerated human domination on the natural system. The products of this era are the railroads, factories and bridges, which lead human lives toward a comfortable lifestyle. As a development process, it led to mass production in agriculture and industries. However, as a consequence, the natural environment rapidly deteriorated through the use of chemical fertilizers and pesticides for mass crop production and emissions in the environment through factories.

The era of Neo-techni from 1930 to 2000 followed the same development process in a more sophisticated way. The major energy achievement of this era was Electricity, Oil and Nuclear power. The innovation and application of aluminum and plastic, as well as the development of automobiles and airplanes for transportation has massively changed the human living style (mostly in the Western world). The impact on global climate became most visible in this era. The innovation of high-tech production of toxic chemicals, emissions into the atmosphere and mass consumption of goods and services were the major characteristics of this era.

However, in the 21st century, the Eco-techic era shows some new scenarios on this trend, where people began to realize the serious consequences of climate change caused by anthropogenic activities. At the lowest level, people have been exploring energy sources that will have the least impact on global climate. Investment in renewable energy such as the use of sunlight, wind and water for energy generation are a few examples of this kind. Similarly, the encouragement and initiation of the general public toward the use of biodegradable, recyclable and reusable products is taking place on a small scale but is increasing (e.g. the concept of LEAD certified buildings, the ban of plastic bag use in Bangladesh and other countries). Likewise, searching for the product that has low impact on local and global climate, searching for natural products, using rechargeable products and increases in using compost manure in organic farming are some good steps of the 21st century to minimize the impact on climate change (Flaming, 1998; 2004). There is a growing trend of knowledge about climate change from the general public level to the national and international level.

2.2 The Major Contributor to the Climate Change and Science

In connecting the historical account of climate science, Karl Marcus Kriesel (1968) notes that the writings of Charles Lous Joseph de Secondat Baron De La Brede et de Montesquieu (1689-1755, French jurist and political philosopher) was the first scientist who examined the effect of the physical environment upon societies and political systems produced by it, have been widely interpreted as being environmentalist or deterministic (Kriessel, 1968:557). Kriessel examines Montesquieu's contribution or role in bringing together the physical environment and the nonphysical societal attributes as customs, manners, morals and tradition, which combined, produce a particular political or governmental system (557). Kriessel cites Platt (1948) in explanation of environmental determinism, where he states:

> *Determinism... refers to the idea to that everything in human life is caused inevitably by previous events or conditions. Extreme environmentalism refers to the idea that everything in human life is caused by the natural environment; and mild environmentalism refers to the idea that human life should be viewed as under the*

> *direct influence of the natural environment (Plat, 1948: 126 as cited by Kriessel, 1968: 557).*

Montesquieu's view, which was written in The Spirit of the Laws, summarizes that physical environment has a direct impact on human society. Montesquieu's major focus was to explain how mankind is influence by the biophysical and socio-cultural and politico-economic environment. Kriessel cites Montesquieu as:

> *Mankind are influenced by various causes: by the climate, by the religion, by the laws, by the maxims of government, by precedents, morals, and customs; whence is formed a general spirit of nations. In proposition as, in every country, and one of these causes acts with more force, the other in the same degree are weakened. Nature and the climate rule all most over the savage; customs govern the Chinese; the laws tyrannize in Japan; moral had formerly all their influence at Sparta; maxims of government, and the ancient simplicity of manners, once prevailed at Rome* (Montesquieu cited by Kriessel, 1968: 560).

These theses of Montesquieu as illustrated by Kriesel (1968) clearly indicate that concern of climate change was firmly rooted in the academic and philosophical arena of the 16th and 17th centuries, which took a long time to verify, and only then with the application of technological tools that did not materialize until the 20th century.

Likewise, James Rodger Fleming (1998) provides a brief chronology of how epistemology of climate science developed. Fleming cites (1998:1-16 and in presentation 1-24):

Aristotle: The same parts of the Earth are not always moist or dry, but they change accordingly as rivers come into existence and dry up... The principle and cause of these changes is that the interior of the earth grows and decays, like the bodies of plants and animals.

Theophrastus (successor of Aristotle): If... the winters are more severe, and more snow falls than formerly.... It follows that the monsoon has greater duration. Is it possible for humans to change the climate? Yes! through deforestation and irrigation.

Jean-Baptiste Dubos (l'Abbé Du Bos 1670-1742): Who noted Genius is not born in every climate. Dubos did not directly explain about the biophysical environment of the Earth but provided the relationship of human and climate.

David Hume (1711-1776): As the science of man is the only solid foundation for the other sciences, so the only solid foundation we can give this science itself must be laid on experience and observation (Hume xxxx). Hume believed that the moderation of the climate had been caused by the gradual advance of cultivation in the nations of Europe (Flaming 1998: 18).

> *"Allowing, therefore, this remark (of Du Bos) to be just that Europe is becoming warmer than formally; how can be account for it? Plainly, by no other method, than by supposing that the land us at present much better cultivated, and that the woods are cleared, which formerly threw a shade upon the earth and kept the rays of the sun from penetrating to it".*

Hume had no tools as we have today to measure the impact of warming climate, but his vision of far sightedness certainly gives future tendency of the climate change the world is facing today.

Thomas Jefferson (April 13, 1743 – July 4, 1826 the third President of the United States1801–1809), (Emphasized data collection Climate could be "improved"..."We want... [an index of climate] for all the States, and the work should be repeated once or twice in a century, to show the effect of clearing and culture towards the changes of climate).

> **Thomas Jefferson Fully Aware of Climate Change**
> A change in our climate however is taking place very sensibly. Both heats and colds are becoming much more moderate within the memory even of the middle-aged. Snows are less frequent and less deep. They do not often lie, below the mountains, more than one, two, or three days, and very rarely a week. They are remembered to have been formerly frequent, deep, and of long continuance. The elderly inform me the earth used to be covered with snow about three months in every year. The rivers, which then seldom failed to freeze over in the course of the winter, scarcely ever do so now. This change has produced an unfortunate fluctuation between heat and cold, in the spring of the year, which is very fatal to fruits. From the year 1741 to 1769, an interval of twenty-eight years, there was no instance of fruit killed by the frost in the neighborhood of Monticello. An intense cold, produced by constant snows, kept the buds locked up till the sun could obtain, in the spring of the year, so fixed an ascendency as to dissolve those snows, and protect the buds, during their development, from every danger of

> returning cold. The accumulated snows of the winter remaining to be dissolved all together in the spring produced those overflowing of our rivers, so frequent then, and so rare now. [3]

President Jefferson's contribution on climate science remained the most influential in history; he initiated research on major, basic science including climatology.

> "To you therefore we address our soliciations, and to lesson to you as musch as possible the ambiguities of our project, I will venture even to sketch the sciences which seem useful and practicable for us, as they occur to me while holding my pen. Botany, chemistry, zoology, anatomy, surgery, medicine, natural philosophy, agriculture, mathematics, astronomy, geography, politics, commerce history, ethics, law, arts, fine arts."
> Jeffersion as cited by Fulling, 1945: 266

Jefferson was highly influenced by the European Enlightenment thinkers such as John Locke (1632-1704); Charles-Louis Secondat Montesquieu (1689-1755); Isaac Newton (1642-1727); François Quesnay (1694-1774); Guillaume-Thomas Raynal (1713–1796); Jean-Jacques Rousseau (1712–1778); Anne-Robert-Jacques Turgot (1727–1781); François-Marie Arouet Voltaire (1694–1778) etc. Thomas Jefferson was also one of the contributors to the Enlightenment era from North America along with Benjamin Franklin, James Otis, John Adams and others, who directly or indirectly contributed to the fostering of scientific research.

John Tyndall (1820–1893): Noted in 1859 that IR absorption by trace gases is "a perfectly unexplored field of inquiry. Elementary gases, oxygen, nitrogen, and hydrogen, are almost transparent to radiant heat, More complex molecules, such as H2O, CO2, O3 and hydrocarbons, even in very small quantities, absorb much more strongly than the atmosphere itself....The atmosphere admits of the entrance of the solar heat, but checks its exit; and the result is a tendency to accumulate heat at the surface of the planet. The aqueous vapor constitutes a local dam, by which the temperature at the

[3]Sources: *Thomas Jefferson on Climate Change (15.10.2008) http://xroads.virginia.edu/HYPER/JEFFERSON/ch07.html; Found it in Dr Richard Keen's Global Warming Quiz, via Roger Pielke, Sr.'s Climate Science. http://omniclimate.wordpress.com/2008/10/15/*thomas-jefferson-*on- http://american-conservativevalues.com/blog/2010/03/thomas-jefferson*-fully-aware-of-climate-change/climate-change/

earth's surface is deepened; the dam, however, finally overflows, and we give to space all that we receive from the sun. Changes in the amount of any of the radiatively active constituents of the atmosphere–water vapor, carbon dioxide, ozone, or hydrocarbons–could have produced all the mutations of climate which the researches of geologists reveal..."

Svante Arrhenius (1859–1927) (in *Philosophical Magazine*, 1896): Depicted a model of CO_2 controlling ice ages and interglacials and discussed that the geometric decline in CO_2 causes a linear decrease in temperature. However, industrial emissions were not yet of concern to him. His climate model is often cited, but it is not continuous with modern results or concerns (also cited by Weart, 2003);

> *"In his 1896 paper, Svante Arrhenius laid the foundation for the modern theory of the greenhouse effect and climate change. The paper is required reading for anyone attempting to model the greenhouse effect of the atmosphere and to estimate the resulting temperature change. Arrhenius demonstrates how to build a radiation and an energy balance model directly from observations. Arrhenius was fortunate to have access to Langley's data, which are some of the best radiometric observations ever undertaken from the surface. The successes of Arrhenius' model are many, even when judged by modern-day data and computer simulations: the suggestion of the diffusivity factor including its correct numerical value; the remarkably accurate simulation of the total emissivity of the atmosphere which seem to agree within 5% of modern-day values; the logarithmic dependence of the C02 radioactive heating effect; and others documented in the text....The main goal of this paper was to estimate the surface temperature increase due to an increase in C02. Towards this goal, Arrhenius developed a detailed and quantitative model for the radiation budget of the atmosphere and the surface"* (Ramanathan and Vogelmann, 1997: 38).

2.3 The Pre-modern Development on Climate Science

The modern climate change science begins from the work of Guy Stewart Callendar (1898-1964), who gave the new evidence on how climate change is taking place. His publication of 1939, illustrates the concentrated on rising temperatures, rising fossil fuel consumption, rising CO_2 concentrations and a detailed understanding of IR–The Callendar Effect–Climatic change brought about by anthropogenic increases in the concentration of atmospheric carbon dioxide, primarily through the processes of combustion.

> *"As man is now changing the composition of the atmosphere at a rate which must be very exceptional on the geological time scale, it is natural to seek for the probable effects of such a change. From the best laboratory observations, it appears that the principal result of increasing atmospheric carbon dioxide...would be a gradual increase in the mean temperature of the colder regions of the earth* (G. S. Callendar, 1939 as cited by Flaming, 1998:107 and Flaming, 2007)

Callendar provided new direction to climate science with the evidence from temperature data collected from various metrological stations. He showed the growing trend of global temperature from 1860 to 1940, which are the first comparative presentation of climate change (see annex or Flaming, 1998: 116-117).

Roger Randall Dougan Revelle (1909-1991) (in *Report of the Environmental Pollution Panel,* President's Science Advisory Committee, 1965): By the year 2000, there will be about 25% more CO_2 in our atmosphere than at present. This will modify the heat balance of the atmosphere to such an extent that marked changes in climate, not controllable through local or even national efforts, could occur.

Jule Charney (known as the Charney report) (National Academy of Sciences, *Carbon Dioxide and Climate: A Scientific Assessment* Report, 1979): The consensus has been that increasing carbon dioxide will lead to a warmer earth with a different distribution of climatic regimes. Doubling CO_2 in models results in 1.5 to 4.5 C warming. Positive feedbacks will increase the warming.

A similar situation was presented in the same year's publication of the Science Journal by Carl Sagan, Owen B. Toon and James B. Pollack in an article titled "Anthropogenic Albedo Changes and the Earth's Climate" where they summarized that:

> *"The human species has been altering the environment over large-geographic areas since the domestication of fire, plants, andanimals. The progression from hunter to farmer to technologisthas increased the variety and pace more than the geographicextent of human impact on the environment. A number of regionsof the earth have experienced significant climatic changes closelyrelated*

in time to anthropogenic environmental changes. Plausiblephysical models suggest a causal connection. The magnitudesof probable anthropogenic global albedo changes over the pastmillennia (and particularly over the past 25 years) are estimated.The results suggest that humans have made substantial contributionsto global climate changes during the past several millennia,and perhaps over the past million years; further such changesare now under way (Sagan, Toon and Pollack 1979 Page 1363).

They also presented how anthropogenic activities changed the global environment. The following figure summarizes the scenario of anthropogenic changes in the environment.

As presented above, the environmental problems and issues were well understood by the scientists even before 1900; however, they were largely overlooked. The exploration of the climate change origin, its history, and the contribution of scientists extensively began its focus in the 1950s. There are a large number of publications regarding the concern of global environment by individual authors, groups of authors or in the name of environment conservation. The formation of the International Union for Conservation of Nature (IUCN-1948), and the International Council for Science (ICSU-1931) are exemplary contributors as groups of scientists (further described in chapter three), who extensively contributed to bring climate science to the public agenda. James Rodger Fleming (1998) provides a good historical account of climate change in his book "Historical Perspectives on Climate Change." Similar attempts have been made by the first chair of the Intergovernmental Panel on Climate Change (IPCC) Bert Bolin (2007), who began to write on the climate change issue in the 1960s, with the analysis of carbon dioxide and atmosphere and ended with his latest book entitled "A History of the Science and Politics of Climate Change of 2007," before he passed away. Other such attempts can be found in the work of Mathew Paterson (1996), through his book "Global Warming and Global Politics" and Spencer R. Weart's book on "the Discovery of Global Warming (2008).

The following are milestones on the way to understanding changes of the greenhouse effect of the earth's atmosphere (until 1967 when the first, full, three-dimensional global circulation model study for doubled CO_2 concentration appeared). The table summarizes the scientist's concern in the specific subject on climate change from 1842 as milestone to 1924 and the theoretical shift.

The following table depicts the similar story in different order.

Anthropogenic changes in the environment.		
Epoch	Some motives for environment change	Some (largely) inadvertent environmental changes and examples of their timing
Hunting and gathering	Preparation of land	Deforestation of temperate regions (North America, Native American Creation of grassland to about A.D. 1600; Eastern Europe, production of steppes – preclassical) Deforestation of tropical regions (Africa, Production of savanna–Since discovery of fire)
Agricultural	Expansion of farmland	Destrification (Sahara Arabia region, beginning about 5000 B.C. to present; India, Pakistan, Sumeria, 2000 B.C. to A.D. 400; Peru About A.D. 1200)
	Generation of energy	Deforestation of temperature regions(china, 2000 B.C. to A.D. 500; Western and Central Europe, A.D. 1000 to 1990; United States, A.D. 1800 to 1900)
	Deforestation of tropics (Africa, indonesia, South America, since origin of agriculture)	
Technological	Urbanizatior	Creation of urban heat islands
	Expansion of farmland	Concentration of surface and water pollutants
	Creation of artificial lakes	Alteration of hydrological cycle by farming and irrigation
	Production of synthetic chemical	Destruction of soil by increased erosion
	Generation of energy	Alteration of composition of atmosphere (carbon dioxide, aerosols, smog)
	Production of raw materials	Destruction of natural plant and animal communities (desertification, deforestation, temperte and tropical regions – changes mainly after about 1800)

Source: Adopted from Sagan; Toon and Pollack, (1979:1363).

Milestones continued in different order:

The above tables show that the concern of climate change is not a new phenomenon. As previously discussed, climate science and concern about it primarily began in the Greek era; however, it took hundreds of years to become a fully accepted science. 1972 was the milestone year for the institutionalization of climate science through the first World Conference on Global environment, which recommended establishing the United National

Table 2.1 Climate change theories as classified by Brooks (1950) as cited by Flaming, 1998

Climate change theories	The theorists and scientists
Changes in elements of the Earth's orbit	Adhémar (1842), Croll (1864, 1875), Drayson (1873), Ekholm (1901), Spitaler (1907), Milankovic (1920, 1930, 1941)
Changes of solar radiation	Dubois (1895), Simpson (1930, 1934, 1939–40), Himpel (1937), Hoyle and Lyttleton (1939)
Lunar-solar tidal influences	Pettersson (1914)
Elevation of land masses – mountain building	Lyell (1830–33), Wright (1890), Ramsay (1909–10, 1924), Brooks (1926, 1949)
Changes in atmospheric circulation	Harmer (1901, 1925), Gregory (1908), Hobbs (1926), Flint and Dorsey (1945)
Changes in oceanic circulation	Croll (1875), Hull (1897), Chamberlin (1899), Brooks (1925), Lasareff (1929)
Changes in continent-ocean distribution	Czerney (1881), Harmer (1901, 1925), Gregory (1908), Brooks (1926), Willis (1932)
Changes in atmospheric composition	Arrhenius (1896), Chamberlin (1897, 1899), Ekholm (1901), Callendar (1938, 1939)
Volcanic dust in the atmosphere	Humphreys (1913, 1920), Abbot and Fowle (1913)
Cosmic dust theory	Hoyle and Lyttleton (1939), Himpel (1947)
Sunspot theory	Czerny (1881), Huntington (1915), Huntington and Visher (1922)
Polar migration and continental drift theory	Kreichgauer (1902), Wegener (1920), Köppen and Wegener (1924)

Source: Adopted from Flaming (1998) with permission on 04/03/2010.

Environment Program. Similarly, the Club of the Rome also published its most authentic report "The Limits to Growth" (1972), which draws global attention to the global climate. There is no direct challenge on the research outcome of the Rome Club. The "Limits to Growth" report states that if the present growth trends in world population, industrialization, pollution, food production and resource depletion continue unchanged, the limits to growth on this planet will be reached sometime within the next one hundred years. The most probable result will be a rather sudden and uncontrollable decline in both population and industrial capacity. This was a second shock after Rachel Carson's book *Silent Spring* (1962), which largely drew the attention of the general public regarding the seriousness of global climate change.

Table 2.2 Milestones on climate science

Scientific Breakthroughs	Scientist(s)	Year	Journal/Book	(Book) Title
Indication that the atmosphere acts like a shield that keeps heat in the system	Jean Baptiste Joseph Fourier	1824 1827	*Annales de Chemie et de Physique* 27: 136–167 *Mémoires de l'Académie Royale des Sciences* 7: 569–604	Remarques générales sur les températures du globe terrestre et des espacesplanétaires Mémoire sur les temperatures du globe terrestre et des espacesplanétaires
First full description of the greenhouse effect	John Tyndall	1863	*Proceedings of the Royal Institute of Great Britain* 3: 158	On the transmission of heat of different qualities through gases of different kinds
Deforestation as a cause for climate change	Eduard Brückner	1890	*Klimaschwankungenseit 1700 nebstBemerkungenüber die Klimaschwankungen der Diluvialzeit.* GeographischeAbhandlungen. Edited by E. D. Penck. Vienna: Hölzl	
First warming rate for doubled CO2 concentration	Svante Arrhenius	1896	*Philosophical Magazine* series 5: 237–276	On the influence of carbonic acid in the air upon the temperature of the ground
First full link from CO2 concentration change via gas absorption of heat radiation to surface temperature change	Guy S. Callendar	1938	*Quarterly Journal of the Royal Meteorological Society* 64: 223–240	The artificial production of carbon dioxide and its influence on temperature
First radioactive transfer calculations of an enhanced greenhouse effect	Gilbert N. Plass	1956	*Tellus* 8: 140–154	The carbon dioxide theory of climate change
First equilibrium model study for doubled CO2 concentration	Syukuro Manabe, Fritz Möller	1967	*Monthly Weather Review* 89: 503–532	On the radiation equilibrium and heat balance of the atmosphere

Source: Adopted from Grassl (2007) with permission on 04/03/2010)

"You have proposed that the discussion about climate change and global warming should be divided into three periods. What are these three periods?
The tenor of scientific discourse about climate change in the decades from the second quarter of the nineteenth century to the present has changed considerably and exhibits three major periods. Each of these periods is characterized by one of three phases of scientific effort and its relationship to society. Briefly put, they are a period of hypothesis (up to about 1945), a period of gathering evidence and testing hypotheses (roughly 1945–1975), and a period of controversy over the application of apparent scientific consensus (from perhaps 1975 to present). Of course, all three of these aspects of science continued throughout recent history: theories have been developed and tested all along, but these periods do seem to follow an emergent dialectic, and the public debate on the relationship between science and society has unmistakably intensified in the most recent decades" (interview with J. Donald Hughe - by Georg Götz-2012:49-50).

2.4 The Major Steps to Tackle the Climate Change

Having growing concerns and evidence of global climate change, UNEP continued its consultation with the scientific and government agencies to reach a mutual understanding. As a follow up to previous conferences and public concern, in *1977, at the request of its Governing Council UNEP convened a meeting of experts on the ozone layer which formulated a draft World Plan of Action on the Ozone Layer. The Plan of Action, adopted by the Governing Council at its eighth session, for research into and assessment of the state of the atmospheric ozone layer and the consequences of its modification was to be implemented by UN bodies, specialized agencies, international, national, intergovernmental and non-governmental organizations and scientific institutions* (UNEP, 1986: 4) and in 1979, the first World Climate Conference was organized by the World Metrological Organization (the details regarding the role of organization is discussed in chapter three). As we noted above, the Charney report was also published in 1979 along with the Sagan; Toon and Pollack (1979) groundbreaking research showing anthropogenic action as a major source of climate change. Following up the same sequence *at its seventh session (Geneva, October 1984), the COOL called the attention of all countries and economic organizations not yet reporting production figures for*

CFCs 11 and 12, as well as for other halocarbons, to the need for reporting pertinent chemical production, release and usage data, including the more detailed data on production and uses needed for socioeconomic analyses (UNEP, 1986: 5).

Similarly, in 1985, from October 9 - 15, a joint UNEP/WMO/ICSU Conference was convened in Villach (Austria) to assess the role of increased carbon dioxide and other radioactive constituents of the atmosphere (collectively known as greenhouse gases and aerosols) on climate changes and associated impacts. The conference concluded that the other greenhouse gases reinforce and accelerate the impact because of CO_2 alone. As a result of the increasing concentrations of greenhouse gases, it is now believed that in the first half of the next century, a rise of global mean temperature could occur that is greater than any in man's history (WMO, 1986). This conference pin pointed the need of an international organization to assess the impact of climate change processes.

> However, the need for an international agency was highlighted by Vladimir Zworykin in 1945, which was largely ignored by the scientists. Zworykin states:
> *The eventual goal to be attained is the international organization of means to study weather phenomena as global phenomena and to channel the world's weather, as far as possible, in such a way as to minimize the damage from catastrophic disturbances, and otherwise to benefit the world to the greatest extent by improved climatic conditions where possible (Vladimir Zworykin 1945 as cited by Flaming in The Wilson Quarterly (1976 and 2007: 65).*

As result of this, the Villach conference and several others, the ENEP and WMO established the Intergovernmental Panel on Climate Change (IPCC) in 1988. Since then, the IPCC has been playing an important role in producing such knowledge.

The scientists, internationals organizations, governments and governmental organizations highly acknowledge the role of the IPCC. The United Nations Educational, Scientific and Cultural Organization: UNESCO (2009: 2) states that:

> *Since the establishment of the Intergovernmental Panel on Climate Change (IPCC), the scientific understanding of climate change has much improved. In their Fourth Report, published in 2007: (IPCC, 2007), the IPCC concluded that:*

Global warming is 'undeniable': 11 of the last 12 years were among the 12 warmest years since records have been kept (in 1850).

Almost all of the observed temperature rise during the second half of the twentieth century is 'very probably' due to human activity.

The increase in temperature projected for 2100 will be between 2 and 4.5 °C, with a doubling of the concentrations of carbon dioxide compared to pre-industrial levels. These 'best estimates' are

averages, in a broader range of 1.1 to 6.4 °C (1.4 to 5.8 °C, in the previous report of 2001).

All scenarios predict a reduction of sea ice in the Artic and Antarctica.

It is highly probable that, in the future, extreme weather events such as heat waves and heavy precipitation will become more frequent and that tropical hurricanes will become more intense.

In 2005, the atmospheric concentration of carbon dioxide (the most important greenhouse gas) largely exceeded the concentrations of the past 650,000 years.

Past and future CO2 emissions will continue to contribute to global warming and rising sea levels for more than a millennium (UNESCO, 2009: 2).

Conclusion

In conclusion, climate science (and change) has a long history, which can be traced from Greek society, began to receive attention during the enlightenment era, slowly in 19th century and accelerated in the 20th century. The growth of international scientific agencies dealing with the global environment began to accelerate from the beginning of the 20th century and still continues. The following figure adopted from Meyer et. al (1997) provides the growing trend of international scientific agencies on global environment.

The figure shows the growth of international scientific organizations, 1880–1990

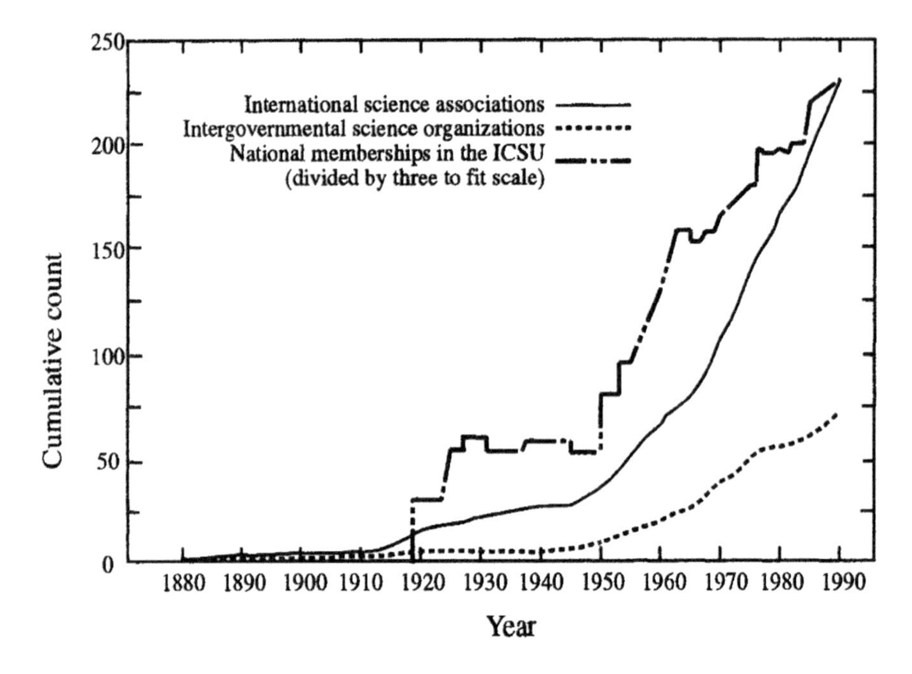

Adopted from Meyer; Frank; Hironaka; Schofer and Tuma (1997: 636)

Meyer, Frank, Hironaka, Schofer and Tuma (1997) have analyzed that trend from 1880 to 1990, which provides a clear indication that the concern regarding global environment is not a new phenomenon. However, the major attention began in the 1950s as we explored in above section.

Since the inception of the IPCC in 1988 and the publication of the first assessment report in 1990, climate change has been seen as common global agenda. The detection of climate change takes several years and is a critical process. Schindler (1999) gives an outline of how complicated the climate change research is.

> *"To a patient scientist, the unfolding greenhouse mystery is far more exciting than the plot of the best mystery novel. But it is slow reading, with new clues sometimes not appearing for several years. Impatience increases when one realizes that it is not the fate of some fictional character, but of our planet and species, which hangs in the balance as the great carbon mystery unfolds at a seemingly glacial pace"* (Schindler, 1999: 6).

Annex 1: Guy Stewart Callendar's graphs on Rising Climate

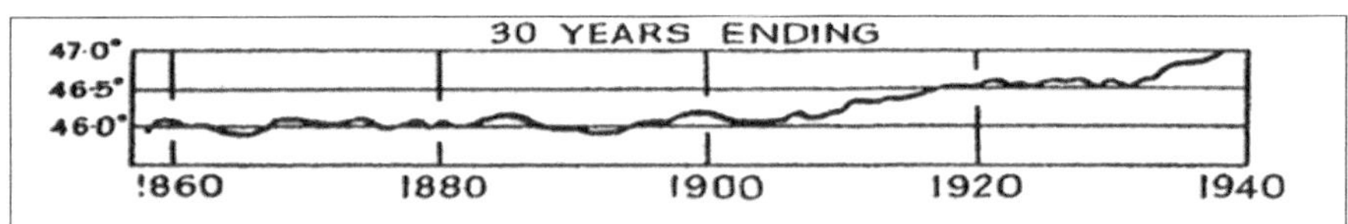

Rising temperatures circa 1858 to 1939. Orginal caption: "The 30 year moving average from the combined means ar Edinburh, Oxford, Greenwich, De Bilt, Bergen. Oslo, Stockholm, Copenhagen, Wilno and Berlin.;; Source: G. S. Callendar, "The Composition of the Atmosphere through the Ages," Meteorol. Mag. 74 (1939)

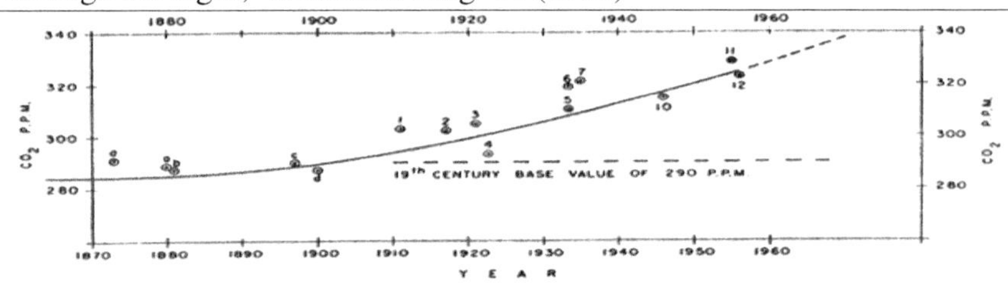

"Amount of CO_2 in the free air of the N. Atlantic region, 1870–1956. Data points indicate individual measurements: the solid line represents the amount from fossil fuel." Source: G. S. Callendar, "On the Amount of Carbon Dioxide in the Atmosphere," Tellus 10 (1958).

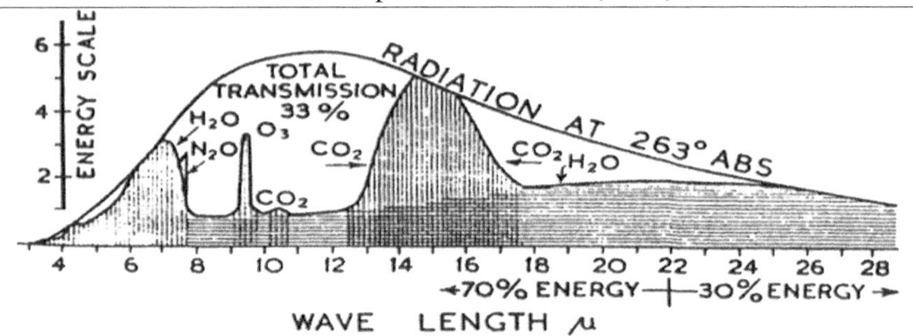

The generalized atmospheric spectrumn in infrared wavelenghts. Source: G. S. Callendar, "Infra-red Absorption by Carbon Dioxide, With Special Reference to Atmospheric Radiation," Quart. J. Roy. Meterorol. Soc. 67 (1941)

Source: Flaming 1998; 116-117).

Endnotes and References

Arrhenius, S., (1896) On the influence of carbonic acid in the air upon the temperature of the ground.*Philosophical Magazine and Journal of Science,* Series 5, Vol. 41, No. 251, April, 237-276.

Bolin, Bert (2007) A History of the Science and Politics of Climate Change, The Role of the Intergovernmental Panel on Climate Change, Cambridge University Press.

Callendar, G. S. (1938) The artificial production of carbon dioxide and its influence on temperature.*Quarterly Journal of the Royal Meteorological Society* 64**:**223-240.

Carbon Dioxide and Climate: A Scientific Assessment Carbon Dioxide and Climate: A Scientific Assessment Report of an Ad Hoc Study Group on Carbon Dioxide and Climate Woods Hole, Massachusetts July 23–27, 1979 to the Climate Research Board Assembly of Mathematical and Physical Sciences National Research Council NATIONAL ACADEMY OF SCIENCES Washington, D.C. 1979: "The Charney Report" (1979). The National Academy of Sciences created a committee of 9 scientists, chaired by pre-eminent MIT atmospheric scientist Jule Charney, to investigate the implications of increasing carbon pollution. The resulting report, Carbon Dioxide & Climate: A Scientific Assessment, is widely cited as being as an impetus for the U.S. Congress, federal agencies, and indeed the international community to take up the issue of climate change. The report authors based their conclusions on a single emissions scenario and five very simple climate models. Total pages: 34. Time to prepare: less than 6 months. Full report is available here http://www.atmos.ucla.edu/~brianpm/download/charney_report.pdf

Carson, Rachel (1962) *Silent Spring*. Boston: Houghton Mifflin Co. and in Rachel Carson, "Silent Spring," in Diane Ravitch, ed., *The American Reader: Words that Moved a Nation* (New York: HarperCollins, 1990), 323-325. http://www.uky.edu/Classes/NRC/381/carson_spring.pdf

David Hume (.........) A Treatise of Human Nature in introduction http://www.creatorix.com.au/philosophy/t06/t06f13.html (accessed on 04/30/2019)

Dunne, Joseph (1997). Back to the Rough Ground: 'Phronesis' and 'Techne' in Modern Philosophy and in Aristotle. Notre Dame, Indiana: University of Notre Dame Press. (ISBN 978-0-2680-0689-1)

Fleming, James R. (2004) Climate dynamics, science dynamics, and technological change, 1804-2004, Conference on International Commission on History Of Meteorology, BarockerBibliothekssaal, Kloster Polling WeilheimerStraße, D-82398 Polling, Germany, http://www.meteohistory.org/2004polling_preprints/docs/polling_program.pdf http://www.meteohistory.org/2004polling_preprints/docs/abstracts/fleming_abstract.pdf (accessed on 04/03/2010)

Fleming, James Rodger (1998), Historical Perspectives on Climate Change, Oxford University Press, New York/Oxford, ISBN: 0-19-507870-5.

Fleming, James Rodger (1998), Historical Perspectives on Climate Change, Oxford University Press, New York/Oxford, ISBN: 0-19-507870-5.

Fleming, James Rodger (2007) The Callendar Effect: The life and work of Guy Stewart Callendar (1898-1964), the scientist who established the carbon dioxide theory of climate change. Boston, American Meteorological Society.

Fleming, James Rodger (2009) Historical Perspectives on Climate Change; Some highlights, with reference of Historical Perspectives on Climate Change and Climate Change and Anthropogenic Greenhouse Warming: A Selection of Key Articles, 1824-1995, with Interpretive Essays, Fixing the Sky: The checkered history of weather and climate control.STS Program, Colby College http://www.lindahall.org/mediafiles/Fleming.pdf (accessed on 04/03/2010)

Fulling, Edmund H. (1945) Thomas Jefferson. His Interest in Plant Life as Revealed in his Writings-II, Bulletin of the Torrey Botanical Club, Vol. 72, No. 3 (May - Jun. 1945), pp. 248-270

Götz, Georg (2012), Global Change, Interviews with Leading Climate Scientists, (SpringerBriefs in Earth System Sciences, DOI: 10.1007/978-3-642-23444-6_1), Springer Heidelberg Dordrecht London New York

Grassl, Hartmut (2007) Guy Stewart Callendar: A Pioneer of Anthropogenic Climate Change Theory, GAIA 16, 222-225, 2007. Max Planck Institute for Meteorology, Hamburg, Germany.

Heidegger, Martin (1993) "The Question Concerning Technology." Basic Writings. Ed. David Farrel Krell. San Francisco: Harper Collins, 311-341

http://www.ipcc.ch/pdf/assessment-report/ar4/wg1/ar4-wg1-chapter1.pdf (accessed on 04/03/2019)

http://www.nap.edu/catalog.php?record_id=12181#toc http://blogs.nwf.org/ (accessed on 04/03/2019) https://secure.ametsoc.org/amsbookstore/reviews/Callendar_GAIA_Review.pdf (accessed on 04/02/2019)

IPCC (2007) full citation is Le Treut, H., R. Somerville, U. Cubasch, Y. Ding, C. Mauritzen, A. Mokssit, T. Peterson and M. Prather, 2007: Historical Overview of Climate Change. In: Climate Change 2007: The Physical Science Basis. Contribution of Working Group I to the Fourth Assessment Report of the Intergovernmental Panel on Climate Change [Solomon, S., D. Qin, M. Manning, Z. Chen, M. Marquis, K.B. Averyt, M. Tignor and H.L. Miller (eds.)]. Cambridge University Press, Cambridge, United Kingdom and New York, NY, USA.

Kriesel, Karl Marcus (1968) Montesquieu: Possibilistic Political Geographer, Annals of the Association of American Geographers, Vol. 58, No. 3 (Sep., 1968), pp.557-574 Published by: Taylor & Francis, Ltd. on behalf of the Association of American Geographers Stable URL: http://www.jstor.org/stable/2561806 (Accessed: 02/04/2010)

Lakatos, Imre (1970) History of Science and Its Rational Reconstructions, Proceedings of the Biennial Meeting of the Philosophy of Science Association, Vol. 1970 (1970), pp. 91-136 Published by: The University. http://www.jstor.org/stable/pdfplus/495757.pdf

Longwell, Chester R. (1954) Origin of the Word Climate, Science, New Series, Vol. 120, No. 3113 (Aug. 27, 1954), p. 355 http://www.jstor.org/stable/pdfplus/1683033.pdf

Meyer, John W.; Frank, David John; Hironaka, Ann; Schofer, Evan and Tuma, Nancy Brandon (1997) The Structuring of a World Environmental Regime, 1870-1990, International Organization, Vol. 51, No. 4 (Autumn, 1997), pp. 623-651. http://www.jstor.org/stable/pdfplus/2703501.pdf

Ramanathan V. and Andrew Vogelmann M. (1997) Greenhouse Effect, Atmospheric Solar Absorption and the Earth's Radiation Budget: From the Arrhenius-Langley Era to the 1990s, Ambio, Vol. 26, No. 1, Arrhenius and the Greenhouse Gases (Feb., 1997), pp. 38–46 http://www.jstor.org/stable/pdfplus/4314548.pdf (accessed on 04.03/2019)

Sagan, Carl; Toon, Owen B.; and Pollack, James B. (1979) Anthropogenic Albedo Changes and the Earth's Climate Science 21 December 1979: Vol. 206. no. 4425, pp. 1363–1368

Schindler, David W. (1999) "The Mysterious Missing Sink" *Nature* 398: 105–106.

Shapin, Steven and Arnold Thackray (1974) Prosopography as a research tool in history of science: The British scientific community, 1700–1900. History of Science 12: 1–28.

Sources: Thomas Jefferson on Climate Change (15.10.2008) http://xroads.virginia.edu/~HYPER/JEFFERSON/ch07.html; Found it in Dr Richard Keen's Global Warming Quiz, via Roger Pielke, Sr.'s Climate Science. http://omniclimate.wordpress.com/2008/10/15/thomas-jeffer son-on- http://american-conservativevalues.com/blog/2010/03/thomas-jefferson-fully-aware-of-climate-change/climate-change/

Tabachnick, David Edward (2006) The Tragic Double Bind of Heidegger's Techne, PhaenEx 1, no. 2 (fall/winter 2006): 94-112; http://www.phaenex.uwindsor.ca/ojs/leddy/index.php/phaenex/article/viewFile/226/233: Tabachnick notes that Generally, the ancient Greek word techne is translated as "craft" or "art" but also "technical knowledge." While sometimes used interchangeably, techne is distinct from episteme, which means "scientific knowledge." Where episteme may be "knowledge for the sake of knowledge," techne is instrumental, oriented towards the deliberate production of something. Furthermore, not only are products wrought via techne different from things produced by nature (physics), but they are also different from things produced by chance (tuche). While something could be made of the etymological meeting of the compound techne and logos (reason) in the modern word technology, the Greek sense of techne already implies the application of reason. Aristotle, for example, defines techne in the Ethics as "a state of capacity to make, involving a true course of reasoning (logos)" (Aristotle 1958, 140a10)". http://www.phaenex.uwindsor.ca/ojs/leddy/index.php/phaenex/article/viewFile/226/233

The Limits to Growth (1972) Abstract established by Eduard Pestel; A Report to the Club of Rome (1972), by Donella H. Meadows, Dennis l. Meadows, Jorgen Randers, William W. Behrens III; http://www.facebook.com/topic.php?uid=9364228327&topic=4478 http://www.bibliotecapleyades.net/sociopolitica/esp_sociopol_clubro me6.htm (accessed on 04/04/2018)

Tuan, Yi-Fu (1991) A View of Geography, Geographical Review, Vol. 81, No. 1 pp. 99-107, in page 99 state "What is the intellectual character and core of geography? An answer, from a broadly humanist viewpoint, that may satisfy the genuinely curious and literate public lies in the definition of the field as the study of the earth as the home of people.

Home is the key, unifying word for all the principal subdivisions of geography, because home, in the large sense, is physical, economic, psycho- logical, and moral; it is the whole physical earth and a specific neighborhood; it is constraint and freedom-place, location, and space" http://www.jstor.org/stable/pdfplus/215179.pdf (accessed on 04/02/2019).

UNEP (1986) UNEP Workshop on Chlorofluorocarbons; Background Factual On Current Production Capacity, Use, Emissions, Trade and Wk&Ent Regulation Of Cfcs Separately By Country And /Or Region: TOPIC I OVERVIEW, Submitted by United Nations Environment Program. www.unep.org/.../ccol8-5-related_activities_to_work_of_cc.86-02-24.doc

UNESCO (2009) The United Nations World Water Development Report 3; Water in a Changing World; Published by the United Nations Educational, Scientific and Cultural Organization, 7 place de Fontenoy, 75352 Paris 07 SP, France. http://unesdoc.unesco.org/images/0018/001818/181893E.pdf (accessed on 04/03/2019)

UNESCO (2009) The United Nations World Water Development Report 3; Water in a Changing World; Published by the United Nations Educational, Scientific and Cultural Organization, 7 place de Fontenoy, 75352 Paris 07 SP, France. http://unesdoc.unesco.org/images/0018/001818/181893E.pdf (accessed on 04/03/2019)

Weart, S. R. (2003). The discovery of global warming, Cambridge, Mass., Harvard University Press.

Weart, Spencer R. (2008) The Discovery of Global Warming, Harvard University Press

World Meteorological Organisation (WMO) (1986) *Report of the International Conference on the assessment of the role of carbon dioxide and of other greenhouse gases in climate variations and associated impacts,* Villach, Austria, 9-15 October 1985, WMO No.661. http://www.icsu-scope.org/downloadpubs/scope29/statement.html (accessed on 04/03/2010).

Young, Damon A. (2009) Bowing To Your Enemies: Courtesy, Budo and Japan, Philosophy East & West Volume 59, Number 2 April 2009 188–215.

Zimmerman, B.J. (1990). Self-regulated learning and academic achievement: An overview. *Educational Psychologist, 25*, 3-17.

3

Role of International Organization in Addressing the Climate Change Issues and Creation of Intergovernmental Panel on Climate Change (IPCC)

The chapter describes the creation of the IPCC as a consequence of the World Climate Program and the Villach expert group, indicating the role of the secretariats of UNEP and the World Meteorological Organization, as well as the input from non-governmental organizations. It looks at its initial composition and procedures.

In the first two chapters of this book we discussed how science evolved in the climate change issue and how science helps or creates the regime. We also discussed the politics and debates regarding the regime theory. International regimes are structures designed to foster international cooperation among participants' countries. Every country needs help solving transnational problems. Examples seen in the current environment include global terrorism, the fight against the transnational drug problem and the fight to minimize HIV and AIDS, none of which can be resolved by the single state. International organizations can create powerful tools to solve a particular problem, which helps to increase their power, access and authority through collaborative efforts, mutual agreements and policy formation. This situation creates a favorable environment for formulating new regimes, where solutions can be contemplated. The IPCC is a good example of international regime formation on climate change because, based on the outcomes of its first four assessment reports, it is creating global agreement on how climate change may hamper the life of the planet. This demonstrates the climate change issue is scientific fact, and the alarming rate of climate change is caused by anthropogenic activities (IPCC, 2007). In this chapter we briefly examine the formation of international organization, their role in the global scenario, the early science-based organizations, as well as the need and the grounds for the IPCC formation.

3.1 The History of the International Organizations

There is a long history of international organization formation. Broadly, this can be traced from Greek civilization (Plato, Aristotle), although they have flourished mostly since the Enlightenment era. However, empirical studies show that the scientific study of international organizations does not begin until the 19th century. The modern form of international organizations began with the treaty in Utrecht, the Netherlands in 1713 and Vienna, Austria in 1815 with the outlining of the peace treaty principles. There is not much historical literature available on how international organizations were formed and how their expansion occurred beyond state borders prior to 1900. However, we do know the internationalization or formalization of organizations beyond state borders was aimed primarily at securing the political and legal security of the state. Another goal of formalizing organizations was to build cooperation between nations and citizens for their welfare and the exchange of business commodities. Traditionally, international organizations have been mostly established by the states to fulfill political goals (Archer 1992).

Historically, the study of international organizations examines the formal organizations and their roles to address the particular issue of areas of international cooperation from the political perspective. According to Archer (1992[1]), the term "international organization" was first used by Professor James Lorimer of England in 1867. Archer does not provide details regarding how and in which context Lorimer used the term. In this respect, Pitman Potter (1945) has done extensive research about the origin of the term, confirming that Lorimer was the first lawyer who used the term in his lecture

[1]According to Clive Archer notes that "Lorimer was a Professor of Law at the University of Edinburgh. The Gazeteer for Scotland gives the following information about him: James Lorimer (1818 – 1890) Lawyer and political philosopher. Born in Aberdalgie (Perth and Kinross). Lorimer became an advocate and was appointed Regius Professor of Public Law at the University of Edinburgh (1862). He was an expert on international law and was noted for two publications; namely 'The Institutes of Law: A treatise of the principles of jurisprudence, as determined by nature' (1872) and 'The Institutes of the Law of Nations: A treatise of the jural relations of separate political communities' (1884). In these he deprecated utilitarianism, and both have been criticized for their elitism and support for imperial colonialism, which was reaching its zenith at that time. I follow what Pitman Potter says about the term, though I would distinguish between International Organizations and International Institutions" (Archer1992).

before the Royal Academy in Edinburgh on May 18, 1867 on the heading of "On the application of the principle of relative or proportional, equity to international organization." According to Potter, Lorimer again used the term in his publication in 1971. Following Lorimer, Thomas Willing Balch used the term in 1874 (Potter 1945:805).

Political scientists examine international organizations in terms of international relations, governance and power dynamics. To examine society, they use various historical perspectives (developed by ancient and classical social thinkers) such as power and authority (rewards and punishment) and the political community (including nationalism). Power and authority notions were developed first by Plato, and further explanations developed by Bodin continued to be built upon by the classical organizational theorists (Weber, Taylor, Simon, etc.). Likewise, another perspective is to see the world in terms of mathematical order (quantitative approach) (Lucas, 1977, 1976). Archer (1992) examines these perspectives in two major classifications i.e. (i) traditional and (ii) revisionist. Traditional perspective considers international organization as a part of institutionalized relationship between states and government. This perspective is state centric, which only focuses on governmental international organizations. The traditional school of thought (which covers both realist and neo-realist school of thought) is developed mostly by the lawyers who primarily study organizations such as League of Nations and United Nations, NATO, European Union etc. who contribute to the formation of international government. They examine international organizations' role in global legal policy formation. Likewise, revisionists also focus on the state centric approach to study international organizations; however, they do not discard the roles of non-governmental international organizations in world politics (Archer, 1991).

There is no empirical evidence to state when international organizations began their formal roles. According to Archer (1992)[2] the rise of modern international organizations began in 1919 at the Versailles Peace Conference. The participants at Versailles were the representatives of victorious powers ready to write a peace treaty, including many national interest groups and

[2]"However, Archer (1992) provides a foundation history of international organization, where he cites Speechaert (1957: xiii) and notes that 509 international organizations were founded during 1693 to 1914, which reached 666 by 1915 to 1944 and 803 by 1945 to 1954. This clearly shows that formal organization must have begun in 17 th century and gradually increased. It also gives a chance to pose the alternative argument that formation of international organizations might not only influenced by politics. It might be religion or social motive which enforced to establish organizations at international level.

international non-governmental organizations (INGOs) wanting to advance public health, the lot of workers, the cause of peace, or the laws of war (Archer 1992:3). This conference was influenced by the previous Hague Conferences in 1899 and 1907, which formed the grounds for the creation of the League of Nations[3] (Bhandari 2011, 2012, 2014, Bhandari and Bhattarai 2017). The League of Nations could not generate a consensus on world politics, as demonstrated by World War II. Therefore, world leaders were looking for a new international organization that could bring the world's nations together to manage permanent peace across the globe. The foundation of the United Nations[4] was the outcome of that effort. There have been debates

[3]"The League of Nations was founded in 1919 as a result of the Treaty of Versailles and the end of World War I. Woodrow Wilson had personally represented the United States at the Versailles peace conference, and he arrived in Paris intent upon establishing a collective security organization that would prevent another world war from ever happening again. The league and its covenant were the ultimate expression of that vision, and President Wilson submitted the treaty to the Senate confident that he could persuade enough of its members to vote for ratification" http://www.nps.gov/archive/elro/glossary/league-of-nations.htm "In order to promote international cooperation and to achieve international peace and security by the acceptance of obligations not to resort to war, by the prescription of open, just and honorable relations between nations, by the firm establishment of the understandings of international law as the actual rule of conduct among Governments, and by the maintenance of justice and a scrupulous respect for all treaty obligations in the dealings of organized peoples with one another, Agree to this Covenant of the League of Nations" Conference Article of League of Nations (Geneva 15, November 1920) http://cyberschoolbus.un.org/unintro/unintro3.htm (accessed on 03/15/2016)

[4]"The name 'United Nations,' coined by United States President Franklin D. Roosevelt, was first used in the 'Declaration by United Nations' of 1 January 1942, during the Second World War, when representatives of 26 nations pledged their Governments to continue fighting together against the Axis Powers. States first established international organizations to cooperate on specific matters. The International Telecommunication Union was founded in 1865 as the International Telegraph Union, and the Universal Postal Union was established in 1874. Both are now United Nations specialized agencies. In 1899, the International Peace Conference was held in The Hague to elaborate instruments for settling crises peacefully, preventing wars and codifying rules of warfare. It adopted the Convention for the Pacific Settlement of International Disputes and established the Permanent Court of Arbitration, which began work in 1902. The forerunner of the United Nations was the League of Nations, an organization conceived in similar circumstances during the First World War and established in 1919 under the Treaty of Versailles "to promote international cooperation and to achieve peace and security." The International Labour Organization was also created under the Treaty of Versailles as an affiliated agency of the League. The League of Nations ceased its activities after failing to prevent the Second World War. In 1945, representatives of 50 countries met in San Francisco at the United Nations Conference on International Organization to draw up the United Nations Charter. Those delegates deliberated on the basis of proposals worked

about the role of the United Nations, its usefulness and its power dynamics (Pangle and Ahrensdorf 1999; Grant 2001). However, the role of the United Nations to guide international society to resolve global geo-politico-socio-economic and most recently, environment problems is unavoidable (Archer 1992). One of the United Nations roles is to also bridge the gap between the states and nongovernmental organizations. However, the trend and field of international organizations has been broadening since the World War I. As Kratochwil (2006) states, international organizations can be conceived of as the investigation of the various organizational forms that populate the international arena. This leads to the conclusion that the study of international organizations does not just cover the international organizations but includes all forms of organizations having international influences or relationships. On the basis of Kratochwil's notion, the study of international organizations covers NGOs, nation-states, international regimes, security alliances, multinational corporations, economic classes and democratic forms of governance, nationalisms, ethnicities and cultures (Kratochwil 1994). In this context, international organizations cannot be the sole concern of political science but become matters for multidisciplinary investigation.

Theoretically, organizational research broadly examines (1) producing units and what factors determine organizational effectiveness or productivity and (2) sets of individuals whose well-being is affected by the terms of organizational membership and whose motivation to continue that membership depends on their assessment of its comparative contribution to their well-being (Kahn, 1990:3). This notion can be applied to investigating the role of international organizations because they follow more complex formalities than domestic formal organizations. The roles of organizations depend upon the motives behind why, how and for what purpose organizations were formed (Bhandari 2011, 2012, 2014, Bhandari and Bhattarai 2017).

To understand the international organizations, it is essential to investigate what criteria make an organization international. The Year Book of International Organizations (1976/1977), published by the Union of International

out by the representatives of China, the Soviet Union, the United Kingdom and the United States at Dumbarton Oaks, United States in August-October 1944. The Charter was signed on 26 June 1945 by the representatives of the 50 countries. Poland, which was not represented at the Conference, signed it later and became one of the original 51 Member States. The United Nations officially came into existence on 24 October 1945, when the Charter had been ratified by China, France, the Soviet Union, the United Kingdom, the United States and by a majority of other signatories. United Nations Day is celebrated on 24 October each year (Extracted from: Basic Facts about the United Nations 2000, Sales No. E.00.I.21. http://www.un.org/aboutun/history.htm). (accessed on 03/15/2016)

Associations (UIA), broadly states eight major criteria: (1) The aims must be genuinely international with the intention to cover at least three states. (2) Membership must be individual or involve collective participation, with full voting rights, from at least three states and must be open to any individual or entity appropriately qualified in the organization's area of operations. Voting must be arranged so that no one national group can control the organization. (3) The constitution must provide for a formal structure giving members the rights to periodically elect governing bodies and officers. Provision should be made for continuity of operation with a permanent headquarters. (4) Officers should not all be the same nationality for more than a given period. (5) There should be a substantial contribution to the budget from at least three states and there should be no attempt to make profits for distribution to members. (6) Those with an organic relationship with other organizations must show they can exist independently and elect their own officials. (7) Evidence of current activities must be available. And there are some negative criteria: (8) size, politics, ideology, fields of activity, geographical location of headquarters, and nomenclature are irrelevant in deciding whether a set-up is an "international organization" or not (as cited by Archer, 1992:33-34). UIA provides a clear picture of how organizations should be categorized and evaluated.

Several authors have tried to define international organizations along the same lines as these suggested by the UIA(1976/1977) (1992); however, they depend on disciplinary orientations. An international organization as permanently expressing a juristic will be distinct from that of its individual members (Pentland 1973). It works as a legal body of bureaucratic structure. In the contemporary world, the impact of international organizations can be found in every sphere of the political, social, economic and environmental arenas. The political function of international organizations is to provide the means of cooperation among states in areas in which cooperation provides advantages for all or a large number of nations. Their social function is to try to reduce social inequality. Their economic function can be to reduce inequality on a global scale, and their environmental function can be to make collaborative efforts to overcome global environmental problems (Young 1999; Bennett 1991).

International organizations can be grouped according to their objectives and their functions. There are three major categories of organizations: intergovernmental organizations (IGOs) e.g. United Nations agencies, United Nations Environment Program (UNEP), Intergovernmental Panel on Climate Change (IPCC), etc., International non-governmental organizations (INGOs) e.g. International Union for Conservation of Nature (IUCN), World Wildlife Fund (WWF), The Flora and Fauna International (FFI), etc. and multinational corporations (MNCs)[5] e.g. World Bank, Regional Development Banks, etc. IGOs are based on a formal instrument of agreement between the governments of nation states, including three or more nation states as parties to the agreement; and possessing a permanent secretariat performing ongoing tasks. INGOs are defined and classified as in the United Nations Economic and Social Council (ECOSOC)[6] definition of INGOs; they should be international NGOs in terms of aims, members, structure, officers, finance, autonomy and activities, all of them taking place in three or more countries. To be considered multinational enterprises, organizations should have the products and services in more than three countries. In this paper, focus will be on INGOs, those established to attain certain goals, and their relationship with certain organizational theories (please see footnote for detail)[7] according to

[5]"Yearbook of International Organizations": The Union of International Association was founded one hundred years ago, in 1907, by Henri La Fontaine (Nobel Peace Prize laureate of 1913), and Paul Otlet, a founding father of what is now called information science. It is a non-profit non-governmental organization registered under Belgian law as an AISBL. It has consultative status with ECOSOC (since 1951) and UNESCO (1952) http://www.uia.be/homepage (accessed on 03/15/2016).

[6]"The first avenue by which non-governmental organizations took a role in formal UN deliberations was through the Economic and Social Council (ECOSOC). 41 NGOs were granted consultative status by the council in 1946; by 1992 more than 700 NGOs had attained consultative status and the number has been steadily increasing ever since to 3,052 organizations today. Article 71 of the UN Charter opened the door providing for suitable arrangements for consultation with non-governmental organizations. The consultative relationship with ECOSOC is governed today by ECOSOC resolution 1996/31, which outlines the eligibility requirements for consultative status, rights and obligations of NGOs in consultative status, procedures for the withdrawal or suspension of consultative status, the role and functions of the ECOSOC Committee on NGOs, and the responsibilities of the UN Secretariat in supporting the consultative relationship. Consultative status is granted by ECOSOC upon recommendation of the ECOSOC Committee on NGOs, which is comprised of 19 Member States." http://www.un.org/esa/coordination/ngo/(accessed on 03/15/2010)

[7]scientific management or management as science, classical school (1910s); human relation (1920s): focuses on attitude; bureaucracy (1940s): order, system, rationality, uniformity, and consistency, lead to equitable treatment for all employees by management; Group dynamics (1940s) individual participation in decision-making; Leadership(1950s): the importance of groups having both social task leaders; Decision theory(1960s)individuals "satisfies" when they make decisions; Socio-technical school(1960s): Called for considering technology and work groups when understanding a work system; Environment and technology system (1960s) mechanistic and organic structures and stated their effectiveness with specific types of environmental conditions and technological types; Systems theory-(1970s): organizations

niche and demands[8]. These INGOs are considered as non-state actors and have significant influence on socio-economic and human services delivery (education, health and human and women's rights), economic development (agriculture, microcredit and infrastructures), environment conservation and world politics. There are many varieties of INGOs, and they have a long history of their products and services delivery. For example, the International Red Cross Society is the one of the oldest INGOs and was established in February 1863 in Geneva, Switzerland, with the purpose of treating war victims and delivering health services. Fauna & Flora International was established in 1903 as the world's first international conservation organization with the purpose of conserving flora and fauna; it was instrumental in establishing much of today's global and local conservation infrastructure, including organizations such as the IUCN, WWF, the Convention on International Trade in Endangered Species of Wild Fauna & Flora (CITES), and conservation instruments such as the Red List of endangered species[9].

3.2 The Regulatory Bounded Treaties, Conventions and Science Behind International Organizations Formation

The Congress of Vienna was held November 1, 1814 through June 8, 1815, and is considered the first internationally binding treaties, where European nations agreed to settle the future boundaries of the continent. This is also considered the foundation of the peace treaty agreement and origin

as open systems with inputs, transformations, outputs, and feedback; systems strive for equilibrium and experience equifinality; Contingency theory (1980s): organization processes and characteristics of the situation; called for fitting the organization's structure to various contingencies.

[8]"International organizations, whether governmental or non-governmental, use any of an extensive range of terms in their official titles. These may include terms such as union, association, office, agency, centre, or alliance. There is a great deal of confusion associated with the meanings to be attached to such terms in practice. It is therefore not usual to attempt to classify an organization on the basis of whether it is a "union," a "confederation," a "committee," or a "league," for example. A "centre" may in fact resemble an "association" more than it resembles most other "centers"; equally an "association" can be more like what is commonly understood to be a "centre." The range of terms can be usefully ordered by relating the organizations in question to the meetings by which they were established or through which they work. This brings out the strengths and limitations of this seemingly obvious approach to classifying organizations" (UIA Classification: Yearbook of International Organizations, Paul Taylor and A J M Groom (London, Frances Pinter, 1977; New York, Nichols Publishing Company, 1978).

[9]http://www.fauna-flora.org/aboutus.php (homepage) (accessed on 03/15/2016)

of international governmental organizations (IGOs). The Vienna Congress created the new web in European politics in terms of balance of power. This was represented by both governments and the general public lobbying to develop a concrete peace keeping policy. The Vienna Congress approved the provision of civil rights for Jews in Article XVI (Reinalda 2009: 17-25). Reinalda (2009) elaborates this also introduced the multilateral conference as instruments, consultation as a process of learning, diplomatic relation regulation and change and also acknowledged the role of the press. The outcome of the Congress was well distributed and published by the journals. From 1815 to 1899, twenty-seven multinational security conferences were held in Europe (Reinalda 2009: 26). The agreement on freedom of navigation in Rhine was begun in 1815, which materialized with the development of the central commission for the navigation in 1919, illustrated in the article of Treaty of Versailles 354-62. Another milestone of that era was the civil rights and anti-slavery citizen's movements and the beginning of the NGO issue on human health. In a conference in Geneva in October 1863, 16 states approved the foundation of private national societies, funded by private donations. The International Committee of the Red Cross (ICRC) was founded in 1863 in Geneva, Switzerland by Henry Dunant, with the primary goal to aid wounded soldiers[10] (Red Cross 2010). This era also gave birth to formal international organizations like the International Telegraphic Union founded in1865, (in its first convention on International Telegraph concluded on May 17, 1865, twenty states signed the agreement). ITU is now the United Nations agency for information and communication technology issues, which has 191 country members, 554 sector members and 144 associate members (ITU 2010). ITU is unique in the sense that NGOs dealing with the communication sector can be members (membership dues apply)[11].

---

[10]The Red Cross Now: The Movement is made up of almost 97 million volunteers, supporters, and staff in 186 countries. It has three main components: (1) The International Committee of the Red Cross (ICRC) (2) The International Federation of Red Cross and Red Crescent Societies and (3) National Red Cross and Red Crescent Societies. Its major goals are: Reduce the number of deaths, injuries and impact from disasters; Reduce the number of deaths, illnesses and impact from diseases and public health emergencies; Increase local community, civil society and Red Cross Red Crescent capacity to address the most urgent situations of vulnerability; Promote respect for diversity and human dignity, and reduce intolerance, discrimination and social exclusion. Source: The International Red Cross and Red Crescent Movement (2010) http://www.ifrc.org/who/movement.asp?navid=03_08&gclid=CISkfekrKACFWV75QodJWHXbQ (accessed 03/10/2016).

[11]The ITU mission: bringing the benefits of ICT to all the world's inhabitants; http://www.itu.int/net/about/mission.aspx(accessed on 03/09/2010)

Similarly, the Vienna Congress contribution was the foundation of the Universal Postal Union. The Union was founded through the conference in Berne on September 15, 1874, which was attended by twenty-two nations; today all countries in the world are members. The UPU is an intergovernmental body and a specialized agency of the United Nations since July 1, 1948. The UPU maintains particularly close ties with United Nations agencies such as:

- International Telecommunications Union (ITU);
- International Civil Aviation Organization (ICAO), established April 4, 1947;
- International Labor Organization (ILO), established in 1919 as part of the Treaty of Versailles;
- World Trade Organization (WTO), established in 1995, one of the second youngest of the international organizations;
- International Air Transport Association (IATA), established in Havana, Cuba in April 1945;
- International Organization for Standardization (ISO), established on February 23, 1947
- World Customs Organization (WCO), active since 1953;
- International Police Organization (INTERPOL), established in 1923;
- United Nations Development Program (UNDP), established in 1965 in the field of postal development;
- United Nations Drug Control Program (UNDCP), established in 1997, the youngest Un agency; and
- United Nations Environment Program (UNEP), established in 1972 to increase awareness for the environment among Posts (IPU, 2010)[12].

3.3 The Fall of the Vienna Congress and the Foundation of the League of Nations

The Vienna Congress and the various security related conferences of Europe could not normalize the power dynamism and competition of European nations and could not stop WWI or WWII. However, that period established the arbitration and international law as normative power in international relations; the Alabama case and the treaty of Washington on arbitration 1871 (between UK and USA); the creation of private institute of international law

[12]The UPU as a specialized agency of the United Nations http://www.upu.int/about_us/en/the_upu_as_a_un_specialized_agency.html (accessed on 03/09/2016)

in 1873 and the international law association of 1873; the establishment of the inter-parliamentary union in 1889 and the activation of the permanent court on arbitration in 1900 in Hague were very important steps toward the establishment of international law and order (Reinalda 2009). More important, this time also helped flourish the role of civil society organization and citizen's movement for human rights and women rights. Despite the peace building efforts, 1899 through 1914 was a time of anxieties and fear of war, with war breaking out in 1914 and ending in 1918.

The second regulatory bounded form of an intergovernmental international organization was the League of Nations, which was founded in 1919–1920 as a result of the Treaty of Versailles in an effort to stop future wars. (The Treaty of Versailles was one of the peace treaties signed at the end of World War I. It ended the state of war between Germany and the Allied Powers. It was signed on June 28, 1919.) By February 23, 1935 the League of Nations had fifty-eight members. The overall goals of League were upholding the new-found Rights of Man such as the rights of non- whites, women, and soldiers; disarmament; preventing war through collective security; settling disputes between countries through negotiation; diplomacy and improving global quality of life (Reinalda 2009)[13]. United States' President Woodrow Wilson was the main proponent of the League but never joined because of the nonalignment motives of American citizens, particularly with Europe. From 1920 to 1940, the League managed to make thirteen agreements in the economic field; fifteen agreements in communication and transportation; six agreements on human health; and nine agreements in the social wellbeing of the citizens of the members' countries. The World Health organization, which was founded in 1922 with the name of Health Organization, is one example of collaborative works resulting from these agreements.

Likewise, the League of Nations had a scientific and knowledge production body, namely the League of Nations Committee on Intellectual Cooperation, sometimes known as The International Committee on Intellectual Cooperation or League of Nations Committee on Intellectual Cooperation. It was founded in 1922 with an office in Geneva, where it could not operate well because of the lack of funding. It moved to Paris in 1926 with French funding with a slightly different name; "International Institute for Intellectual Cooperation" and began to function properly. The members of this committee were

[13]Reinalda, Bob (2009).Routledge History of International Organizations, From 1815 to the Present Day, Routledge, member of the Taylor & Francis Group, an Informa Business, UK, USA.

the most distinguished figures of the twentieth century, including Henri Bergson (French philosopher; Nobel Prize Winner on literature in 1927); Albert Einstein (Physician, Nobel Prize Winner on Physics in 1921); Marie Curie (physicist and chemist, Nobel Prize winner in both fields); Béla Bartók (one of the greatest composers of the 20th century); Thomas Mann (German novelist, short story writer, social critic, philanthropist, essayist and 1929 Nobel Prize laureate); Salvador de Madariaga (Spanish diplomat, writer, historian and pacifist); and Paul Valéry (French poet, essayist and philosopher). The role of the International Institute for Intellectual Cooperation was to promote international cultural/intellectual exchange between scientists, researchers, teachers, artists and other intellectuals. This organization continued its work until 1946, when its role was taken over by The United Nations Educational, Scientific and Cultural Organization (UNESCO) (Reinalda, 2009; UNESCO 2010;)[14].

During the League of Nations era, knowledge producing NGOs were flourishing through the academic institutions. Prior to the League's formation, there was the International Association of Academies (IAA), which continued its knowledge production from 1899 to1914, and from 1919 to 1931, there was International Research Council (IRC). The works of the previous council was taken over by the International Council for Science (ICSU) in 1931. The ICSU has a dual membership system such as the National Scientific Members and International Scientific Unions. The ICSU deals with the scientific issues that cannot be solved alone by the national or international scientific unions. The ICSU works collaboratively with UN agencies and other national and international scientific agencies. The major international scientific ICSU member organizations include European Science Foundation (ESF), Interacademy Council (IAC)

Interacademy Medical Panel (IAMP), Interacademy Panel (IAP), International Association of Universities (IAU), International Council for Engineering and Technology (ICET),

International Council for Philosophy and Humanistic Studies (CIPSH), International Group of Funding Agencies for Climate Change Research (IGFA), International Social Sciences Council (ISSC), International Union

[14]Creation - The United Nations Educational, Scientific and Cultural Organiza tion (UNESCO) - located http://www.nation sencyclopedia.com/United-Nations-Related-Agencies/The-United-Nations-Educational-Scientific-and-Cultural-Organization-UNESCO-CREATION.html#ixzz0hiqtjH34 and http://en.wikipedia.org/ wiki/International_Committee _on_Intellectual_Cooperation (accessed on 03/09/2016)

of Technical Associations and Organizations (UATI), Organization for Economic Cooperation and Development (OECD), International Union for Conservation of Nature (IUCN) and World Federation of Engineering Organizations (WFEO) (ICSU 2010)[15].

Despite the many efforts to avoid war through various agreements among the League of Nations member countries, it could not obtain its goal of peacekeeping, and World War II occurred because of power struggles between European nations. From the humanitarian perspective, there was no victory for any nation; the entire world was defeated by the war. However, the League of Nations' failure demonstrated war could not achieve peace for all and provided an experimental ground for the collective security, which was adopted in the United Nations charters.

3.4 The Fall of the League of Nations and Foundation of the United Nations (UN)

Just as President Woodrow Wilson proposed for the League of the Nations, on January 1, 1942 President Franklin D. Roosevelt coined the name "United Nations," where twenty-six World War victims' countries were present. In 1945, 50 countries' representatives met in San Francisco at the United Nations Conference on International Organization, to draw up the United Nations Charter, and all 50 countries signed the charter on June 26 1945, which framed the United Nations. Poland signed the charter later and became the 51st original charter signing state. In October 24, 1945, the United Nations officially came into existence, when the Charter was ratified by China, France, the Soviet Union, the United Kingdom, and the United States and by a majority of other signatories. Currently, almost all 192 countries of the world are members of the UN. Only Kosovo, which was declared independent from Serbia in 2008, Taiwan (China claims it is a part of China) and the Vatican City, which was founded in 1929, have not felt it necessary to join the UN.

The UN has four major purposes: (1) to maintain international peace and security; (2) to develop friendly relations among nations; (3) to cooperate in solving international problems and in promoting respect for

[15]ICSU (2010). Overview: ICSU has two categories of full Members,National Members and International Scientific Unions; there are currently 30 Scientific Union Members and 119 National Scientific Members covering 137 countries, In addition, ICSU has 21 International Scientific Associates. http://www.icsu.org/4_icsumembers/OVERVIEW.php (accessed on 03/09/2010

human rights; and (4) to be a center for harmonizing the actions of nations. In the contemporary world, the UN is visible in every aspect of social, economic, environmental and political issue including peacekeeping and development through its forty plus affiliated organizations (UN, 2010). In addition to closely working with member governments, the UN also has a tradition of establishing new agencies to address emerging issues; however, sometimes it also works collaboratively with other IGOs or INGOs on the global or the transboundary issues and with NGOs if an issue is related to a particular nation. The UN also has a body to collaborate with NGOs and INGOs through the Economic and Social Council (ECOSOC) and the Non-Governmental Liaison Service (NGLS). The UN also provides observer status to IGOs who can participate in the sessions and the work of the General Assembly, some of which have permanent mission offices in New York. The international organizations who have permanent mission offices include the African Union; Caribbean Community (CARICOM); Central American Integration System; Commonwealth Secretariat; Cooperation Council for the Arab States of the Gulf; European Union; International Criminal Police Organization (INTERPOL); International Union for the Conservation of Nature and Natural Resources (IUCN); League of Arab States; International Red Cross; regional development banks, etc. There are 25 intergovernmental organizations of this kind and 50 other organizations that do not have mission offices in New York but have observer status in the UN system (UN 2010)[16].

In additional to the directly affiliated agencies, the UN also collaborates with all kinds of governmental, nongovernmental organizations, academic institutions, private business associates, public sector organizations, national and international organizations, etc. through the Global Compact Network Forum (UN 2010).

The UN Global Compact Participants

The UN Global Compact is the world's largest corporate citizenship and sustainability initiative. Since its official launch on 26 July 2000, the initiative has grown to more than 7700 participants, including over 5300 businesses in 130 countries around the world. It is a network-based initiative with the Global Compact Office and six UN agencies at its core. The Global Compact involves all relevant social actors: companies, whose actions it seeks to influence; governments, labor, civil society organizations, and the

[16]UN (2010). Permanent observer of United Nations, http://www.un.org/en/members/ intergovorg.shtml (accessed on 03/11/2016)

United Nations, the world's only truly global political forum, as an authoritative convener and facilitator. Source: The overview of the UN Global Compact (UN 2010)[17]

This indicates that in its 65 years, the UN's involvement and the role for the healthy planet became omnipresence in every aspect of the contemporary world including global environmental change.

3.5 Environmental Change: Public Concern and Actions

"Up to our own day American history has been in a large degree the history of the colonization of the Great West.... [The frontier produced] a man of coarseness and strength...acuteness and inquisitiveness, [of] that practical and inventive turn of mind... [full of] restless and nervous energy...that buoyancy and exuberance which comes with freedom.... The paths of the pioneers have widened into broad highways. The forest clearing has expanded into affluent commonwealths. Let us see to it that the ideals of the pioneer in his log cabin shall enlarge into the spiritual life of a democracy where civic power shall dominate and utilize individual achievement for the common good" – Frederick Jackson Turner[18] in 1893.

The concern about the global environment change is not a new phenomenon. The first assertions that humans are responsible for the Earth came through the book by George Perkins Marsh published in 1864 entitled Man and Nature. In the 1874 revised edition, Marsh changed the title to "The Earth as Modified by Human Action: Man, and Nature" where he stated that "We are not passive inhabitants of Earth. . . . We give Earth its shape and form. We are responsible for Earth"[19]. This was most likely the first book on ecological problems, which created a path for scientific research on anthropogenic cause on global environment. When humans were recognized as being responsible

[17]The overview of the UN Global Compact (UN 2010); http://www.unglobalcompact.org/ParticipantsAndStakeholders/index.html (accessed on 03/11/2016)

[18]Historian Frederick Jackson Turner: in an 1893 speech to the American Historical Association. http://www.radford.edu/~wkovarik/envhist/5progressive.html; from the Environmental History Timeline (accessed on 5/11/2016)

[19]Lienhard, John H. (. . . . .) Engine of our ingenuity; GEORGE PERKINS MARSH No. 595 Marsh, G.P., Man and Nature . Cambridge: The Harvard University Press, (1965). (This is an annotated reprint of the original 1864 edition.) http://www.uh.edu/engines/epi595.htm (accessed on 03/11/2016)

for global environmental change, people began to think about "how such issues could be solved." It was also determined environmental problems were not endemic, or a single nation's problem. As a matter of fact, different countries began to work together through mutually biding and non-binding treaties of cooperation and collaboration (Weart, 2008)[20].

There have been several agreements to address the environmental issues. According to R. B. Mitchell (2009)[21], there are 1538 Bilateral Environmental Agreements; 1000 Multilateral Environmental Agreements; 259 Other (non-multi, non-bi) Environmental Agreements; 213 Bilateral Environmental Non-Binding Instruments (non-agreements); 228 Multilateral Environmental Non-Binding Instruments (non-agreements) and 100 Other (non-multi, non-bi) Environmental Non-binding Instruments (non-agreements), listed so far. The first such agreement was proposed by the Swiss in 1872 as an international commission to protect migratory birds. The first Convention for the Preservation of Animals, Birds and Fish in Africa was signed in 1900 in London by the European colonial powers with the intent to protect African game species, particularly to limit the export of ivory, which was leading to severe hunting pressure on the African elephant[22]*(see annex 1 for detailed list of the environment related major agreements from 1872 to 2009).*

As a consequence, Fauna and Flora International (FFI) was founded in 1903, which is the world's first international conservation organization. The fauna and floras major focus in the beginning was to protect the wildlife in Africa and still continues[23]. There is long gap of INGO foundation after the inception of FFI. After forty-five years, the International Union for Conservation of Nature (IUCN)[24] was founded at Fontainebleau, France in 1948, with the support of the newly formed United Nations Educational, Scientific and

---

[20]Weart, Spencer R. (2008). The Discovery of Global Warming: Revised and Expanded Edition (New Histories of Science, Technology, and Medicine), Harvard University Press.

[21]http://iea.uoregon.edu/page.php?query=home-contents.php and Data from Ronald B. Mitchell. 2002-2010. International Environmental Agreements Database Project (Version 2010.2).Available at: http://iea.uoregon.edu/ (accessed on 11 March 2019).

[22]United Nations Environment Programme (UNEP) (1992) The world environment 1972-1992: Two decades of challenge, ed. M. K. Tolba, O. A. El-Kholy, E. El-Hinnawi, M. W. Holdgate, D. F. McMichael, and R. E. Munn, 529-67. New York: Chapman & Hall.

[23]Fauna & Flora International (2010). About us http://www.fauna-flora.org/aboutus.php (accessed on 03/11/2016)

[24]"The first Director General of UNESCO, (Sir Julian Huxley), wishing to give UNESCO a more scientific base, sponsored a congress to establish a new environmental institution to help serve this purpose, which ultimately helped to establish the IUCN" (Christoffersen, Leif E. (1997) IUCN: A Bridge-Builder for Nature Conservation,Green Globe Yearbook 1997).

Cultural Organization (UNESCO). There were 18 governments, 7 international organizations, and 107 national nature conservation organizations with a group of individual scientists and lawyers as participants. The participants all agreed to form the institution and signed a "constitutive act" with the name of International Union for the Protection of Nature (IUPN), now IUCN, with its headquarters located in the Lake Geneva area in Gland, Switzerland. It was founded during the same period in which the international community created the United Nations and its agencies (IUCN, 2010; (Bhandari 2011, 2012, 2014, Bhandari and Bhattarai 2017).

IUCN is typically listed as an NGO in Switzerland and USA, though it occasionally describes itself as a GONGO. It has observer status at the United Nations and consultative status with UN Economic and Social Council (ECOSOC), FAO and UNESCO (MacDonald, 2005:2)[25]. In the United States, IUCN's legal status is as an International Organization, designated by Executive Order No. 12986 (January 18, 1996)[26], and IUCN is supported in the United States by charitable organization (IUCN-US) established under 501(c) 3 statuses. 501(c) is a provision of the United States Internal Revenue Code (26 U.S.C. §501(c)), listing 26 types of non-profit organizations exempt from some federal income taxes.

> ***"What is IUCN?*** *IUCN, the International Union for Conservation of Nature, helps the world find pragmatic solutions to our most pressing environment and development challenges. It supports scientific research, manages field projects all over the world and brings governments, non-government organizations, United Nations agencies, companies and local communities together to develop and implement policy, laws and best practice. IUCN is the world's oldest and largest global environmental network - a democratic membership union with more than 1,000 government and NGO member organizations, and almost 11,600 volunteer scientists in more than 154 countries. IUCN's work is supported by over 1,000 professional staff in 60 offices and hundreds of partners*

[25]MacDonald, Kenneth Iain (2005) IUCN: A History of Constraint: Text of an Address given to the Permanent workshop of the Centre for Philosophy of Law Higher Institute for Philosophy of the Catholic University of Louvain (UCL), Louvain-la-neuve. http://perso.cpdr.ucl.ac.be/maesschalck/MacDonaldInstitutional_Reflexivity_and_IUCN-17.02.03.pdf (accessed 03/11/2016)

[26][Federal Register: January 22, 1996 (Volume 61, Number 14)][Presidential Documents](Page 1691-1693]From the Federal Register Online via GPO Access [wais.access.gpo.gov][DOCID:fr22ja96-114] http://www.epa.gov/fedrgstr/eo/eo12986.htm

in public, NGO and private sectors around the world. The Union's headquarters are located in Gland, near Geneva, in Switzerland" (source: http://www.iucn.org *accessed on 03/11/2016).*

The IUCN has been playing a major role in bringing science and conservation together through its hybrid membership system. IUCN is one of the major international organizations developing strategies and policy for global environment (World Conservation Strategy, 1980); after the publication of Our Common Future known as Brundtland report in 1987, the 1983 General Assembly passed Resolution 38/161 in 1983, IUCN published Caring for the Earth: A Strategy for sustainable living in 1991, which is considered one of the major milestones for the sustainable development policy formation.

IUCN capitalized on the subsequent burst of environmental activity in governments around the world, particularly the establishment of departments or ministries of environment. It was also allowed to play a key role in the preparations for the first United Nations Conference on the Human Environment in Stockholm in 1972. This conference led directly to the creation of the United Nations Environment Program (UNEP); with the intention of strengthening the environmental dimensions of the UN. But IUCN staff prepared background papers and acted as consultants and, as governments developed reports for the conference, they turned to people who were associated with IUCN (MacDonald, 2005:8). The IUCN was the key player for Convention on the International Trade in Endangered Species (CITES); on the conservation of wetlands of international importance (RAMSAR); and on the conservation of the World Heritage (World Heritage Convention) (MacDonald, 2005:8; Bhandari 2012[27]). Particularly from the 1972 Stockholm Conference, IUCN has been positioning its stand in most of the global conventions or conferences held in the world including Rio 1992, Durban 2002, Bali 2007 and Copenhagen 2009. IUCN does not work against any government or agency; it plays the collaborative role to develop mutual understanding to address global environmental issues including climate change. As noted above, IUCN was one of the INGOs who played a role to establish the ENEP and works closely with the UN agencies including IPCC and The World Meteorological Organization (WMO)[28] to address the climate change issue. World Wildlife

[27]Bhandari, Medani P. (2012). Environmental Performance and Vulnerability to Climate Change: A Case Study of India, Nepal, Bangladesh, and Pakistan. Climate Change and Disaster Risk Management. Series: Climate Change Management, pp. 149-167. Springer, New York / Heidelberg, ISBN 978-3-642-31109-3.

[28]WMO (2010). WMO in Brief: The World Meteorological Organization (WMO) is a specialized agency of the United Nations. It originated from the International

Fund for Nature (WWF) was formed within the IUCN in 1961 to raise the fund but later became a different organization. At present, most of the big international conservation organizations are member of IUCN, including intergovernmental, NGOs around the world, who are working to address the environmental issue. One of the IUCN's major programs are to address the climate change, including policy lobbying to program implementation in the ground (IUCN, 2010; Bhandari 2012).

3.6 The United States: A Pioneer to Address the Environment

Because the United States was the main player in establishing the League of Nations, followed by the United Nations, it also has pioneered facilitating the public and private sector's ability to address the environmental and climate change issues.

The National Oceanic and Atmospheric Administration (NOAA) traces back to the United States' Coast Survey established in 1807, the United States Weather Bureau established in 1870, and the United States Commission of Fish and Fisheries established in 1871[29].One of the major missions of NOAA is to conduct atmospheric and climate research. The vision statement states:

> *Climate Research: "NOAA Research, the research and development arm of NOAA, conducts research to provide the nation with better weather forecasts, earlier warnings for natural disasters, and an overall greater understanding of our oceans, climate, and atmosphere. This research helps prepare the nation for the new challenges of tomorrow as society and natural surroundings continue to change...."NOAA scientists continually strive to understand and describe climate variability and change to enhance society's capacity to anticipate and respond to climate change....Integration of research across existing disciplines is a central theme of NOAA's future climate research. The links among the land, ocean, polar ice, atmosphere, and biosphere must be*

Meteorological Organization (IMO), which was founded in 1873. Established in 1950, WMO became the specialized agency of the United Nations in 1951 for meteorology (weather and climate), operational hydrology and related geophysical sciences. http://www.wmo.int/pages/about/index_en.html (accessed on 03/11/2016).

[29]NOAA (2010). History "NOAA: 200 Years of Science, Service, and Stewardship" (Lammon, E. & Lopez, A. (Eds.). (2007). London: Faircount LLC.)

> *further explored, bolstering our nascent understanding of the complex interrelationships that comprise the global climate system"* (NOAA, 2010).[30]

Similarly, the first in the United States and the largest grassroots environmental organization was the Sierra Club, founded on May 28, 1892. The first president of the Sierra Club, Mr. John Muir, was one of the major players to establish the world's first National Park, "Yellowstone National Park," in 1872. It was the same year the Swiss Government proposed the conservation of migratory birds in the world. The Sierra Club first started the conservation campaign to defeat a proposed reduction in the boundaries of Yosemite National Park. By 1920, the Club became a very powerful organization in the United States and opposed a plan to build dams in the Park. Since its inception, the Yellowstone model has been copied in national parks throughout the world. At the union label, the American Forestry Association was founded in 1875, Appalachian Mountain Club was founded in 1876, the Smoke Prevention Association of America was founded in 1907, and the Environment Defense fund was established in 1867 (Brulle, 2000)[31]. Similarly, at the government level, broader environment conservation was also initiated in the United States.

However, the sole body to tackle environmental and climate change came in 1970 under the name of Environment Protection Agency (EPA) through the firm influence of Rachel Carson's Silent Spring, which was published as a column in the New York Times and as a book in 1962 (Lewis, 1985)[32]. Thebook'smission and "what we do" section states:

> *The EPA leads the nation's environmental science, research, education and assessment efforts. The mission of the Environmental*

[30]NOAA (2010) History of NOAA: The foundation of NOAA was through the Act of February 10, 1807 (chapter VIII; 2 Stat. 413), signed by President Thomas Jefferson. http://celebrating200years.noaa.gov/about.html http://celebrating200years.noaa.gov/about/HConRes147.pdf http://celebrating200years.noaa.gov/visions/atmospheric/welcome.html (accessed on 03/11/2016)

[31]Brulle RJ. (2000). Agency, Democracy, and Nature: The U.S. Environmental Movement from a Critical Theory Perspective. Cambridge, MA: MIT Press

[32]Lewis, Jack (1985) The Birth of EPA, [EPA Journal – November 1985] http://www.epa.gov/history/topics/epa/15c.htm (accessed on 03/11/2016)

Protection Agency is to protect human health and the environment. Since 1970, EPA has been working for a cleaner, healthier environment for the American people (EPA, 2010)[33].

There have been very positive impacts of the United States' environment management system on the rest of the world. This system is based on the National Environmental Policy Act of 1969 (NEPA); the Clean Air Act 1970 (CAA); the Clean Water Act 1977 (CWA); the Endangered Species Act 1973 (ESA); and the International Environment Protection Act of 1983[34], as well as others acts. These acts forced the formulation of the policy for environmental reform in the United States. U.S. policies also insisted first to the Western World and then to the rest of the world to examine their position and incorporate similar types of policies for environment management. The German Environmental Action Program of 1970 is another similar early program (Rehbinder 1976). I believe these acts play significant roles in addressing the global environmental crisis.

3.7 Back to the United Nations: The UNEP and WMO, the Founder of IPCC

As we noted in the second section of this chapter, United Nations brought together some of the exiting international organizations founded prior to its inception and established several other with specific roles and responsibilities. However, there are overlaps in the programs and policies among the UN agencies. For example, WHO, UNDP, UNEP and FAO all have the environmental program and policies? However, UNEP, WMO[35], UNFCC are the specific agencies responsible for addressing climate change and global environment. WMO is the core scientific body of the UN. The major role of the WMO is to facilitate, help or establish the meteorological observations stations and provide the training and research platform to member countries

---

[33]EPA (2010). Our Mission and What We Do, EPA, USA. http://www.epa.gov/epahome/whatwedo.htm (accessed on 03/11/2016)

[34]http://www.epa.gov/epahome/laws.htm

[35]Ibid (see 18) and the vision of WMO is to provide world leadership in expertise and international cooperation in weather, climate, hydrology and water resources and related environmental issues and thereby contribute to the safety and well-being of people throughout the world and to the economic benefit of all nations. http://www.wmo.int/pages/about/mission_en.html (accessed on 4/11/2015).

(there are 189 members), whereas UNEP has several roles including climate research and facilitations.

The UNEP was founded by the UN General Assembly resolution 2997 (XXVII) of December 15, 1972. This was recommended by the first UN Conference on Human Environment, held in Stockholm, Sweden, from June 5-16, 1972, where 113 countries representatives, 19 inter-governmental agencies, about 400 IGOs' and NGOs' leaders were participants. The Stockholm Conference is considered to be the foundation of the modern political and public awareness of global environmental problems. The UNEP has played a pivotal role in achieving its primary mission.

> The mission of the UNEP is *"to provide leadership and encourage partnership in caring for the environment by inspiring, informing, and enabling nations and peoples to improve their quality of life without compromising that of future generations" (UNEP: Organization Profile 2009, p2).*

And further, according to the UNEP Organization Profile (2009: 3)[36]

> *UNEP is the United Nations system's designated entity for addressing environmental issues at the global and regional level. Its mandate is to coordinate the development of environmental policy consensus by keeping the global environment under review and bringing emerging issues to the attention of governments and the international community for action. The mandate and objectives of UNEP emanate from:*
>
> 1. *UN General Assembly resolution 2997 (XXVII) of 15 December 1972;*[37]

---

[36]UNEP (2009). Organization Profile, UNEP, Nairobi Kenya http://www.unep.org/PDF/UNEPOrganizationProfile.pdf (accessed on 05/11/2017).

[37]UN General Assembly resolution 2997 (XXVII) of 15 December 1972; Mandates UNEP to "promote international cooperation in the field of the environment and to recommend, as appropriate, policies to this end, and to provide general policy guidance for the direction and coordination programs within the UN system". The Assembly further decided that the Executive Director of UNEP would be entrusted with, inter alia, the responsibility to "coordinate, under the guidance of the Governing Council, environmental programs within the UN system, to keep their implementation under review and to assess their effectiveness, and to advise, as appropriate of environmental and under the guidance of the Governing Council, intergovernmental bodies of the UN system on the formulation and implementation of environmental programs". http://www.nyo.unep.org/emg2.htm (accessed on 03/11/2010)

2. *Agenda 21, adopted at the UN Conference on Environment and Development (UNCED:the Earth Summit) in 1992;*
3. *the Nairobi Declaration on the Role and Mandate of UNEP, adopted by the UNEP Governing Council in 1997;*
4. *The Malmö Ministerial Declaration and the UN Millennium Declaration, adopted in 2000; and*
5. *Recommendations related to international environmental governance approved by the 2002 World Summit on Sustainable Development and the 2005 World Summit.*

The role of UNEP is to implement the mandate given by the governing council. At these conferences, such as UNCED (Rio), the Nairobi declaration and Malmo (Sweden), the WSSD assigned more responsibility to UNEP for coordinating with the governments and other organizations working for the environmental field. The UNEP has major roles in the environmental movement, which is highly appreciated by world leaders. For example, the Malmo declaration number 22, on page 5 notes that:

> *Governments and UNEP have to play a major role in the preparation for the 2002 review of UNCED at the regional and global levels and ensure that the environmental dimension of sustainable development is fully considered on the basis of a broad assessment of the state of the global environment. The preparations for the conference should be accelerated (UNEP 2000)*[38].

Similarly, in the Nairobi declaration:

> *"The role of UNEP, as the principal UN body in the field of the environment, should be further enhanced. Taking into account its catalytic role, and in conformity with Agenda 21 and the Nairobi Declaration on the Role and Mandate of UNEP, adopted on 7 February 1997, UNEP is to be the leading global environmental authority that sets the global environmental agenda, promotes the coherent implementation of the environmental dimension of sustainable development within the UN system, and serves as an*

[38]Governing Council of the United Nations Environment Program, Decisions Adopted by Global Ministerial Environment Forum Sixth special session, Malmö, Sweden, 29-31 May 2000. http://www.unep.ch/natcom/assets/milestones/malmo_declaration.PDF(accessed on 03/11/2016)

> *authoritative advocate for the global environment" (UNEP-New York, 2010)*[39].

And in UNCED declaration:

> *"UNEP played a pivotal role in coordinating the UN system's preparations for UNCED, held in Rio de Janeiro in June 1992. The Designated Official on Environmental Matters (DOEM) regularly reviewed the collective environmental work of UN bodies and agencies in preparation for UNCED and was involved in discussions on post-UNCED institutional arrangements. UNCED in adopting* Agenda 21 *, reaffirmed UNEP's coordinating role, stating that, "in the follow-up to the Conference, there will be a need for an enhanced role for UNEP and its Governing Council. The Governing Council should, within its mandate, continue to play its role with regard to policy guidance and coordination in the field of the environment, taking into account the development perspective". Agenda 21 further stipulated that UNEP should concentrate, inter alia, on "promoting international cooperation in the field of environment and recommending, as appropriate, policies to this end" (UNEP-New York, 2010)*[40].

Regarding the role of UNEP, former UN secretary general *Kofi A. Annan* states:

> *UNEP has served as an expert 'watchdog', monitoring the state of ecosystems and species worldwide. It has been, and remains, the environmental conscience of the United Nations. UNEP has played an instrumental role in the adoption of international environmental conventions and treaties aimed at preserving the ozone layer, conserving biological diversity, coping with climate change, protecting the oceans and seas, controlling the movement of toxic wastes and controlling the trade in endangered wildlife species (Annan, 1997)*[41].

---

[39]UNEP-New York office webpage http://www.nyo.unep.org/emg2.htm(accessed on 03/11/2017)

[40]Ibid (29) http://www.nyo.unep.org/emg2.htm

[41]Annan, Kofi (1997) UNEP: An indispensable Contribution, Our Planet, The Way Ahead, Vol 9, N.1. http://www.unep.org/ourplanet/imgversn/91/contents.html (accessed on 03/11/2016)

UNEP is small in terms of budget, human resources and program function in comparison to other UN agencies e.g. UNDP, FAO, UNESCO, etc. However, as an intergovernmental agency it has been playing a critical role in the formation of environmental institutions. The prefunding role for the foundation of Intergovernmental Panel on Climate Change (IPCC) was an example of this kind.

3.8 The Intergovernmental Panel on Climate Change (IPCC) Establishment

> *Post-World War II advances in basic atmospheric science that led to greatly increased understanding of the mechanisms of the large-scale circulation of the atmosphere; Initiation of a number of new geophysical observations (especially the Mauna Loa measurements of atmospheric carbon dioxide) during the 1957 International Geophysical Year; Recognition of the potential meteorological observing capabilities of Earth-orbiting satellites; The advent of digital computers; and The willingness of countries, even in the developing Cold War environment, to use the institutions of the United Nations System for cooperation in addressing important global problems* (John W. Zillman[42] 2009).

The grounds for foundation of the IPCC goes back to several scientific conferences held on the climate change and global environmental issues in various times and locations. The most important such conference was the First World Climate Conference held in Geneva from February 12-23, 1979 sponsored by the WMO, in collaboration with UNESCO, FAO, WHO, UNEP, ICSU and other scientific partners. More than 300 experts from over 50 countries participated in this conference (Weart 2004). The conference concluded that the climate change was a serious problem declaring that (a) to take full advantage of man's present knowledge of climate (b) to take steps to improve significantly that knowledge; and (c) "*to foresee and prevent potential man-made changes in climate that might be averse to the*

[42]Chairman of the International Organizing Committee for World Climate Conference-3; former President of WMO (1995-2003) and former President of the International Council of Academies of Engineering and Technological Sciences (2005); John Zillman (2009) A history of climate activities, WMO Bulletin Volume 58(3)

well-being of humanity" (emphasis added) (UNEP, 1990)[43]. This conference was the major breakthrough in advancing climate change knowledge, which also founded the World Climate Program (WCP), a major body of Climate Research under WMO.

> *"World Climate Program is the lead agency and coordinator, the purpose of which is to provide an authoritative international scientific voice on climate change and to assist countries in applying climate information and knowledge to sustainable development and the implementation of Agenda 21. It was started in 1979 as a successor to Global Atmospheric Research Program (GARP) with the major objectives of determining to what extent climate can be predicted and the extent of man's influence on climate. The four major components of WCP are the World Climate Data and Monitoring Program (WCDMP), the World Climate Applications and Services Program (WCASP), the World Climate Impact Assessment and Response Strategies Program (WCIRP) and the World Climate Research Program (WCRP)" (Baum, 1997; WMO, 2010)*[44].

UNEP, WMO and ICSU, through the WCP, coordinated a series of meetings and workshops between 1980 and 1985. The first international assessment was about the CO_2 issue; an expert meeting held in Villach, Austria, in November 1980 (World Climate Program, 1981; Agrawala 1998)[45]. In October 1982, the WCP (WMO/UNEP/ICSU) meeting in Geneva recommended that continuing assessments of $C0_2$, believed to be responsible for global warming, be held every five years, starting from the first meeting in

[43]UNEP (1990). How policy-makers are responding to global climate change. United Nations Environment Programme Information Unit for Climate Change Fact Sheet 201. Nairobi, Kenya

[44]Baum, Steve (1997) Glossary of Oceanography and the Related Geosciences with References, Texas Center for Climate Studies,Texas A&M University. http://oceanz.tamu.edu/~baum/paleo/paleogloss/node1.html http://www.wmo.int/pages/prog/wcp/wcdmp/wcdmp_home_en.html http://wcrp.wmo.int/wcrp-index.html (accessed on 03/11/2016).

[45]World Climate Program (1981). On the assessment of the role of CO2 on climate variations and their impact. Report of a WMO/UNEP/ICSU meeting of experts in Villach, Austria, November 1980, Geneva, WMO. Agrawala, Shardul (1998). Context and Early Origins of the Intergovernmental Panel on Climate Change, Climatic Change 39: 605-620.

1980. Following that meeting, an Interim Assessment was prepared (Morrissey 1998)[46]. These series of meetings concluded that climate change was a serious problem caused by the anthropogenic activities and also decided to arrange another international conference in Villach, Austria in 1985.

As a consequence, the joint UNEP/WMO/ICSU Conference was convened in Villach (Austria) from October 9-15, 1985 and attended by scientists from 29 countries, reached the following conclusions and made these recommendations:

1. *Many important economic and social decisions are being made today on long-term projectsmajor water resource management activities such as irrigation and hydro-power, drought relief, agricultural land use, structural designs and coastal engineering projects, and energy planningall based on the assumption that past climatic data, without modification, are a reliable guide to the future. This is no longer a good assumption since the increasing concentrations of greenhouse gases are expected to cause a significant warming of the global climate in the next century. It is a matter of urgency to refine estimates of future climate conditions to improve these decisions.*
2. *Climate change and sea level rises due to greenhouse gases are closely linked with other major environmental issues, such as acid deposition and threats to the Earth's ozone shield, mostly due to changes in the composition of the atmosphere by man's activities. Reduction of coal and oil use and energy conservation undertaken to reduce acid deposition will also reduce emissions of greenhouse gases, a reduction in the release of chloro-fluorocarbons (CFCs) will help protect the ozone layer and will also slow the rate of climate change.*
3. *While some warming of climate now appears inevitable due to past actions, the rate and degree of future warming could be profoundly affected by governmental policies on energy*

[46]Morrissey, Wayne A.(1998) Global Climate Change: A Concise History of Negotiations and Chronology of Major Activities Preceding the 1992 U.N. Framework Convention, UNT Libraries Government Documents Department, The University of North Texas Libraries http://digital.library.unt.edu/ark:/67531/metacrs527/m1/1/high_res_d/ (accessed on 03/11/2016).

conservation, use of fossil fuels, and the emission of some greenhouse gases (WMO 1986)[47].

The conference also made the recommendation stating that the UNEP, WMO and ICSU should establish a small task force on green- house gases, or take other measures, to:

i. *Help ensure that appropriate agencies and bodies follow up the recommendations of Villach 1985.*
ii. *Ensure periodic assessments are undertaken of the state of scientific understanding and its practical implications.*
iii. *Provide advice on further mechanisms and actions required at the national or international levels.*
iv. *Encourage research in developing countries to improve energy efficiency and conservation.*
v. *Initiate; if deemed necessary, consideration of a global convention* (WMO 1986).

The collaborative research between UNEP, WMO and ICSU was broken because of the international policy of the United Nation system. ICSU as a nongovernmental scientific organization was not responsive to any government (Agrawala 1998); however, the scientists and the scientific organization members of ICSU had been playing important roles to foster the climate change research.

Following the Villach conference, the World Meteorological Congress held in Geneva in May 1987 attested to the outcome of the Villach Conference. Prior to the conference, "Our Common Future," the World Commission on Environment and Development (the Brundtland Commission Report), was already published. The report, which showed the seriousness of the global environment, states that:

Failures to manage the environment and to sustain development threaten to overwhelm all countries. Environment and development

[47]WMO (1986). Report of the International Conference on the assessment of the role of carbon dioxide and of other greenhouse gases in climate variations and associated impacts, Villach, Austria, 9-15 October 1985, WMO No.661. SCOPE 29 - The Greenhouse Effect, Climatic Change, and Ecosystems, Statement by the UNEP/WMO/ICSU International Conference on The Assessment Of The Role Of Carbon Dioxide And Of Other Greenhouse Gases In Climate Variations And Associated Impacts Villach, Austria, 9-15 OCTOBER 1985 http://www.icsu-scope.org/downloadpubs/scope29/statement.html (accessed on 03/11/2016)

are not separate challenges; they are inexorably linked. Development cannot subsist upon a deteriorating environmental resource base; the environment cannot be protected when growth leaves out of account the costs of environmental destruction. These problems cannot be treated separately by fragmented institutions and policies. They are linked in a complex system of cause and effect (emphasis added: no 40; Our Common Future 1987)[48].

The report was also based on the Villach findings in highlighting global warming as a major threat to sustainable development (WCED 1987; Zillman 2009). The WMO executive council authorized the Secretary General to discuss the matter with the UNEP executive director and to form the Intergovernmental Panel on Climate Change (IPCC). The 70th UN general assembly of December 6, 1988, also recognizes the Villach outcome. The resolution states:

Recalling also the conclusions of the meeting held at Villach, Austria, in 1985, which, inter alia, recommended a program on climate change to be promoted by Governments and the scientific community with the collaboration of the World Meteorological Organization, the United Nations Environment Program and the International Council of Scientific Unions (UN 1988)[49].

The issue of the global climate change has been in discussion in the National and international forum; however, until 1988, there was no intergovernmental international authority who could collaborate with the world research centers and produce the global report on climate change. To fulfill this gap, the UN General Assembly Resolution 43/53 on December 6, 1988; under the title of "Protection of global climate for present and future generations of mankind" in resolution number (5) states:

"Endorses the action of the World Meteorological Organization and the United Nations Environment Program in jointly establishing an Intergovernmental Panel on Climate Change to provide

[48]Our Common Future (1987). Chapter 1: A Threatened Future UNDocuments Gathering a Body of Global Agreementsfrom A/42/427. Report of the World Commission on Environment and Development, http://www.un-documents.net/ocf-01.htm (accessed on 03/15/2017).

[49]UN general assembly resolution A/RES/43/53: of 70th plenary meeting of 6 December 1988 http://www.un.org/documents/ga/res/43/a43r053.htm (accessed on 03/11/2010).

internationally coordinated scientific assessments of the magnitude, timing and potential environmental and socio-economic impact of climate change and realistic response strategies, and expresses appreciation for the work already initiated by the Panel (UN 1988).

Further, in resolution 43/53 number (10) the UN general assembly gives the following mandate to the Executive Directors of the WMO and UNEP:

Requests the Secretary-General of the World Meteorological Organization and the Executive Director of the United Nations Environment Program, through the Intergovernmental Panel on Climate Change, immediately to initiate action leading, as soon as possible, to a comprehensive review and recommendations with respect to:

1. *The state of knowledge of the science of climate and climatic change;*
2. *Programs and studies on the social and economic impact of climate change, including global warming;*
3. *Possible response strategies to delay limit or mitigate the impact of adverse climate change;*
4. *The identification and possible strengthening of relevant existing international legal instruments having a bearing on climate;*
5. *Elements for inclusion in a possible future international convention on climate;*

The IPCC's immediate task was to prepare a comprehensive review and recommendations with respect to the state of knowledge of the science of climate change; social and economic impact of climate change, possible response strategies and elements for inclusion in a possible future international convention on climate (IPCC 2010)[50]. The role of the IPCC is:

To assess on a comprehensive, objective, open and transparent basis the scientific, technical and socio-economic information relevant to understanding the scientific basis of risk of human-induced climate change, its potential impacts and options for adaptation

[50]IPCC (2010). The History of IPCC, http://www.ipcc.ch/organization/organization_history.htm (accessed on 03/11/2010)

and mitigation. Review by experts and governments are an essential part of the IPCC process. The Panel does not conduct new research, monitor climate-related data or recommend policies. It is open to all member countries of WMO and UNEP" (IPCC 2004)[51].

At the outset, the IPCC elected Bert Bolin, a professor of meteorology from Sweden, as the first chair, a position in which he served until 1997, covering the first two IPCC assessments. Bolin's book (2007) is a thorough insider's view of the processes followed by the IPCC, using his perspectives. Although it focuses primarily on the substantive issues, it includes considerable detail on the procedural elements of the IPCC work.

In its initial organization, the respective secretariats played a key role. The IPCC set up three working groups dealing respectively with science, impacts and responses on the basis of a suggestion, according to Bolin, of Mostapha Tolba, the Executive Secretary of UNEP. Tolba, having been a negotiator at UNCED, was well aware of how multilateral negotiations took place. In the early period, the secretariats of UNEP and WMO were active in ensuring the IPCC would be set up to be effective. However, in doing so they decided to maintain a remarkably low profile, restricting their activities to providing the administrative services necessary for the IPCC to function, and essentially leaving the governance to the members themselves.

The three working groups were intended to draw on slightly different scientific constituencies because impact and responses would require factoring in research outside the physical sciences and would touch on political issues. Working Group, I would be dominated by climate scientists, while Working Groups II and III would have a wider participation, including, as time went on, economists and other social scientists. Each working group issued its own report and there was no common synthesis. As has been the case throughout, the Working Group I report was considered most important because it established the scientific basis for political negotiations. It also included a summary for policymakers that reflected a consensus of the different drafters.

As a given mandate the IPCC adopted its first assessment report on August 30, 1990 in Sundsvall, Sweden; in 1992, it published the Supplementary Reports; in 1994, the Special Report; in 1995, the Second IPCC Assessment Report; in 2001, the Third IPCC Assessment Report; and in 2007, the Fourth IPCC Assessment Report. The IPCCflyer for fourth assessment report states that for the 2007 report there were2500+ Scientific Expert

[51]http://www1.ipcc.ch/pdf/10th-anniversary/anniversary-brochure.pdf (accessed on 03/11/2017)

Reviewers, 800+ Contributing Authors and 450+ Lead Authors from 130+ Countries. This demonstrates that the IPCC has a major role in climate change knowledge production.

Conclusion

In this chapter, briefly discussed the origin of international organizations with the specific focus on international governmental organizations. The history of international organizations is moderately long; however, the environmental discourses are relatively new. The role of the international organizations in the beginning of 20th century, were mostly surrounded around the security issues. However, the Congress of Vienna, the League of Nations and the United Nations also paved the way for scientific development to address the global environment. At the government level, the United States and Switzerland were the pioneers for environment conservation. In terms of research and development, the United States is the leading country in natural resource management throughout history. Similarly, the historical account also shows that INGOs and NGOs also played critical roles in raising the awareness on global environment. Of the UN agencies, the flora and fauna International, the International Union for Conservation of Nature (IUCN), World Wildlife Fund, ICSU and several others played important roles in obtaining government commitments to address the global environment issues. As a result, thousands of international binding or non-binding convention treaties are in force. The foundation of IPCC is one of the best examples of a governmental scientific body providing the science base knowledge to global communities. The following chapter provides the more details on IPCC formalization and its working procedures.

Annex-1

The Major Environmental Agreements:

- 1872: The Swiss government proposed an international commission to protect migratory birds.
- 1900: Convention for the Preservation of Animals, Birds and Fish in Africa, which was signed in London by the European colonial powers with the intent to protect African game species, particularly to limit the export of ivory which was leading to severe hunting pressure on the African elephant.

(Continued)

- 1900: European littoral states sign treaty to regulate transportation of toxic substances on the Rhine River.
- 1906: Convention concerning the Equitable Distribution of the Waters of Rio Grande for Irrigation (US-Mexico water treaty)
- 1909: Canada-US Boundary Waters Treaty
- 1911: The North Pacific Fur Seal Commission was established by USA, Canada, USSR and Japan to regulate harvest of seals in North Pacific.
- 1918: US-Canada negotiate and sign the Migratory Bird Treaty Act, designed to protect bird species–particularly waterfowl–that seasonally migrate between the two nations.
- 1923: Convention for the preservation of the habitual fishery of the Northern Pacific Ocean and the Bering Sea
- 1931: First international convention to discuss the regulation of commercial whaling, eventually led (in 1946) to the International Convention for the Regulation of Whaling and the establishment of the International Whaling Commission, a permanent body responsible for negotiating & setting policy re- the harvest and preservation of whales.
- 1940: Convention on Nature Protection and Wildlife Conservation in the Western Hemisphere.
- 1946: Founding of the United Nations and World Bank, centerpieces for an international effort to promote world peace and post-war reconstruction. These institutions subsequently played leading roles in international environmental cooperation, through the World Bank, UN-IMCO, FAO, UNDP, WHO, and later, UNEP and UNESCO.
- 1946: International Convention for the regulation of Whaling
- 1950: International Convention to Protect Birds
- 1951: International Plant Protection Convention
- 1954: International Convention for the Prevention of Pollution of the Sea by Oil, signed in London, the culmination of 28 years of negotiations by Western European and North American nations.
- 1958: International Maritime Consultative Organization (UN-IMCO) established, assuming principal responsibility for negotiating international agreements on ocean pollution.
- 1959: Antarctic treaty
- 1963: Treaty Banning Nuclear Weapon Tests in the Atmosphere, in outer space and under water
- 1964: Agreed measures on the conservation of Antarctic flora and fauna
- 1962/69: Amendments to the International Convention on Oil Pollution.
- 1971: RAMSAR Convention on Wetlands of International Importance Especially as Waterfowl Habitat.
- 1972: UNESCO-sponsored Convention for the Protection of World Cultural and Natural Heritage, which designates World Heritage Sites.
- 1972: Oslo Convention for the Prevention of Marine Pollution by Dumping from Ships and Aircraft.
- 1972: Great Lakes Water Quality Agreement (US - Canada).

(*Continued*)

- Key agreements on marine pollution in the North Sea and East Atlantic.
- 1972: Stockholm-United Nations Conference on the Human Environment.

Outputs:

- Stockholm Declaration of the UN Conference on the Human Environment: 26 principles, intended as a foundation for future developments in international environmental cooperation.
  Action Plan for the Human Environment: consisting of 109 recommendations for governmental and intergovernmental action across the full range of environmental policy issues, ranging from species conservation, forests and atmospheric and marine pollution, to development policy, technology transfer and impact of environment on trade.
  Resolved to establish United Nations Environment Program (UNEP) and the Environment Fund.
- 1972: London Convention for the Prevention of Marine Pollution by Dumping of Wastes and Other Matter (restricts toxic & nuclear waste dumping at sea).
- 1973: International Convention for the Prevention of Pollution from Ships (MARPOL) (restricts release/dumping of oil, garbage, sewage, ballast waters, etc.).
- 1974: Paris Convention for the Prevention of Marine Pollution from Land-based Sources intended to control land-based pollution to the North Sea.
- 1974: Helsinki Convention on the Protection of the Marine Environment of the Baltic Sea.
- 1975/80: Mediterranean Action Plans: control marine and land-based pollution.
- 1973: Convention on the International Trade of Endangered Species (CITES).
- 1974/84: First and Second UN Population Conferences: contentious events that, nonetheless, helped to focus attention and coordinate support for implementation of family planning programs in many countries.
- 1979: Convention on Long-Range, Transboundary Air Pollution (LRTAP). Negotiated between Canada, the US and European countries primarily in response to concerns about acid rain, this was the first major international effort to regulate air pollution.
- 1980: World Conservation Strategy. Coordinated by IUCN/WWF/UNEP, this was a major effort sponsored by non-government agencies to promote national conservation programs in LDCs.
- 1982: UN Conference on the Law of the Sea (UNCLOS). Established 200 mile territorial jurisdictions over coastal waters.
- 1982: Whaling moratorium adopted by IWC.
- 1983: International Tropical Timber Agreement (formation of ITTO: Int. Tropical Timber Organization)
- 1985: Helsinki Protocol on the Reduction of Sulphur Emissions.
- 1985: Vienna Convention for the Protection of the Ozone Layer, established initial targets for gradual reductions in CFC production.
- 1987: Montreal Protocol (London Amendments, 1990) on Substances that Deplete the Ozone Layer, established specific time-tables for reductions and phase-out of CFC's by the turn-of-the-century, and established financial mechanism (Ozone Fund) to assist LDCs and former Soviet Bloc nations in phase-out.

(*Continued*)

- 1987: Our Common Future published (Report of the World Commission on Environment & Development/Brundtland Commission)
- 1988: Intergovernmental Panel on Climate Change formed by UNEP & WMO.
- 1989: Basel Convent on Control of Transboundary Movements of Hazardous Wastes and Their Disposal.
- 1990: Kingston Protocol on Specially Protected Areas and Wildlife in the Caribbean
- 1991: Protocol on Environmental Protection of Antarctica, established a moratorium on mineral and related exploration and development for 50 years.
- 1991: Canada-US Air Quality Agreement, reducing emissions that cause acid rain
- 1991: European Union: Major progress on international environmental efforts, with 280 items of environmental legislation ranging across a range of policy areas, including toxics, water quality, waste management, air pollution, wildlife protection and noise pollution.
- 1992: Rio-United Nations Conference on Environment & Development.

Outputs:

- Rio Declaration: statement of key principles for environment & development
  Agenda 21: detailed list of recommendations
  Statement of Forest Principles (scaled down from Forest Convention)
  Biodiversity Convention (signed by 153 countries, but not US)
  Climate Change Convention
  Global Environment Facility (GEF)
  Established UN Commission on Sustainable Development to review progress of Rio efforts
- 1993: North American Commission for Environmental Cooperation, established as side agreement to North American Free Trade Agreement with the intent of addressing environmental problems and arbitrating related conflicts that arise through international trade between Mexico, US and Canada.
- 1994: UN Convention to Combat Desertification (particularly in Africa).
- 1994: 3rd International Population Conference, Cairo, established broad consensus over need to make women's issues–health, education, employment, rights & empowerment–as central to concerns of family planning, fertility management and social development.
- 1995: Beijing International Conference on Women & Development
- 1996: Protocol to the 1972 London Convention on the Prevention of Marine Pollution by Dumping of Wastes and Other Matter
- 1997: International agreement to reduce the production, storage and use of land mines.
- 1997: Kyoto Protocol on the Reduction of Greenhouse Gases: Established first binding, numerical targets for reducing greenhouse gases.
- 1998: Rotterdam Convention on Prior Informed Consent for Trade in Hazardous Chemicals and Pesticides
- 1999: World Trade Organization meeting in Seattle crashed by environmental protests.

(*Continued*)

- 1999: Canada-US Pacific Salmon Treaty renewed
- 2000: Ozone Annex to the 1991 Canada-US Air Quality Agreement, reducing emissions that cause smog (especially NO-x)
- 2001: Cartegena (Bio-Safety-GMO) Protocol to UN Convention on Biological Diversity
- 2001: UN-Stockholm Convention on Persistent Organic Pollutants (DDT, PCBs, dioxin, furans)
- 2001: Bonn Framework Agreement for the Kyoto Protocol of the UN Convention on Climate Change
- 2002: Rio + 10: UN World Summit for Sustainable Development, Johannesburg
- 2005: Kyoto Protocol of 1997 comes into force
- 2005 Bali Strategic Plan for Technology Support and Capacity Building adopted by UNEP Governing Council mandating national level support to developing countries
- 2005 Millennium Ecosystem Assessment highlights the importance of ecosystems to human well-being, and the extent of ecosystem decline
- 2005 World Summit agrees to explore a more coherent institutional framework system for international environmental governance
- 2006: Asia-Pacific Partnership on Clean Development and Climate
- 2007: UNFCC-Bali Conference on Climate Change: post-Kyoto road-map
- 2009: COP 15 Copenhagen

Sources: Various including UNEP 1992; Reinalda 2009: Mitchell 2010, and http://www.mta.ca/faculty/socsci/geograph/genv4111/International%20laws.pdf *(accessed on 03/11/2010)*

References

Agoumi, Ali (2003). Vulnerability of North African Countries to Climatic Changes, Adaptation and Implementation Strategies for Climate Change, International Institute for Sustainable Development, Canada (funded project by USAID) http://www.cckn.net/pdf/north_africa.pdf (accessed on 03/06/2010)

Archer, Clive (1992). International Organizations (2nd Edition), Routledge, USA.

Barnett, Michael (1997a) The Politics of Indifference at the United Nations and Genocide in Rwanda and Bosnia. In This Time We Knew: Western Responses to Genocide in Bosnia, edited by Thomas Cushman and Stjepan Mestrovic, 128-62. New York: New York University Press.

Bhandari, Medani P. (2017). The Pedagogical Development of the International Organization and Organizational Sociology Theories and Perspectives, in Douglass Capogrossi (Ed.). Educational Transformation: The University as Catalyst for Human Advancement. Xlibris Corporation, USA (in press).

Bhandari, Medani P. and Bhattarai, Keshav (2017). Institutional Architecture for Sustainable Development: A Case Study from India, Nepal, Bangladesh, and Pakistan, SocioEconomic Challenges, Volume 1, Issue 3, 2017, ARMG Publishing, (6-21)

Bhandari, Medani P. (2012). Environmental Performance and Vulnerability to Climate Change: A Case Study of India, Nepal, Bangladesh, and Pakistan. Climate Change and Disaster Risk Management. Series: Climate Change Management, pp. 149-167. Springer, New York / Heidelberg, ISBN 978-3-642-31109-3.

Bhandari, Medani P. (2012a). Exploring the International Union for the Conservation of Nature (IUCN's) National Program Development in Biodiversity Conservation: A Comparative Study of India, Pakistan, Nepal, and Bangladesh. Sociology – Dissertations. 73. Available at http://surface.syr.edu/soc_etd/73.

Brechin, Steven R. and Bhandari, Medani P. (2011). Perceptions of climate change worldwide,WIREs Climate Change 2011, Volume 2:871–885.

Grant, Charles (2001). A European View of ESDP, Centre for European Policy Studies working paper (April 2001).

Ivanova, Maria (2007). Moving Forward by Looking Back: Learning from UNEP's History, in Global Environmental Governance: Perspectives on the Current Debate (Lydia Swart and Estelle Perry, eds.), New York: Center for UN Reform Education, 2007.

Kratochwil, Friedrich (1994). 'Citizenship: On the Border of Order'. Alternatives, 19

Kratochwil, Friedrich (2006). Constructing a New Orthodoxy?: Wendt's Social Theory of International Politics and the Constructivist Challenge in Stefano GUZZINI (ed), Anna LEANDER (ed), Constructivism and International Relations: Alexander Wendt and his Critics, New York, Routledge, 2006, 21-47

Lucas, Robert E. (1976). "Econometric Policy Evaluation: A Critique." Journal of Monetary Economics, 2, Supplement (1976), Carnegie-Rochester Conference Series, Vol. 1.

Lucas, Robert E. (1977). "Understanding Business Cycles." Journal of Monetary Economics, Supplement, Carnegie-Rochester Conference Series, Vol. 5.

Pentland C. (1973). International Theory and European Integration, Free Press, NY.

Pangle, T. L.&Ahrensdorf, P. J. (1999). Justice among nations: on the moral basis of power and peace, University Press of Kansas.

Potter, Pitman B.(1922). An introduction to the study of international organization, The Century Co publisher.

Potter Pitman B. (1945). Origin of the Term International Organization, The American Journal of International Law, Vol. 39, No. 4 (Oct., 1945), pp. 803-806,

Reinalda, Bob (2009).Routledge History of International Organizations, From 1815 to the Present Day, Routledge, member of the Taylor & Francis Group, an Informa Business, UK, USA.

UNEP (1992). The world environment 1972-1992: Two decades of challenge, ed. M. K. Tolba, O. A. El-Kholy, E. El-Hinnawi, M. W. Holdgate, D. F. McMichael, and R. E. Munn, 529-67. New York: Chapman & Hall.

UNEP (2009). Organization Profile, United Nation Environment Program-UNEP, Nairobi, Kenya. http://www.unep.org/PDF/UNEPOrganization Profile.pdf (accessed on 03/08/2010)

Weart, Spencer (2004). The Discovery of Global Warming, Harvard University Press, paperback edition +online edition.

Young, O.R. (1999) The Effectiveness of International Environmental Regimes, MIT Press, Cambridge, MA.

4

Getting the Facts Right[*]– IPCC – Formalization and its Report Procedures[†]

Previous chapter showed the historical scenarios of how international organizations became concern on climate change science and how climate change became global agenda. The formation of IPCC is the outcome of various conferences, meetings, workshops of many international organizations (mostly UN), and also numerous efforts of scientific communities.

4.1 Introduction: Science, Politics and Regimes

Media commentators and political leaders who are opposed to dealing with climate change have often questioned the validity and reliability of the scientific predictions about the causes and consequences of global warming. To do so, they have often attacked the scientists as academic, or biased. The recent "climate-gate" scandal, based on hacking of the e-mails of some scientists in the United Kingdom has helped fuel skepticism among critics about the value of the scientific arguments (Revkin, 2009). The evidence, however, shows that the scientific consensus arrived at by the Intergovernmental Panel on Climate Change (IPCC) is a solid one, given the composition of the panel, and an innovative means of connecting science with politics.

Determining the facts of climate change obviously involves the relationship of science and politics. The relationship is complex, since the criteria for truth are different in the two spheres. Albert Einstein is said to have remarked

[*]This chapter is modified from Getting the Facts Right and part of it was published in the Yale Conference Proceedings. The initial ideas belong to Mathison, who insist me to complete this book, originally, we planned to author together. Due to his business he encouragement me to complete.

[†]Thanks, are in order to Professor Mathiason's graduate assistants who coded the participation lists, including Sarah Yagoda, Uwe Gneiting, Katherine Aston and Pierpaolo Capalbo.

when asked "why is it that when the mind of man has stretched so far as to discover the structure of the atom we have been unable to devise the political means to keep the atom from destroying us?' And he replied:' That is simple, my friend. It is because politics is more difficult than physics." (Clark, 1955).

Dealing with climate change involves creating an international regime, a set of arrangements among States and other stakeholders designed to solve a global problem that cannot be solved by individual nation-states. And, because it involves the physical world, it must engage science as well as politics. This analysis shows how the IPCC bridges the two fields, by getting the facts right so the policies can be effective.

Regime theory is increasingly being used to explain the process of negotiating the international agreements to solve global problems. Developed to help explain the creation of regimes like the law of the sea, regimes were defined as "sets of implicit or explicit principles, norms, rules, and decision-making procedures around which actors' expectations converge in a given area of international relations." (Krasner 1983, 2). While this was not originally designed to describe the negotiation process, it serves this well since, in practice, agreements must be reached sequentially on principles, norms, rules and procedures.[1]

The sequence of steps to create a regime starts with an agreement on the facts. Making this possible for managing climate change has been the continuing contribution of the International Panel on Climate Change (IPCC). Beyond this, however, the IPCC has established a key precedent for how to achieve a multi-stakeholder approach to dealing with global problems.

The process of deciding on the principles, which, as Krasner put it, "... are beliefs of fact, causation, and rectitude," (Krasner 1983, p. 2). can be very long. Deciding on the shape of a problem is a requisite to deciding what to do about it. If there is no agreement on the principles, it is almost impossible to agree on the other elements of the regime. State obligations that would be set out in norms and rules depend on the nature of the problem and its facts, and decision-making procedures similarly depend on these facts, if only to verify whether States are in compliance. Much of the debates during the negotiation have to do with which facts to believe and how to apply them to remedies.

In the case of climate change, agreeing that there was a problem, and what caused it, took about twenty years, as has been very well documented by Paterson (1996, Chapter 1) and Bolin (2007). One problem is that there

[1]This was discussed extensively in a paper presented to the International Studies Association in 2007 (Mathiason, 2007).

are natural changes in climate, so any data on change has to be seen in that context. A second problem is that climatic changes are gradual, so many of the scientific conclusions must be based on a projection into the future, using models that can be, and have been, questioned. Before non-scientists can be convinced that there is a problem, scientists must achieve their own consensus.

Although many climate scientists had begun to accept the hypothesis that human behavior could alter the climate over the first two-thirds of the 20th Century, the idea was given an initial boost at the first United Nations Conference on the Human Environment in Stockholm in 1972. While pollution rather than climate change was the main focus of the conference, at least one recommendation suggested a need to look at climate.

Stockholm Conference (SC) recommendation 70 read:

> It is recommended that Governments be mindful of activities in which there is an appreciable risk of effects on climate, and to this end:
> Carefully evaluate the likelihood and magnitude of climatic effects and disseminate their findings to the maximum extent feasible before embarking on such activities;

And SC Recommendation 79 set in motion a larger process:

It is recommended:

> (a) That approximately 10 baseline stations be set up, with the consent of the States involved, in areas remote from all sources of pollution in order to monitor long-term global trends in atmospheric constituents and properties which may cause changes in meteorological properties, including climatic changes;
> (b) That a much larger network of not less than 100 stations be set up, with the consent of the States involved, for monitoring properties and constituents of the atmosphere on a regional basis and especially changes in the distribution and concentration of contaminants;
> (c) That these programmes be guided and coordinated by the World Meteorological Organization;
> (d) That the World Meteorological Organization, in cooperation with the International Council of Scientific Unions (ICSU), continue to carry out the Global Atmospheric Research Programme (GARP), and if necessary establish new programmes to understand better the general circulation of the atmosphere and the causes of climatic changes whether these causes are natural or the result of man's activities.

4.2 Science and International Organizations

That public international organizations would become facilitators of scientific endeavors was one of the changes in the relationship of science and politics that grew in the late 20th century (Zurn 1998; Andresen, Skodvin, Underdal and Wettestad, 2000; Miller, 2001). The traditional international face of science was through non-governmental or quasi-governmental organizations like the International Conference of Scientific Unions (ICSU) founded in 1931 in France or the International Union for the Conservation of Nature (IUCN) that founded in 1948 also in France, but now headquartered in Switzerland) that brought together national scientific unions in different disciplines. As Miller (2001, 495) notes that "The relationship between science and politics has become increasingly sophisticated over the past half-century. Three features of this growing sophistication stand out". These include (1) an increasingly complex institutional landscape where elements of science and politics are mixed; (2) the construction of objective knowledge and authoritative orderings of society increasingly requires nuanced arrangements that orchestrate activities in the worlds of both science and politics; and (3) the discourses, material artifacts, and institutions that increasingly populate all three domains are hybrids, complex mixtures of facts and values.

While they could partner with governments at the international level, they were not formally part of the international governance structure in the same way as international public organizations of the United Nations System. Connecting these non-governmental institutions with the World Meteorological Organization, which linked national public meteorological services that had a formal interest in climate, a clear science-politics link could be secured. The WMO saw itself as a technical agency somewhat divorced from the political issues that affected other organizations, but was by being intergovernmental, connected to politics. Its Secretary-General was inevitably from a public meteorological entity.

The World Climate Programme, established in 1979, was one result of the collaboration, and its activities led to a slow, but increasing consensus among the climate scientists, led by meteorologists, that much of the observed climate change was caused by human behavior. The program ran a series of workshops and seminars that gradually mobilized this consensus culminating in an international conference at Villach, Austria in 1985 on "Assessment of the Role of Carbon Dioxide and Other Greenhouse Gases in Climate Variations and Associated Impacts". The Villach Conference was the first to express a scientific consensus that based on projections, there would be a significant increase in global temperatures and it led to a movement of the climate change debate to the political sphere.[2]

4.3 The IPCC's Origins and Role

The governments participating in the World Meteorological Organization and UNEP Governing Council recognized the importance of obtaining a scientific consensus on climate. These two bodies agreed to the establishment of an Intergovernmental Panel on Climate Change in 1988. Paterson (1996, 40) reports that the United States instigated the establishment of the IPCC. The recommendation was endorsed by the General Assembly in its resolution 43/177 of 6 December 1988 on Protection of Global Climate for Present and Future Generations of Mankind. The resolution stated: "*Endorses* the action of the World Meteorological Organization and the United Nations Environment Programme in jointly establishing an Intergovernmental Panel on Climate Change to provide internationally coordinated scientific assessments of the magnitude, timing and potential environmental and socio-economic impact of climate change and realistic response strategies, and expresses appreciation for the work already initiated by the Panel" (IPCC 2020-history).

The IPCC began its work almost immediately, in order to prepare an assessment for the Second World Climate Conference in November 1990.

The role of the IPCC is "to assess on a comprehensive, objective, open and transparent basis the scientific, technical and socio-economic information relevant to understanding the scientific basis of risk of human-induced climate change, its potential impacts and options for adaptation and mitigation. Review by experts and governments is an essential part of the IPCC process. The Panel does not conduct new research, monitor climate-related data or recommend policies. It is open to all member countries of WMO and UNEP". (IPCC Brochure 2007)

4.4 IPCC Reports: Procedural Agreements

The institutional and procedural agreements set at the outset have continued over the next 21 years. The mandate has remained the same, the structure of working groups also generally the same and the procedures being followed to

[2]1986 Report of the International Conference on the Assessment of the Role of Carbon Dioxide and of Other Greenhouse Gases in Climate Variations and Associated Impacts, Villach, Austria, 9–15 October 1985 World Meteorological Organization, [Paris]

complete the assessments the same. A change is that the three working groups are now called, respectively, science, adaptation and mitigation.

The procedure for developing, drafting and approving an assessment has been remarkably consistent over the four assessments that have been completed (1990, 1995, 2001 and 2007, 2014 and 2018- on Global warming of 1.5°C). The current procedures were worked out in detail in 1999 and amended in 2003 in an appendix to the Principles Governing IPCC Work. Figure 1 shows the process (as described by the IPCC). It reflects a combination of the procedures used in science with those used in multilateral governance. The first stage involves an effort to survey scientific research systematically to answer questions that are particularly relevant for policy. Bolin (2007, 165) reports that the nine major papers leading up to the Third Assessment in 2001 contained over 25,000 references to papers in the relevant scientific and technical literature. There is exponential increase of such reference in 2007 and 2014 reports. Thousands of experts are working for the next report, which will come in 2021.

In the second phase, the initial draft is subject to an intensive peer-review by other scientists to assess the quality of the drafts. This is a procedure applied to determining whether research can be published in scientific journals. On the basis of the review, the drafts are revised to address comments and concerns. The process of selecting both the drafters and reviewers has always been a matter, but has largely been driven by the same criteria that would be applied to selecting reviewers for a journal.

The third phase derives from intergovernmental practice, where a draft is submitted to a governmentally-influenced review. In other United Nations contexts this is done through expert groups in which proposals are put on the table and considered by a group that is mindful of both factual and political considerations. On this basis, a final draft is prepared that can be accepted my most governments. The sequence of phases leads to as addressing or eliminating issues that might impede a consensus about the facts.

The final phase is adoption by the IPCC (or in some cases, by the working group itself) with predominance given to governmental input. Once this is done, the report has intergovernmental standing and can be used in other contexts. This was important for the final negotiations of the United Nations Framework Convention on Climate Change (First Assessment Report), the Kyoto Protocol (Second Assessment Report) and the Bali Roadmap (Fourth Assessment Report). Fifth Assessment Report (AR5) was completed in 2014, which noted “Warming of the climate system is unequivocal, and since the 1950s, many of the observed changes are unprecedented over decades to

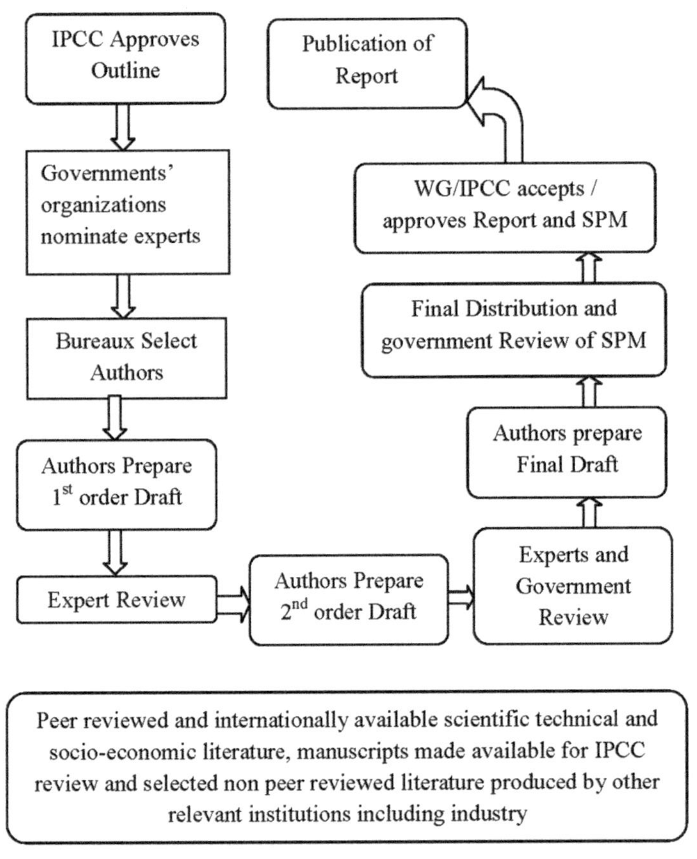

Figure 4.1 IPCC Assessment Preparation Process

Source: http://www1.ipcc.ch/ipccreports/index.htm as of 3 November 2019.

millennia" (IPCC 2014); Without new policies to mitigate climate change, projections suggest an increase in global mean temperature in 2100 of 3.7 to 4.8°C, relative to pre-industrial levels (median values; the range is 2.5 to 7.8°C including climate uncertainty) (IPCC 204).

The procedures for decision-making in the IPCC have been formalized since the beginning. The current Principles Governing IPCC Work were originally approved in 1998 and were amended in 2003 and 2006, usually to respond to concerns and to improve the process. A key element in decision-making is consensus. As the Principles (No. 10) state:

> In taking decisions, and approving, adopting and accepting reports, the Panel, its Working Groups and any Task Forces shall use all

> best endeavours to reach consensus. If consensus is judged by the relevant body not possible: (a) for decisions on procedural issues, these shall be decided according to the General Regulations of the WMO; (b) for approval, adoption and acceptance of reports, differing views shall be explained and, upon request, recorded. Differing views on matters of a scientific, technical or socio-economic nature shall, as appropriate in the context, be represented in the scientific, technical or socio-economic document concerned. Differences of views on matters of policy or procedure shall, as appropriate in the context, be recorded in the Report of the Session.

The purpose of this is to eliminate the possibility that the findings could be discredited on the grounds that alternative points of view were not considered.

Who are the IPCC?

The public face of the IPCC is its overall chair and to a lesser extent the working group chairs, who together with vice-chairs and regional representatives constitute the IPCC Bureau. The Bureau in the IPCC, as in most intergovernmental bodies, functions primarily to make decisions about procedural matters. There have been three chairs of the IPCC. The first, as noted, was Bert Bolin, from Sweden, who organized the first two appraisals. The second was Robert Watson, from the United States, who organized the third appraisal. The third is Rajendra Pachauri, from India, who organized the fourth appraisal and is organizing the fifth.

Early in Bolin's chairmanship and on his proposal (Bolin, 2007, 82), the IPCC decided to ensure balance in the working groups by having co-chairs. Using the usual United Nations approach to geographical balance, one of the co-chairs was from a developing, country and the other from a developed country. Over time, there has been some movement among these figures. For example, Watson was originally the chair of Working Group III in the FAR, eventually moving up to Chair of the IPCC.

However, the IPCC is composed of a much larger group of persons. In order to prepare the drafts, there is a complex system of lead authors, coordinating lead authors, contributing authors, expert reviewers, review editors and government focal points. As set out in the procedures found in Appendix A of the Principles (4.2.1), the compilation of lists is a very open-ended process:

> At the request of Working Group/Task Force Bureau Co-Chairs through their respective Working Group /Task Force Bureau, and

> the IPCC Secretariat, governments, and participating organisations and the Working Group/Task Force Bureaux should identify appropriate experts for each area in the Report who can act as potential Coordinating Lead Authors, Lead Authors, Contributing Authors, expert reviewers or Review Editors. To facilitate the identification of experts and later review by governments, governments should also designate their respective Focal Points. IPCC Bureau Members and Members of the Task Force Bureau should contribute where necessary to identifying appropriate Coordinating Lead Authors, Lead Authors, Contributing Authors, expert reviewers, and Review Editors in cooperation with the Government Focal Points within their region to ensure an appropriate representation of experts from developing and developed countries and countries with economies in transition. These should be assembled into lists available to all IPCC Members and maintained by the IPCC Secretariat.

The process is intended to be inclusive, but with some accountability, and if it is successful, the participants will clearly represent the range of scientific opinion and knowledge. The full composition of the IPCC has not been analyzed before. For this paper, we analyzed the formal lists of participants for 2001 and 2007 in terms of country of nationality, type of organization for which they worked and profession.[3]

Over the last two assessments, scientists from 106 different countries have been members of the IPCC working groups. As would be expected, scientists from the industrialized countries heavily dominate, although that has been changing slowly. Table 1 shows the distribution of the members of the IPCC for the Third (2001) and Fourth (2007) assessments by country group. These are the main caucusing groups in climate change negotiations. JUSCANNZ consists of Japan, the United States, Canada, Norway, Australia and New Zealand, however because of its predominance; the United States is shown separately from the other members in the tables. The other developed countries category includes the Russian Federation and other former states of

[3]To obtain the list of participants we used the secondary documentary sources published by the IPCC itself, mostly online. We compared and compiled the list of IPCC experts from the 2001 and 2007 publications' appendixes where list of experts is listed and put them into an Excel spreadsheet for coding and analysis. Coding was done in each case by two coders using standard inter-coder reliability checks. When information was missing, we used on-line search engines. In addition to the secondary data, we also collected field data (through interviews) from India, Nepal, Bangladesh, and Pakistan, to get to know whether environmental experts of South Asia were actually aware of the contribution of IPCC's role.

Table 4.1 Membership in the IPCC by Country Group and Assessment

Country Group	*2001*	*2007*
United States	36.9%	25.6%
Other JUSCANNZ	16.2%	18.0%
European Union	28.5%	32.3%
Other Developed	3.6%	6.0%
G-77 and China	14.8%	18.2%
TOTAL NUMBER	1429	1268

the Soviet Union that have not joined the Group of 77 as well as South Korea, Mexico and Switzerland. The main change that can be noted is the reduction in the proportion of scientists from the United States. The geographical distribution is improving over time. In 2007, scientists from 91 countries were involved, compared with 79 countries in 2001. The member of scientist from developing countries are increasing because in most of the Conference of Parties, known as COP, United Nations Framework Convention on Climate Change participants has been raising the questions on involvement of developing countries scientists in IPCC working modality (see Regime and Climate change chapter).

This summary table, (which is just an example), however, masks some significant differences by working group. Working Group I on science is more likely to depend on the well-funded scientific research programs, either in universities or government, that are only found in the industrialized countries. As can be seen from Table 2, the dominance of these countries is more pronounced. Although the representation of scientists from developing countries has almost doubled, this is from a low base and reflects a slight drop in the proportion of United States scientists that are involved. Similar trend can be seen in 2014 reports, however, the participation of development countries scientists increased sharply (https://www.ucsusa.org/resources/ipcc-who-are-they-Published Jul 16, 2008 Updated Oct 11, 2018).

Table 4.2 Working Group I by Country Group and Assessment

Country Group/Assessment	2001	2007
United States	38.7%	33.4%
Other JUSCANNZ	16.7%	17.6%
European Union	34.1%	34.1%
Other Developed	4.0%	5.7%
G-77 and China	6.5%	9.2%
Total number	618	619

In contrast, Working Group II, which deals with adaptation, has had a growing relative participation of scientists from developing countries, although much of that is due to a reduction in the participation of scientists from the United States and a general reduction in the number of individuals participating in the Working Group. In contrast, the geographical composition of Working Group III, dealing with mitigation, has been stable.

One factor that is noticeable in the figures is that Working Group I has a much larger participation than the other two. This may reflect the idea that the results of Working Group I, whose conclusions form the basis for the work of the other two Groups, needs to be comprehensive. It may also reflect a concern that the voluminous and highly technical hard science behind the conclusions needs a firm consensus that can only be credible if it involves a large number of scientists.

The IPCC is formally intergovernmental, but the participants can come from a variety of sources. While all must be accepted by the governments who oversee the Panel, slightly less than half of the participants work for the government. Most of the rest come from universities. The only change over time has been a slight decline in the proportion coming from government and a corresponding increase in personnel from universities (Table 5). This primarily took place in Working Group II, as can be seen from Table 6. Perhaps more significantly, the composition of Working Group III dealing with mitigation has always the lowest percentage of government employees,

Table 4.3 Working Group II by Country Group and Assessment

Country Group/Assessment	2001	2007
US	38.4%	15.8%
Other JUSCANNZ	16.2%	18.9%
EU	23.6%	31.1%
Other Developed	2.7%	7.6%
G-77 and China	19.1%	26.6%
Total Number	593	380

Table 4.4 Working Group III by Country Group and Assessment

Country Group/Assessment	2001	2007
US	27.9%	21.2%
Other JUSCANNZ	14.4%	17.5%
EU	26.1%	29.7%
Other Developed	4.5%	4.5%
G-77 and China	27.0%	27.1%
Total Number	222	269

Table 4.5 Place of employment by Assessment

Place of employment/Assessment	2001	2007
Government	48.2%	41.4%
Universities	37.0%	43.2%
Non-governmental	8.2%	6.6%
Private Sector	4.6%	4.6%
International Organization	2.0%	4.2%

Table 4.6 Composition of Working Groups by Place of Employment

	Working Group I		Working Group II		Working Group III	
Place of employment/ Assessment	2001	2007	2001	2007	2001	2007
Government	51.6%	53.3%	53.4%	30.8%	25.0%	29.0%
Universities	37.4%	40.4%	34.7%	50.8%	42.3%	39.0%
Non-governmental	8.4%	4.2%	5.9%	7.6%	13.6%	10.8%
Private Sector	1.9%	0.8%	4.2%	3.2%	12.7%	15.2%
International Organization	0.6%	1.3%	1.7%	7.6%	6.4%	5.9%

but perhaps more significantly it has a sizeable (one quarter) proportion of persons who are from either non-governmental organization, the private sector or international organizations.

There are also geographical differences in where the participant scientists are employed. In the United States, there has been a consistent balance between government and university employment, while in the G-77 countries, with the exception of Asia, the balance is tipped to universities. The exception is due to Chinese participants, most of whom work for government institutions. Over time, for most country groups the proportion of participants from universities has increased.

In both 2001 and 2007, almost all of the participants in Working Group I were physical scientists. However, in 2007, a quarter of the participants in Working Group II were social scientists and in Working Group III, over half of the participants were social scientists. This reflected a concern noted by Bolin that to deal with the policies necessary for adaptation and mitigation, it was necessary to engage economists, sociologists and other social scientists to explore the implications of the scenarios generated by the physical scientists in Working Group I.

There is some overlap between the two assessments. Of the total number of individuals who have been involved, 18 percent participated in both the 2001 and 2007 assessments. Most were involved in only one, providing a

Table 4.7 Composition of Working Group I by Country, Place of Employment and Assessment

Assessment	2001			2007		
Country Group/ Place of Employment	Government	Universities	Other	Government	Universities	Other
US	47.8%	38.9%	13.2%	40.1%	47.2%	12.7%
Other JUSCANNZ	62.8%	31.6%	5.6%	50.0%	39.9%	10.1%
EU	40.6%	36.9%	22.5%	38.9%	41.3%	19.8%
Other Developed	43.7%	49.7%	6.6%	66.7%	16.7%	9.2%
Asian Developing	57.7%	27.9%	14.4%	47.7%	34.2%	18.0%
Africa	35.3%	51.0%	13.7%	34.4%	48.4%	17.2%
Latin America and Caribbean Developing	37.5%	39.3%	23.2%	30.4%	48.2%	21.4%
TOTAL	46.5%	39.3%	14.2%	44.2%	39.8%	19.0%

Table 4.8 Proportion of Scientists Who were in both 2001 and 2007 Assessments

Year	WG I	WG II	WG III	TOTAL
2001	33.5%	21.5%	37.8%	29.2%
2007	32.6%	32.6%	31.5%	32.4%
TOTAL	19.8%	14.9%	20.8%	18.1%

remarkable level of renewal. In any of the two assessments, between 22 and 38 percent of the working groups included persons who were ultimately in both, as is shown in Table 8. (Note that the total for each working group includes both those scientists who are only in one assessment and those in both. The latter group are double counted in the assessment years but counted only once in the total).

There are few differences in this by country group. Only 16 and 17 percent of the European Union and United States participants respectively carried over from 2001 to 2007. Up to a quarter of the participants from Group of 77 countries carried over.

The mix between carry-over and new scientists strengthens the notion that the consensus is solid. There is no evident "old boys" network in the IPCC, but rather that the IPCC has benefited from new research and new researchers.

Over time, the importance of the assessments to governments in helping drive the negotiation process has clearly increased. The Fifth Assessment is expected by 2014 and initial scoping discussions have already been held. Many of these took place at a special expert group meeting in Venice in July 2009. Table 9 shows the composition of the participants by country group

Table 4.9 Participation in July 2009 Expert Meeting on Scoping for the Fifth Assessment

Country Group/ Institutional type	Govern-ment	Universi-ties	Non-Governmental	Private Sector	Number Country Group	Percent Country Group
United States	19.4%	58.3%	19.4%	2.8%	36	20.0%
Other JUSCANNZ	33.3%	62.5%	0.0%	4.2%	24	13.3%
European Union	31.4%	45.1%	21.6%	2.0%	51	28.3%
Other Developed	62.5%	37.5%	0.0%	0	16	8.9%
G77 and China	49.1%	35.8%	15.1%	0	53	29.4%
TOTAL	37.2%	46.7%	10.5%	1.7%	180	100%

and type of organization. The largest group of participants was composed of academics with university affiliations, reflecting the trend noted above. There were, however, differences. The Group of 77 participants were more heavily from governments, as were the participants from other developed countries. The United States had the lowest relative participation of governmental experts. This suggests that different country groups had different constituencies for the assessment. In addition, there were 17 representatives from different international organizations.

The Fifth Assessment will build on the work of the previous four. The various participants in the working groups will be selected during 2010, but the scoping suggests that while many will be new, many will also carry over from previous assessments. This process of renewal coupled with consistency will ensure that the Fifth Assessment will be as credible as its predecessors.

The relatively low representation of experts from developing countries has been a concern since the outset of the IPCC and is one of the issues that will be addressed in the Fifth Assessment. The reasons for this have more to do with the de-linkage of researchers from developing countries from epistemic communities that are based in developed countries. The problems inherent in this are reflected in research carried out by Bhandari in the summer of 2009 as part of a study of the International Union for the Conservation of Nature (IUCN).

In the interviews with environmental experts from the South Asia during the summer of 2009, they were asked about the role of IPCC in their respective countries. The research participants were mostly from the IUCN

Non-governmental Organizations members, IUCN's staff and University Faculties. The criteria of the selection of the participants in this research was (1) they must have at least masters degree or above academic qualification, (2) most have working experiences in the environment field more than three years, (3) most have involve environment and climate change policy at local level (national and international was the preference). Forty-five interviews were in total (5 from Nepal, 15 from India, 10 from Bangladesh and 10 from Pakistan).

Out of 45 only 25 were actually aware about the role of IPCC (9 from India, 4 from Nepal, 6 from Bangladesh and 6 from Pakistan). They all (N=45) had heard about IPCC, but only 25 of them had idea how it works and what it produces for whom. Mostly the research participants from India and Bangladesh said that the IPCC is a knowledge producer like academic publications that has minimal impact on the global climate change debate. They accept that IPCC produces very good knowledge but does not have any mechanism to enforce to the government and stakeholders to apply that knowledge. Further out of 25, fifteen of them said that, until IPCC obtained the Nobel Peace Prize for 2007, none of them gave any attention about its role. They stated that the role of IPCC is to produce the facts and figures on climate change to the governments through the UNFCCC.

On a question about the relevance of the produced knowledge they had mixed reactions. They said that they considered it another western hegemonic body of global environment governance which shows problems but does not successfully provide the option to solve the problem. In the experts selection process, they stated that experts were mainly chosen from the governments and one only had a chance to be an expert if someone had good relationship with the government. While they did not question the expertise of experts, they did not consider them to be good representatives of the developing world.

The issue of how to involve researchers from non-governmental organizations remains a problem, although much of the criticism expressed, like that in the media in developed countries, was not based on a factual understanding of the composition and working of the IPCC.

4.5 Multi-stakeholder Governance in the Future

The IPCC process has been successful in ensuring that the climate change negotiation process generally follows the facts. Despite the arguments made by critics, the IPCC consensus is both wide and deep. The process has been an ingenious combination of academic review processes and intergovernmental consensus-building practices. The composition of the participants is equally heterogeneous and, while maintaining a formal governmental oversight, ensures that the various disciplines concerned with climate change are represented. It is a truly multi-stakeholder process.

The net effect of the structure and process is to present a credible consensus on the facts so that any changes to the predictions will be acceptable and taken into account. The IPCC has, as a matter of policy, stayed away from making policy recommendations, although the work of its working groups on adaptation and mitigation, by drawing out alternative scenarios, can affect policy discussions. While there is a consensus among the IPCC participants, there have been deniers outside the process. Bolin (2007) notes that it is the practice of the IPCC to note any dissents from specific findings in the body of reports, to ensure transparency. With regard to many of the critics, he further notes that the most prominent of them do not come from the disciplines that specialize in the findings that have been agreed.

There is no doubt that climate change, and policy issues surrounding it, will continue over the next decades. The resolution of disputes, verification of commitments and monitoring of changes will be helped by the now well-articulated network of professionals who can reach consensus on facts.

This model can be applied to other international policy areas where science and politics need to be linked, perhaps showing that politics can be easier if science is used. The details elaboration of developing world scientist's involvement in the IPCC processes is explained in chapter titled Developing Country Scientists and the IPCC.

References

Andresen, S., T. Skodvin, A. Underdal and J. Wettestad (2000), *Science and Politics in International Environmental Regimes. Between integrity and involvement.* Manchester: Manchester University Press.

Bolin, Bert (2007), *A History of the Science and Politics of Climate Change: The Role of the Intergovernmental Panel on Climate Change*, Cambridge: Cambridge University Press.

Clark, Grenville (1955), "Einstein Quoted on Politics", Letters to the Times, *New York Times*, April 22, 1955

Intergovernmental Panel on Climate Change, *Principles Governing IPCC Work*, http://www.ipcc.ch/pdf/ipcc-principles/ipcc-principles.pdf and Appendix A, *Procedures for the Preparation, Review, Acceptance,*

Adoption, Approval and Publication of IPCC Reports, http://www.ipcc.ch/pdf/ipcc-principles/ipcc-principles-appendix-a.pdf

Intergovernmental Panel on Climate Change (2001), *Tenth Anniversary Brochure*, http://www1.ipcc.ch/pdf/10th-anniversary/anniversary-brochure.pdf

IPCC (2013), Observed Changes in the Climate System, in: Summary for Policymakers (finalized version), in: IPCC AR5 WG1 2013, p. 2

SPM.3 (2014), Trends in stocks and flows of greenhouse gases and their drivers, in: Summary for Policymakers, p.8 (archived 2 July 2014), in IPCC AR5 WG3 2014

IPCC, (2018), Summary for Policymakers. In: Global warming of 1.5°C. An IPCC Special Report on the impacts of global warming of 1.5°C above pre-industrial levels and related global greenhouse gas emission pathways, in the context of strengthening the global response to the threat of climate change, sustainable development, and efforts to eradicate poverty [V. Masson-Delmotte, P. Zhai, H. O. Pörtner, D. Roberts, J. Skea, P. R. Shukla, A. Pirani, W. Moufouma-Okia, C. Péan, R. Pidcock, S. Connors, J. B. R. Matthews, Y. Chen, X. Zhou, M. I. Gomis, E. Lonnoy, T. Maycock, M. Tignor, T. Waterfield (eds.)]. World Meteorological Organization, Geneva, Switzerland, 32 pp https://report.ipcc.ch/sr15/pdf/sr15_spm_final.pdf (Drafting Authors: Myles Allen (UK), Mustafa Babiker (Sudan), Yang Chen (China), Heleen de Coninck (Netherlands/EU), Sarah Connors (UK), Renée van Diemen (Netherlands), Opha Pauline Dube (Botswana), Kristie L. Ebi (USA), Francois Engelbrecht (South Africa), Marion Ferrat (UK/France), James Ford (UK/Canada), Piers Forster (UK), Sabine Fuss (Germany), Tania Guillén Bolaños (Germany/ Nicaragua), Jordan Harold (UK), Ove Hoegh-Guldberg (Australia), Jean-Charles Hourcade (France), Daniel Huppmann (Austria), Daniela Jacob (Germany), Kejun Jiang (China), Tom Gabriel Johansen (Norway), Mikiko Kainuma (Japan), Kiane de Kleijne (Netherlands/EU), Elmar Kriegler (Germany), Debora Ley (Guatemala/ Mexico), Diana Liverman (USA), Natalie Mahowald (USA), Valérie Masson-Delmotte (France), J. B. Robin Matthews (UK), Richard Millar (UK), Katja Mintenbeck (Germany), Angela Morelli (Norway/Italy), Wilfran Moufouma-Okia (France/ Congo), Luis Mundaca (Sweden/Chile), Maike Nicolai (Germany), Chukwumerije Okereke (UK/Nigeria), Minal Pathak (India), Anthony Payne (UK), Roz Pidcock (UK), Anna Pirani (Italy), Elvira Poloczanska (UK/Australia), Hans-Otto Pörtner (Germany), Aromar

Revi (India), Keywan Riahi (Austria), Debra C. Roberts (South Africa), Joeri Rogelj (Austria/Belgium), Joyashree Roy (India), Sonia I. Seneviratne (Switzerland), Priyadarshi R. Shukla (India), James Skea (UK), Raphael Slade (UK), Drew Shindell (USA), Chandni Singh (India), William Solecki (USA), Linda Steg (Netherlands), Michael Taylor (Jamaica), Petra Tschakert (Australia/Austria), Henri Waisman (France), Rachel Warren (UK), Panmao Zhai (China), Kirsten Zickfeld (Canada)].

Krasner, Stephen D. (ed. 1983) *International Regimes*. Ithaca, NY: Cornell University Press.

Mathiason, John (2007), "Reviving Functionalism and Regime Theory to Explain the Role of International Secretariats," paper prepared for the panel on The Agency and Influence of International Organizations at the 2007 Convention of the International Studies Association, Chicago, February 28, 2007

Miller, Clark (2001) Hybrid Management: Boundary Organizations, Science Policy, and Environmental Governance in the Climate Regime, *Science Technology Human Values* 26; 478

Paterson, Matthew (1996) *Global Warming and Global Politics*, London: Routledge

Revkin, Andrew C. (2009), "Hacked E-Mail Data Prompts Calls for Changes in Climate Research," *New York Times*, November 27, 2009

Zürn, Michael (1998) "Review: The Rise of International Environmental Politics: A Review of Current Research", World Politics, Vol. 50, No. 4 (Jul., 1998), pp. 617–649

5

Developing Country Scientists and the IPCC

The IPCC has been dominated, as has science, by developed country participants. Original research by the authors has indicated some of the factors behind this and will analyze how participation can be increased. This chapter analyzes the developed and developing nations' proportional representation in the IPCC assessment process.

5.1 The Complexity of United Nations System

As the principle of the United Nation system, each member country should have an equal voice and equal voting rights in the public concern issues except the five veto-powered countries; the United States of America, United Kingdom, Russian Federation, France and China. This power centric system does not have the developing nations' representation except emerging neo-power China. The Security Council consists of 15 nations with 10 coming in by election for a two-year term. However, the Council has a voting system to resolve any major issue; if anyone from the five veto nations rejects the resolution, it cannot be passed. This is an issue India, Japan, Brazil and Germany have been raising; they want to be included as a permanent member of the Security Council. In terms of humanitarian and environmental programs, the UN agencies' headquarters are mostly based in the developed world except UNEP in Nairobi, Kenya. However, programs planning, and development programs are mostly focused on the developing world.

In relation to the internal bureaucracy, the UN is predominantly dominated by the developed world workers except for the secretary general. The same story applies to the composition and representation of the IPCC, whereas most of the authors and reviewers have been dominated by the developed world. In this chapter, we analyze the authors' and reviewers' composition and representation mostly from the third and forth assessment reports. In addition to that, we examine and analyze the IPCC survey results about the developing world participation presented in the thirty-first session of the IPCC in Bali, October 26-29, 2009. The IPCC survey covers 38 nations, DC (Developing Countries) 18; EIT (Economy in Transition) 4; and 16 Developed Countries and with the summary of the total (including all countries) authors and reviewers from second assessment to the forth assessment report respectively. This analysis shows that the developing world participation in the IPCC climate change assessment process is significantly low. In contrast, the overall scenario of climate change presented in the first to the forth assessment reports of the IPCC shows that the vulnerability and the risk factor and impact of global warming is more severe in the developing world than in the developed world (Bolin, 2007; IPCC, 2007; German-watch: Global Climate Risk Index, 2010) with few exception. In terms of the emission supplier into the atmosphere, both the industrialized and developing worlds are equally responsible. This chapter also combines these two scenarios. This clearly indicates the IPCC needs to insist on boosting the research capacity in the developing world and needs to involve more authors, reviewers and governments in the IPCC assessment processes.

5.2 The Representation of Developing World Authors, Reviewers and Governments in the IPCC Assessment Processes

This issue of marginalized representation from developing and economy-in-transition countries in the climate change assessment processes has been discussed in the international forum, within the IPCC community and in academic discussions (Runci 2007). Paul J. Runci states that:

> *The need to expand developing countries' scientific capacity and level of participation in the international scientific discourse on climate change takes on greater urgency in the light of the increasing vulnerabilities, growing populations, and persistent resource limitations these areas now face. Enhanced indigenous scientific capacity will be essential for environmental and climate assessment at the regional and sub-regional scales and will be invaluable in providing policy makers with sound information on which to base climate change response strategies (*Runci, 2007: *225).*

The concern and awareness of the global climate change have been increased in both developing and developed worlds. Local people from a specific location know more about the changing scenarios of the local climate than the remotely based scientists. The developing world has a serious concern regarding local and global environmental change (Brechin and Kempton, 1994). In fact, the 2009 World Bank Public Opinion Surveys, Global Public Opinion Organization, British Broadcasting Corporation and Globe-scan, poll results show that the developing world's publics are more concern about the climate change issues than the developed world. Figure 1 (taken from World Bank poll of 2009) illustrates a comparative overview of "public seriousness to the climate change" both from the developed and the developing world. The figure shows that among the 15 surveyed countries where Unites States, France, Japan and Russia and Turkey (economy in transition) from the developed world and China, Egypt, Indonesia India, Iran, Vietnam, Senegal, Kenya, Bangladesh and Mexico from the developing world; the climate change vulnerable countries are more serious.

Figure 1 shows that 90 percent of Mexicans say climate change is very serious problem, followed by 7 percent indicating somewhat serious; in Bangladesh, among the poorest nations, 85 percent believe it is very

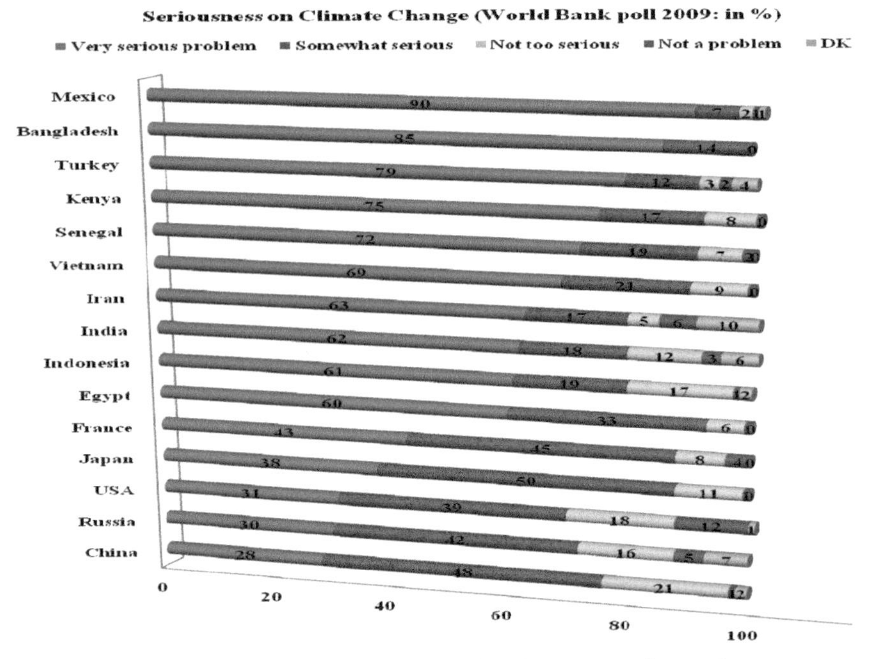

Figure 5.1 The comparative overview of seriousness on climate change.

Source: The World Bank (2009)

serious and 14 percent somewhat serious. Following the same trend, Turkey (economy in transition) 79 percent claimed it is very serious and 12 percent somewhat serious; in the developing nation of Kenya, 17 percent say very serious and 17 percent somewhat serious. However, in the emerging power nation China, only 28 percent think climate change is serious, followed by 48 percent somewhat serious. Likewise, another emerging power nation, India's poll remains in the middle level where 62 percent say it is serious, 18 percent somewhat serious and 12 percent say climate change is not too serious. From the developed world, only 30 percent of Russians say climate change is a very serious problem, 42 percent somewhat serious, 16 percent not too serious and 5 percent claim it is not a problem. This is followed by the United States where 31 percent say it is very serious, 39 percent somewhat, 18 percent not too serious and 12% not a problem. The figure illustrates that the developing world is more concerned about climate change issues. Similar results can be found on the Germanwatch Global Climate Risk Index (2010). The most severely affected countries of adverse climate are all from the developing world. The key message of the Germanwatch Global Climate Risk Index (2010) is:

> *"According to the Germanwatch Global Climate Risk Index, Bangladesh, Myanmar and Honduras were the countries most affected by extreme weather events from 1990 to 2008; All of the ten most affected countries (1990-2008) were developing countries in the low-income or lower-middle income country group; In total, 600,000 people died as a direct consequence from more than 11,000 extreme weather events, and losses of 1.7 trillion USD occurred; Myanmar, Yemen and Viet Nam were most severely affected in the year 2008;* ***Anthropogenic climate change is expected to lead to further increases in precipitation extremes, both increases in heavy precipitation and increases in drought****"* (emphasis added, Harmeling, Sven (2010) Global Climate Risk Index, 2010: 5).

The situation illustrated by the Global Climate Change Index (2010) is summarized through figure 2.

Figure 2 illustrates the total number of adverse climate events that occurred from 1990 through 2008 in the top ten developing countries at risk of global climate change. During the 18-year period, China faced a total of 558 extreme weather events, followed by India with 325, Bangladesh

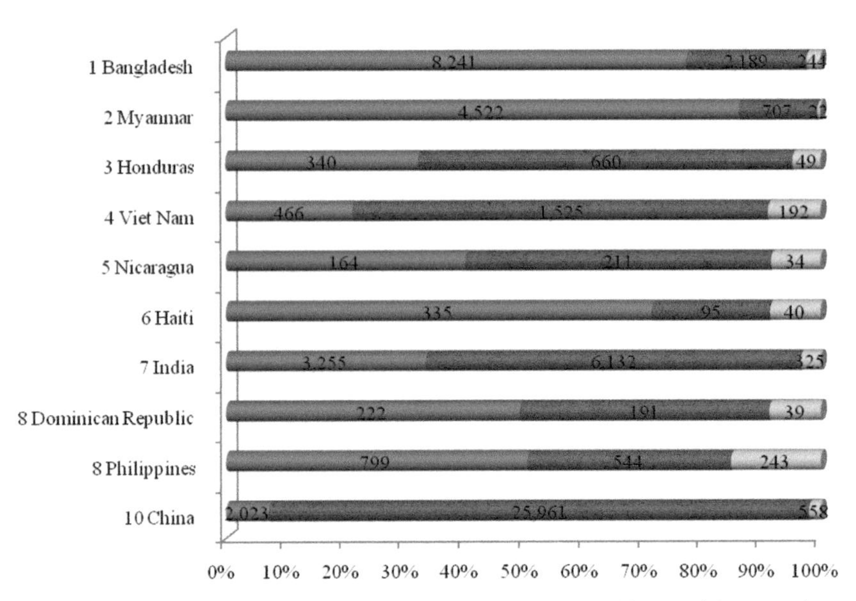

Figure 5.2 The extreme weather events in the top ten Global Climate Risky countries.
Source: The global Climate change Risk index (2010)

with 244, the Philippines having 243, followed by Vietnam with 192, totaling 1746 events. In the same time frame, China faced 2,023 human loss and US$25,961 million in property loss; Philippines 799 human loss and US$544 million in property loss; Dominican Republic 222 human loss and US$191million in property losses, India 3,255 human loss and US$6,132 in property losses, Haiti 335 human loss and US$95 million in property losses, Nicaragua 164 human loss and US$211 million in property loss; Viet Nam 466 human loss and US$1,525 million in property loss ; Honduras 340 human loss and US$660 million in property losses; Myanmar 4,522 human loss and US$707 in property losses and Bangladesh faced 8,241 human loss and US$2,189 million in property losses, totaling 20,367 human and US$38,215 million property loss. The annual global human loss resulting from adverse climate events was 34,782, and in 18 years, the world faced a total of 626,079 human deaths and 1,881,537.48 millions of property loss

(Global Climate Risk Index, 2010). This scenario shows an alarming threat of extreme weather events. However, in terms of the geography, population and economic condition in the Global Climate Risk Index, Bangladesh is the most vulnerable followed by Myanmar, Honduras, Vietnam, Nicaragua, Haiti, India, Dominican Republic, Philippines and China. Surprisingly, the IPCC assessment reports have not analyzed this extreme risk of global climate impact and have not raised the priority for involving scientists from most of the victims' nations. Linking to the IPCC, the authors and reviewers are dominated by the developed world. Figure 3 shows how poorly the developing world is involved in the IPCC climate change assessment process.

Figure 3 shows the geographical distribution of the IPCC assessment process in the third (TAR) and fourth reports (AR4). In the TAR, there was

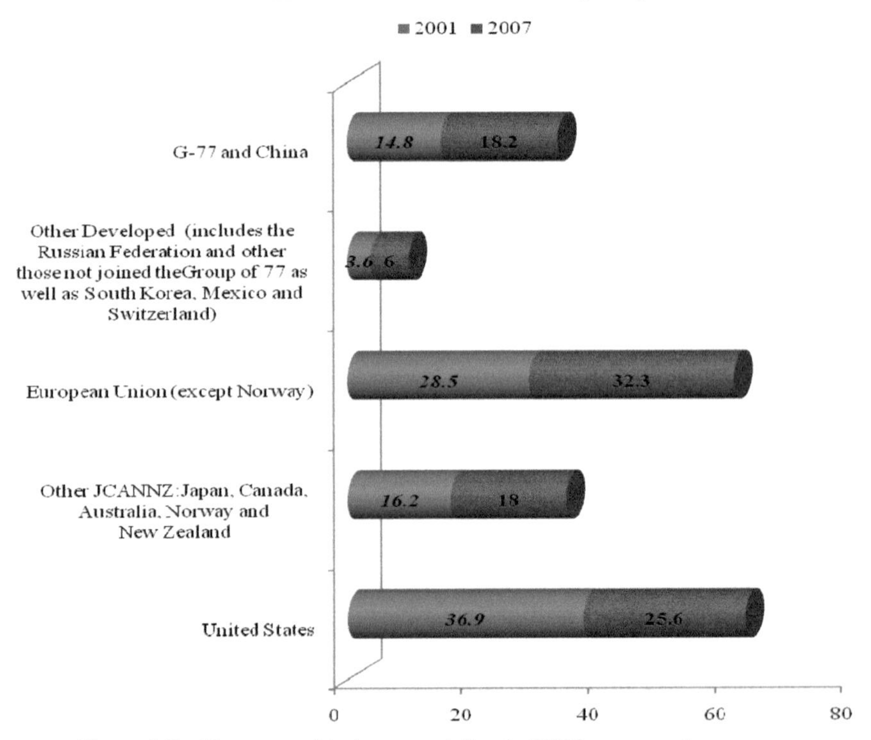

Figure 5.3 The geographical representation in IPCC assessment processes.
Source: The annex of IPCC 2001 and 2007 report.

36.9 percent involvement by the United States in the IPCC process, which dropped by 11.3 points having only 25.6 percent in the AR4; however, the involvement of Japan, Canada, Australia, Norway and New Zealand slightly increased from 16.2 percent in the TAR to 18 percent in the AR4 (increase of 1.8 points). Similarly, the European Union is in the lead in the AR4 with a participation of 32.3 percent, whereas in TAR, it was 28.5 percent (increased by 3.8 points). The participation of other developed nations including the Russian Federation and others that had not joined the Group of 77, as well as South Korea, Mexico and Switzerland is also increased from 3.6 percent in the TAR to 6 percent in the AR4 (increased by 2.4 points). Likewise, the participation of the developed world in the G77 (G77 consists of 130 member countries) and China also increased from 14.8 percent in TAR to 18.2 percent in AR4 (increased by 3.4).

The involvement of the developing world increased from 14.8 percent to18.2 percent in TAR to AR4 respectively. However, it is still insignificant in terms of global coverage, population, public concern and Global Climate Risk of the developing world. For example, Asia alone holds the largest population of 4,117 million, with the highly populated countries such as China 1,331 million; India 1,171 million; Indonesia 243 million; Pakistan 181 million and Bangladesh 162 million. Similarly, Latin America/Caribbean comprises a total population of 580 million, where Brazil has the largest population of 191 million, Mexico has 109.6 million and Africa has 999 million, where Nigeria is the most populated country with 152 million people. Based on 2009's Population Reference Bureau (2009), of a total world population of 6,810 million, the Asian, African and Latin America/Caribbean nations combined have 5,696 million people, which is 83.64 percent of the total world population estimate. North America has 341 million people (United States 307 million and Canada 34 million, respectively), which is only 5 percent of the world population; and Europe has a total population of 738 million including Russia's 141.8 million (Russia is also poorly represented in the IPCC process). Excluding Russia, Europe holds an 8.75 percent of the total world population. The Oceana has a total of 36 million people where Australia alone has 21.9 million people and New Zealand has only 4.3 million people (Population Reference Bureau 2009), which is about 0.53 percent of the world population. North America, Europe and Oceana combined have 14.28 percent of the world's population (Russia excluded). The representation of the world scientists in the IPCC processes was exactly the opposite in TAR's 14.8 percent representation from the developing world that went up slightly in AR4. The following section elaborates how the three working groups were composed in TAR and AR4.

5.3 Developing and Develop World Representation by Working Groups

Over the last two assessments, scientists from 106 different countries have been members of the IPCC working groups. As would be expected, scientists from the industrialized countries heavily dominate, although that has been changing slowly. In this chapter, we have analyzed the distribution of the members of the IPCC for the Third (2001) and Fourth (2007) assessments by country group. These are the main caucusing groups in climate change negotiations. This grouping includes the United States; "Other JCANNZ" (includes Japan, Canada, Australia, Norway and New Zealand); European Union (except Norway); Other Developed (includes the Russian Federation and those that did not join the Group of 77 as well as South Korea, Mexico and Switzerland), G77 (130 member countries) and China. The main change that can be noted is the reduction in the proportion of scientists from the United States. The geographical distribution is improving over time. In 2007, scientists from 91 countries were involved, compared with 79 countries in 2001. However, the following summary figures masks some significant differences by working group. Working Group, I on science is more likely to depend on the well-funded scientific research programs, either in universities or government, that are found dominated by the industrialized countries.

As can be seen from figure 4, the dominance of these countries is more pronounced. Although the representation of scientists from developing countries has almost doubled, this is from a low base and reflects a slight drop in the proportion of United States scientists that are involved.

In contrast, Working Group 2 (figure 5), which deals with adaptation, has had a growing relative participation of scientists from developing countries, although much of that is a result of a reduction in the participation of scientists from the United States and a general reduction in the number of individuals participating in the working group. In contrast, the geographical composition of Working Group 3 (figure 6), dealing with mitigation, has been stable.

One factor that is noticeable in the figures is that WG 1 (figure 4) has a much larger participation than the other two. This may reflect the idea that the results of WG 1, whose conclusions form the basis for the work of the

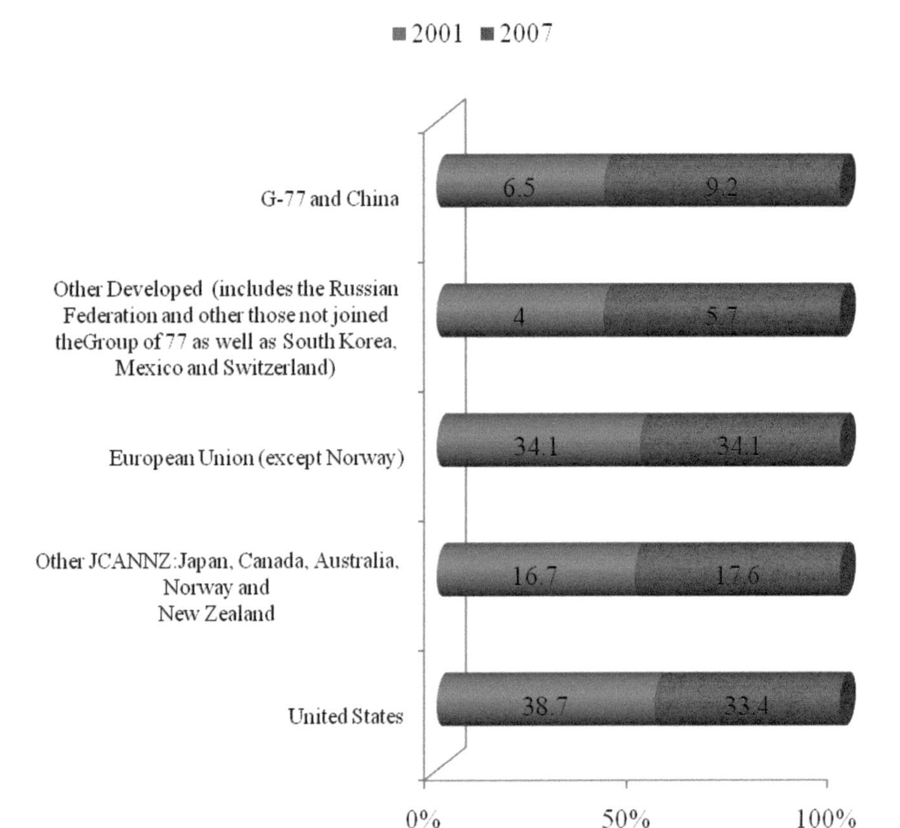

Figure 5.4 Working group 1 country group distribution.

other two groups, needs to be comprehensive. It may also reflect a concern that the voluminous and highly technical hard science behind the conclusions needs a firm consensus that can only be credible if it involves a large number of scientists.

The overall above analysis shows that representation of the developing world in the IPCC assessments process is minimal in terms of population and geographical variation. However, in the coming fifth assessment, which probably will come in 2014, the IPCC seems serious in increasing the participation from the developing world. The press release on the assessment policy procedure entitled "The role of the IPCC and key elements of the IPCC

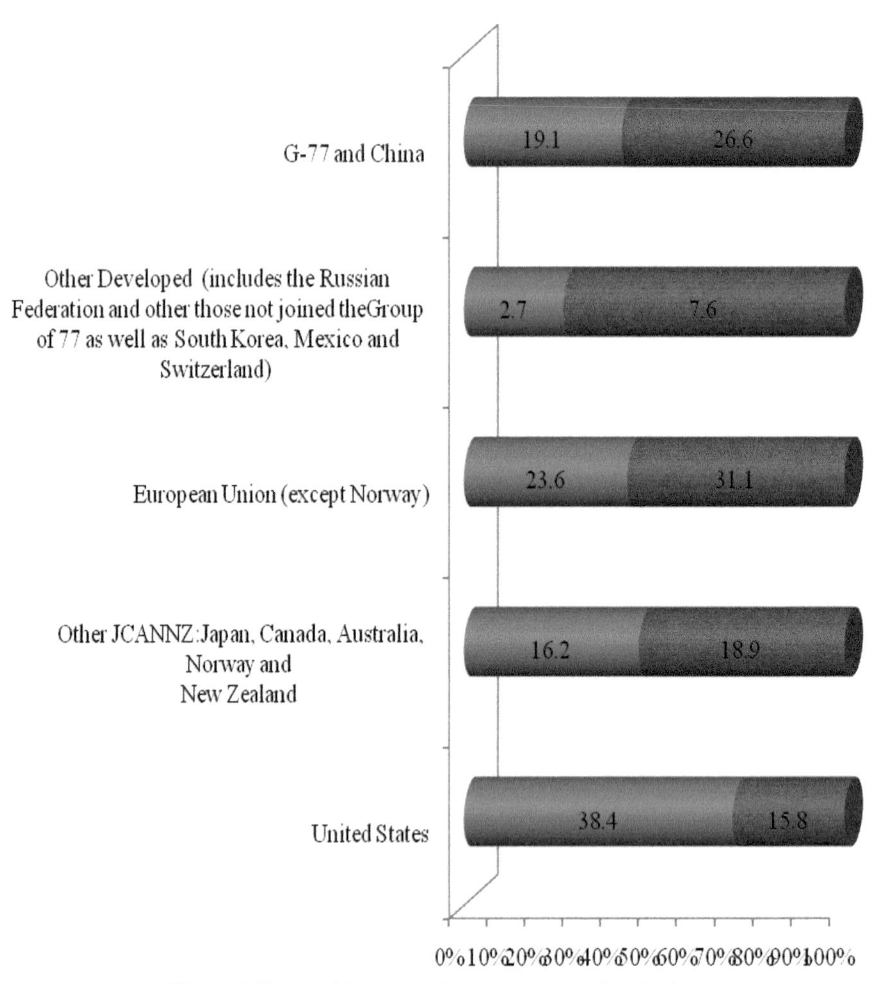

Figure 5.5 Working group 2 country group distribution

assessment process" published in Geneva, February 4, 2010, gives such an indication.

The press release states:

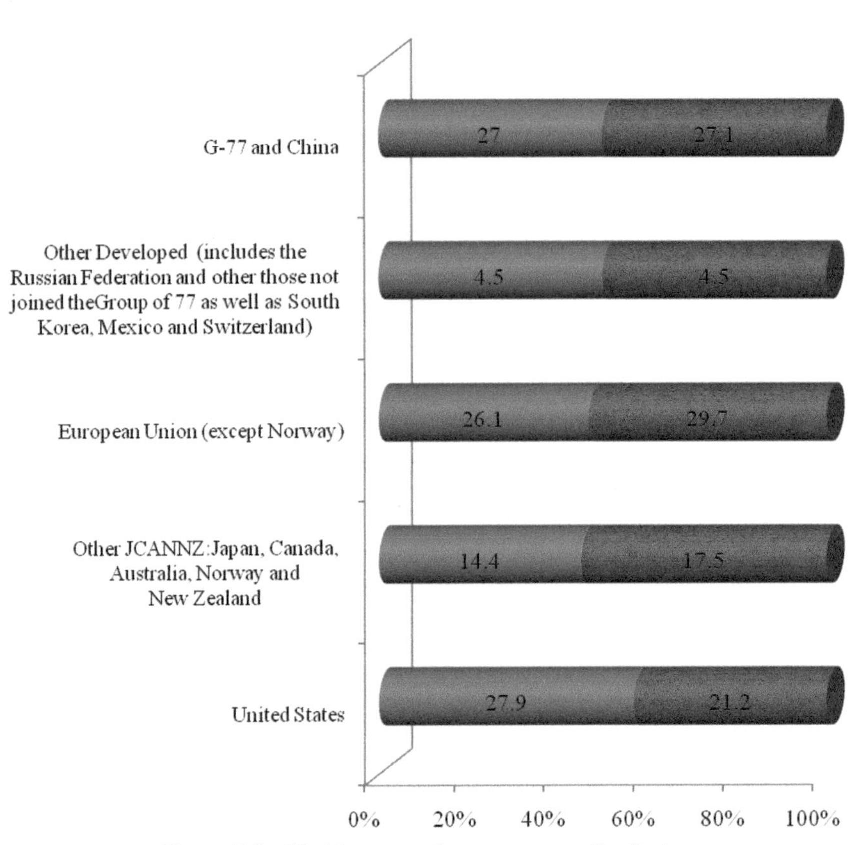

Figure 5.6 Working group 3 country group distribution

- *Author teams that reflect a wide range of expertise and views and work on a voluntary basis; Author teams for the chapters of IPCC reports should represent a range of views and expertise; as well as appropriate geographical representation. The IPCC also aims for gender balance. This is achieved through a wide nomination and selection process, as is currently ongoing for the Fifth Assessment Report (AR5).*
- *More than 450 Lead Authors and more than 800 Contributing Authors (CAs) have contributed to the Fourth Assessment Report (AR4). They were selected from around 2000 nominations.*

- *All experts contributing to the report work for the IPCC on a voluntary basis and are compensated by the IPCC only for their travel expenses to the necessary meetings, including that of the Chair of the IPCC and all of the elected leadership* (IPCC 2010).

In addition to this press release, the IPCC had also conducted its own research on developing (DC), developed and economy in transition countries (EIT) on country participation with the title of "Improving Participation of Developing/EIT Countries in the IPCC". The survey was conducted from 18 Developing Countries (those include Central African Republic, Chile, China, Cook Islands, Ecuador, Islamic Republic of Iran, Liberia, Libya, Myanmar, Republic of Korea, Republic of Mauritius, Rwanda, Sao Tome e Principe, Senegal, Sri Lanka, Suriname, Syrian Arab Republic and Uganda), four from EIT (those include Armenia, Croatia, Kazakhstan, and Uzbekistan) and 16 from DC (including Belgium, Bulgaria, Canada, Czech Republic, Germany, Hungary, Latvia, Lithuania, New Zealand, Norway, Poland, Romania, Slovenia, Sweden, The Netherlands and United Kingdom). However, the report does not state the criteria for choosing these countries.

The survey mandate was attested:

> "*At the 30th Session of IPCC in Antalya, within the framework of the topic "Future Work of IPCC" the IPCC took a decision (Decision 7): "The Panel charges the IPCC Vice-Chairs to carry out over the next six months an assessment of the current shortcomings in involving an adequate number of developing/EIT country scientists and to propose approaches to address this issue"* The objectives of survey was "*To provide an objective basis for this reflection, an extensive analysis of the number and origin of coordinating lead authors, lead authors and reviewers to the second, third, and fourth assessment reports was conducted by the IPCC Secretariat, in consultation with the IPCC Vice-chairs* (IPCC 2009, P1).

The questions were divided into 10 major themes, such as overall representation; review process; government review; regional meetings; literature; grey literature; language; scientific capacity and the expertise; data availability; computational capacity and scenarios. The overall findings show that the IPCC is highly constituted by the developed world scientists. More surprisingly, some countries were not involved or not invited in the process. The following are some of the examples of those findings.

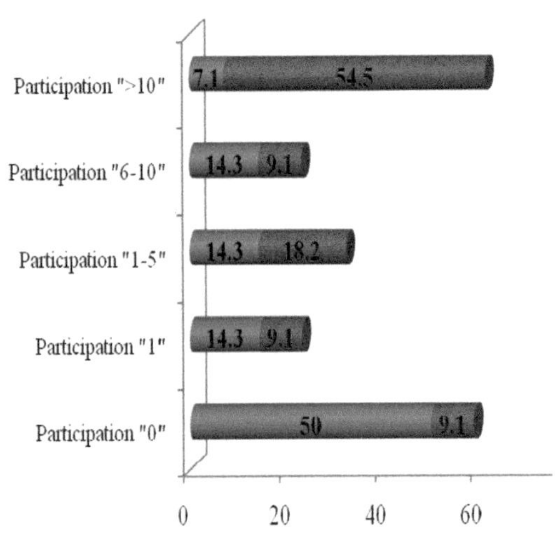

Figure 5.7 Experts nominated by the Focal Point by country as authors in AR4
Source: IPCC (2009).

5.4 IPCC Questionnaire Results and Analysis on Number of Experts in the Past Assessment Reports

The first question was to get the number of authors nominated by the Focal Point of surveyed countries, where 14 DC/EIT and 11 developed countries had responded.

Figures 7 and 8 indicate how poorly authors were chosen by the countries' focal points (the government authorized agencies if applicable, otherwise foreign ministry of respected countries), and representativeness in the IPCC processes. For example, in Figure 7, 50 percent of developing nations have zero nominations, only 14.3 percent have one person nomination, and so on. In contrast, from the developed nations 54.5 percent have more than 10 people nominated, whereas from the developing nations there were only 7.1 percent nations nominated for more than 10 people. Similarly, in the case of developed country's participation illustrated in the figure 8, 31.1 percent of nations were well represented, followed by the 12 percent mix represented and 25 percent represented to some extent. 31.3 percent were

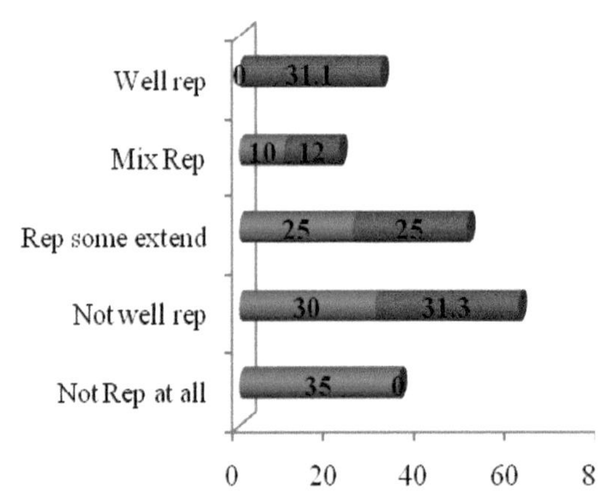

Figure 5.8 Authors representation in the IPCC AR4

Source: IPCC (2009).

not well represented and 0 percent in the not represented at all category. It is the opposite in the case of DCs where 35 percent were not represented at all, 30 percent not well represented, 25 percent represented to some extent, 10 percent mix represented and 0 percent in the well represented categories. These two figures clearly indicate that the developing world which holds the largest population and has been facing the adverse climate change risk, is only partially participating in the IPCC process. This scenario is also repeated in the reviewer case as well.

5.5 The Developed and Developing World Participation in the Review Processes

As the IPCC processes indicates, the report reviewers have better coverage in terms of countries and number of scientist involvement. The following figure 9 illustrates the increment of authors and reviewers from the second to the fourth assessment reports.

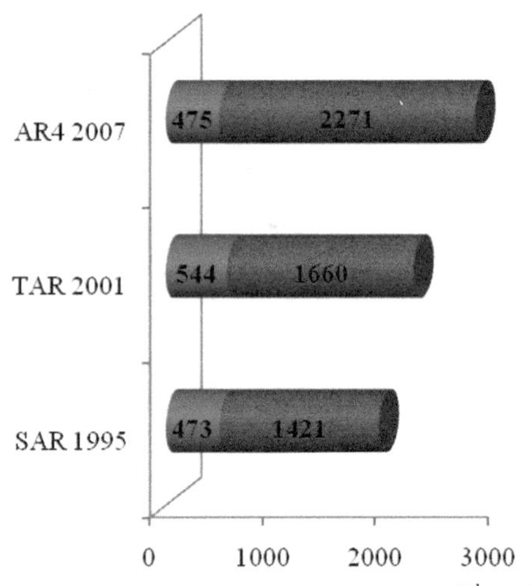

Figure 5.9 The increment of authors and reviewers from 2nd to 4th assessment reports
Source: IPCC (2009).

Figure 9 indicates that, in terms of authors, there is no significant change between the 2nd and 4th assessment report (SAR and SR4); however, in the 3rd assessment (TAR), there 69 more authors than in the 4th. There is no information on why the number of authors dropped. In the case of reviewers, there is significant incremental change. For example, from SAR to TAR, the reviewers' number increased by 239, and from TAR to the AR4 the number more than doubled, increasing by 611, reaching a total of 2271 reviewers. Does this add the participation from developing countries? The simple answer is no. Figure 10 illustrates the numbers of reviewer participation in the assessment processes are highly correlated with the authors nominated by the IPCC focal points (government agencies; see figure 7). Even the overall the number of reviewers increased near to double from SAR to AR4, but there is no developing world participation increment in the IPCC assessment processes. The following figure in scale "0" to ">10" also illustrates a similar result.

Figure 11 shows that there is minimal developing country group participation in the IPCC assessment processes. The figure illustrates that 60 percent of the surveyed DCs and EITs have "0" participation in review process.

Authos and reviewers participation in 2nd, 3rd and 4th assessment synthesis report (total of WG1, WG2 and WG3 in %)

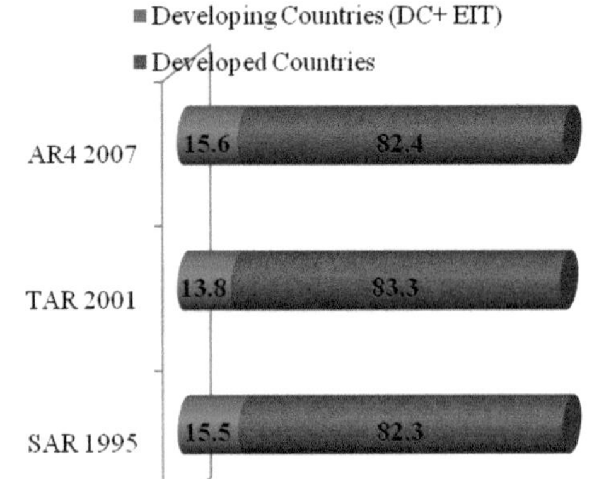

Figure 5.10 The authors and reviewers participation from 2nd to 4th assessment reports

Source: IPCC (2009).

Reviewers participating the in the 4th assessment report and the 3rd assessment report (in %)

DC (Developing Countries) and EIT (Economy In Transition)

Developed Countries

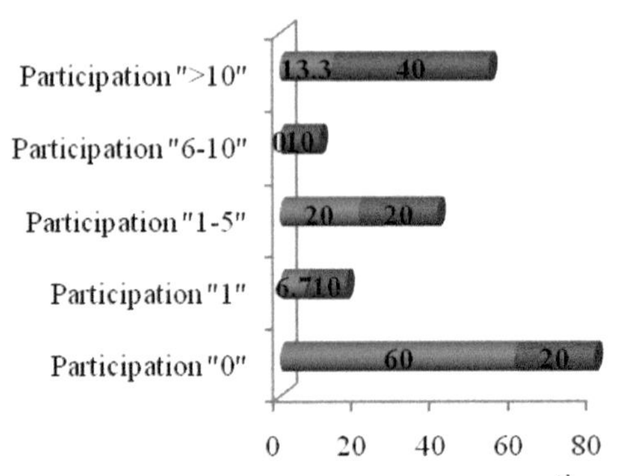

Figure 5.11 Reviewer's participation by country group in the 4th and the 3rd assessment report

Source: IPCC (2009).

Government's involvement in the IPCC 4th assessment report and the 3rd report review process (in %)

■ DC (Developing Countries) and EIT (Economy In Transition)
■ Developed Countries

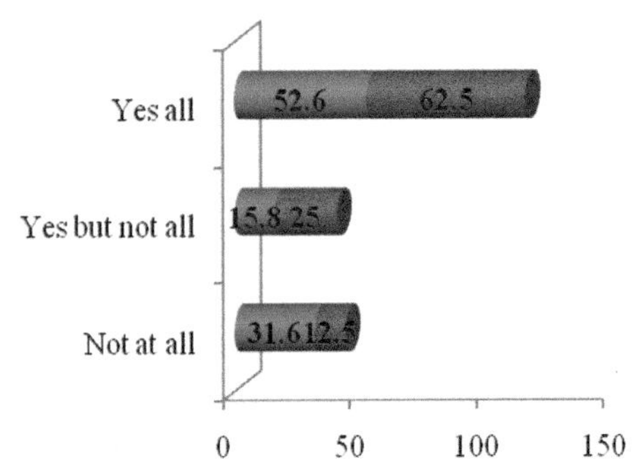

Figure 5.12 Government involvement in reviewing process
Source: IPCC (2009).

Similarly, in the case of the "6-10" category, DC/EIT participation was also "zero." In the case of the developed world 40 percent of countries have more than 10 participants, followed by 10 percent in the 6 to 10 category and 20 percent in the 1 to 5 category and only 20 percent in "0" category. This affirms that the developing world is poorly represented in the IPCC assessment processes. What about the government participation? Figure 12 illustrates there is a better situation there, where 31.6 percent of developing countries' governments were not involved in any assessments, 15.8 percent partially involved and 52.6 percent involved in all assessment process. In the case of developed countries' governments 62.5 percent were involved in reviewing all assessment reports, followed by 25 percent partially and 12.5 percent of governments were not involved at all. This indicates that developing world governments also give priority to the assessment process but not at the same level of developed countries' governments.

Conclusion

These findings show that there is low representation of experts from developing countries. The reasons for this have more to do with the lack of researchers linked to the IPCC from developing countries from epistemic communities that are based in developed countries. The problems inherent in this are reflected in research carried out by Bhandari in the summer of 2009 as part of a study of the International Union for the Conservation of Nature (IUCN).

In the interviews with environmental experts from South Asia during the summer of 2009, they were asked about the role of the IPCC in their respective countries. The research participants were mostly from the IUCN NGO members, IUCN's staff and university faculties. The criteria of the selection of the participants in this research was (1) they must have at least a masters degree or above academic qualification, (2) must have working experiences in the environment field of more than three years, (3) must have been involved in environment and climate change policy at local level (national and international was the preference). Forty-five total interviews were conducted (5 from Nepal, 15 from India, 10 from Bangladesh and 10 from Pakistan).

Out of 45, only 25 were actually aware of the role of the IPCC (9 from India, 4 from Nepal, 6 from Bangladesh and 6 from Pakistan). They all (N=45) had heard about the IPCC, but only 25 of them had an idea of how it works and what it produces for whom. Mostly the research participants from India and Bangladesh said that the IPCC is a knowledge producer like academic publications that have minimal impact on the global climate change debate. They accept that the IPCC produces very good knowledge, but it does not have a mechanism to enforce the government and stakeholders to apply that knowledge. Further, out of the 25, 15 of them said that, until the IPCC obtained the 2007 Nobel Peace Prize, none of them paid attention to its role. They stated that the role of the IPCC is to produce the facts and figures on climate change to the governments through the UNFCCC.

On a question about the relevance of the produced knowledge, they had mixed reactions. They said they considered it another western hegemonic body of global environment governance that identifies problems but does not successfully provide the solution to the problem. In the experts selection process, they stated experts were mainly chosen from the governments and one only had a chance to be an expert if someone had good relationship with the government. Although they did not question the expertise of experts, they did not consider them to be good representatives of the developing world.

The issue of how to involve researchers from NGOs remains a problem, although much of the criticism expressed, like that in the media in developed countries, was not based on a factual understanding of the composition and workings of the IPCC.

5.6 The IPCC's Efforts to Involve Developing Nations

In this chapter, we saw that developing world participation slightly increased from SAR to AR4; however, it is still minimal. We also saw (though public poll data) that in recent years, the developing world is more concerned regarding the climate change issue (World Bank, 2009). The developing world people are aware of global climate change and hold a good deal of knowledge (Brechin and Campton, 1994). The climate risk index (Germanwatch 2010), shows that the top ten climate risk countries are from the developing world and hold the largest proportion of the world population. The IPCC itself has also affirmed that the developing world is more vulnerable than the developed world (with few exceptions) (IPCC, 2007). Most important, the IPCC has been facing criticism from both the developed and developing worlds regarding the minimal involvement of the developing world in its assessment processes. As a result, the IPCC appointed a group to identify a way to involve more authors, reviewers and governments on the Thirtieth Session of the IPCC (Antalya, Turkey, April 21 – 23, 2009). " The aim of the survey is to explore the most relevant issues for improving the involvement of experts from developing / EIT countries in the IPCC work: as authors and reviewers in IPCC products (AR, SR, TP, etc.); as participants in IPCC expert meetings; and for the outreach activities related to IPCC" (IPCC, 2009). The group lead by the vice chairs conducted the survey and submitted the results with the recommendations in the thirty-first session of the IPCC in Bali (October 26-29, 2009).

The summary of the IPCC survey findings matches with what we have explored in the above section of this chapter. The IPCC report summarizes that:

> *"While the balance improved from the SAR to the TAR, the share of authors from developing countries did not increase between the TAR and AR4....The overall participation of developing and EIT countries to the review of reports is significantly smaller (more than 80* percent *of reviewers were coming from developed countries;* (***our calculation shows around 83*** percent, ***see figure 10***). *The stagnation of DC/EIT participation between the TAR and AR4 justifies that measures be taken to improve their participation)* (IPCC, 2009: 2).

And the major recommendations were:

- *Invite the IPCC secretariat to consider ways to enhance the awareness of the Focal Points regarding IPCC activities and possible funding for participation of scientists from their countries, for example by setting up a communication and outreach initiative. Also, it would be important that the Focal Points and Bureau members be encouraged to nominate more experts from developing/EIT countries.*
- *Ensure that procedures for the nomination and selection of authors and reviewers are conceived in a way that facilitates the identification and selection of suitable experts from developing countries/EIT (e.g. easy multi-criteria searching into a single database of experts, for all types of IPCC products including AR, SR, and TP, as well as expert meetings). Ensure an adequate balance in the selection of CLAs, LAs and REs, by taking care that the nominations of suitable experts are considered, with specific attention for poorly represented regions and developing countries/EIT.*
- *Consider ways to increase the provision of financial support for the attendance of more experts from developing and EIT countries to IPCC meetings (increase the number of journeys supported by the Trust Fund), noting that past experience suggests that funding journeys is not sufficient to increase participation in itself.*
- *Organize more regional meetings in developing regions. These may in particular contribute to the identification of grey and non-English literature, drawing it to the attention of the relevant author teams.*
- *Encourage the participation of experts from developing countries/EIT in the outreach process of IPCC products.*
- *The IPCC Bureau in general (and the WG Bureaux, in particular) should encourage more involvement of young experts from the developing countries/EIT in the IPCC process.*
- *The Panel invites the Task Group set up in Plenary XXX, regarding the catalytic role of the IPCC regarding scenario development, to explore possibilities to facilitate DC/EIT country scientist participation. This may in particular include an invitation to the Focal Points in developing/EIT countries/EIT to encourage their experts to contribute scenarios to the library of new socioeconomic scenarios anticipated in the report "Towards New Scenarios"* (IPCC, 2008; IPCC, 2009).

In general, these recommendations provide the reasons of why developing world participation is limited and what necessary steps need to be taken.

However, to materialize this recommendation it needs firm commitment from the governments, multi-national and international agencies including the UN and others. There has been some efforts for the capacity building of developing countries through a global project Assessments of Impacts and Adaptations to Climate Change (AIACC) initiated and developed in collaboration with the UNEP/WMO Intergovernmental Panel on Climate Change (IPCC), which has been funded by the Global Environment Facility, for the purpose of advancing scientific understanding of climate change vulnerabilities and adaptation options in developing countries. Currently they are conducting 24 regional studies with the involvement of 46 countries (5 projects in Latin America, 11 projects in Africa, 5 projects in Asia and 3 projects in small island states), with the following expected outcomes.

- *"Publication of peer-reviewed scientific articles by participating scientists that significantly expand understanding of developing country vulnerabilities and adaptation options*
- *Increased numbers of developing country researchers, particularly young researchers, who are actively engaged in global change research*
- *Increased participation of developing country scientists in future assessments of the IPCC*
- *Wider understanding of climate change issues among stakeholder groups in developing countries*
- *Use of information generated by AIACC studies in National Communications and for planning adaptation actions"* (AIACC, 2010).

These are some of the positive steps to address the global climate change issue, though it still needs a large project that can cover the other developing courtiers. More importantly, the developed world itself; United Nation and its agencies; multi-national organizations, etc. need to work together with the developing countries to enhance and strengthen their capacity to bring them into the main stream of climate change, because they are the major victims of adverse global climate.

5.7 UNFCC Appropriately Illustrates the Developing World'S Situation

"Developing countries are the most vulnerable to climate change impacts because they have fewer resources to adapt socially, technologically and financially. Climate change is anticipated to have far reaching effects on the sustainable development of developing

countries including their ability to attain the United Nations Millennium Development Goals by 2015 (UN 2007). Many developing countries' governments have given adaptation action a high, even urgent, priority.....Developing countries have very different individual circumstances and the specific impacts of climate change on a country depend on the climate it experiences as well as its geographical, social, cultural, economic and political situations. As a result, countries require a diversity of adaptation measures very much depending on individual circumstances" (UNFCC, 2007: 5-6).

The developing world comprises 83.64% of the world population and is the most vulnerable to climate change impacts. Without having their input and knowledge, the IPCC's assessments always will be incomplete and questionable.

Endnotes and References

IPCC assessment reports are available online from second assessment 1995 at http://www1.ipcc.ch/ipccreports/assessments-reports.htm (accessed on 03/15/2010) Bolin, Bert (2007), A History of the Science and Politics of Climate Change: The Role of the Intergovernmental Panel on Climate Change, Cambridge: Cambridge University Press. Rosenzweig, C., P. Neofotis, M. Vicarelli, and X. Xing (eds.) (2008), Intergovernmental Panel on Climate Change (IPCC) Observed Climate Change Impacts Database Version 1.0. Palisades, NY: Socioeconomic Data and Applications Center (SEDAC), Columbia University. Available at http://sedac.ciesin.columbia.edu/ddc/observed/. (accessed on 03/15/2010). Harmeling, Sven (2010), Global Climate Risk Index 2010, Who Is Most Vulnerable? Weather-Related Loss Events since 1990 and How Copenhagen Needs To Respond, German Watch. http://www.germanwatch.org/klima/cri2010.pdf (accessed on 03/15/2010).

Runci, Paul J. (2007), Expanding the Participation of Developing Country Scientists in International Climate Change Research, Environmental Practice 9: 225-227 Cambridge University Press. http://journals.cambridge.org/action/displayFulltext?type=1&fid=1820864&jid=&volumeId=&issueId=04&aid=1820860&bodyId=&membershipNumber=&societyETOCSession=.

Brechin, S. R., & Kempton, W. (1994), Global environmentalism: A challenge to the postmaterialism thesis? Social Science Quarterly, 75, 245-269.

The World Bank (2009), Public attitudes toward climate change: findings from a multi-country poll, (December 3, 2009), The World Bank. http://siteresources.worldbank.org/INTWDR2010/Resources/Background-report.pdf http://www.worldpublicopinion.org/pipa/articles/btenvironmentra/649.php?nid=&id=&pnt=649&lb=bte BBC climate change poll shows rising concerns (December 7, 2009) http://news.bbc.co.uk/2/hi/8396512.stm (accessed on 03/15/2010)

Harmeling, Sven (2010), Global Climate Risk Index 2010, Who Is Most Vulnerable? Weather-Related Loss Events since 1990 and How Copenhagen Needs to Respond, German Watch (page 5). http://www.germanwatch.org/klima/cri2010.pdf (accessed on 03/15/2010)

Population Reference Bureau (2009), World Population Data Sheet, Population Reference Bureau, USA; "press release of (Aug. 12, 2009) states "Global population numbers are on track to reach 7 billion in 2011, just 12 years after reaching 6 billion in 1999. Virtually all of the growth is in developing countries. And the growth of the world's youth population (ages 15 to 24) is shifting into the poorest of those countries". http://www.prb.org/Publications/Datasheets/2009/2009wpds.aspx and http://www.prb.org/pdf09/09wpds_eng.pdf (accessed on 03/18/2010)

IPCC (2010), The role of the IPCC and key elements of the IPCC assessment process, press release at Geneva, on February 4, 2010. http://www.ipcc.ch/pdf/press/role_ipcc_key_elements_assessment_process_04022010.pdf (accessed on 03/18/2010).

IPCC (2009), Improving participation of developing/EIT countries in the IPCC, Summary and recommendations, Thirty-First Session of the IPCC, Bali, 26-29 October 2009, IPCC-XXXI/Doc.11 (14.X.2009) Agenda Item: 9.1, (Submitted by Hoesung Lee and Jean-Pascal van Ypersele, IPCC Vice-chairs) http://www.ipcc.ch/meetings/session31/doc11.pdf (accessed on 03/18/2010).

Richard Moss, Mustafa Babiker, Sander Brinkman, Eduardo Calvo, Tim Carter, Jae Edmonds, Ismail Elgizouli, Seita Emori, Lin Erda, Kathy Hibbard, Roger Jones, Mikiko Kainuma, Jessica Kelleher, Jean Francois Lamarque, Martin Manning, Ben Matthews, Jerry Meehl, Leo Meyer, John Mitchell, Nebojsa Nakicenovic, Brian O'Neill, Ramon Pichs, Keywan Riahi, Steven Rose, Paul Runci, Ron Stouffer, Detlef van Vuuren, John Weyant, Tom Wilbanks, Jean Pascal van Ypersele, and Monika Zurek, (2008), Towards New Scenarios for Analysis of Emissions, Climate Change, Impacts, and Response Strategies. Technical Summary. Intergovernmental Panel on Climate Change, Geneva, 25 pp. The preface states: "The expert meeting included presentations focused on needs for scenarios as seen from a policy-making perspective, a review of past IPCC scenarios, overviews of evolving plans in the research community, needs and opportunities for scenarios on two different time scales ("near term"—to 2035, and "longterm"— to 2100, extended to 2300 for some applications), and a review of options for the benchmark scenarios, referred to in the report as "Representative Concentration Pathways" (RCPs)" (IPCC 2008: preface). http://www.ipcc.ch/pdf/supporting-material/expert-meeting-ts-scenarios.pdf (accessed on 03/18/2010).

AIACC (2010), Project on Assessments of Impacts and Adaptations to Climate Change (AIACC) in Multiple Regions and Sectors, http://www.aiaccproject.org/about/about.html (accessed 03/18/2010).

6

Climate Change Deniers and the IPCC

"Societies are tragically vulnerable when the men and women who compose them lack character. A nation or a culture cannot endure for long unless it is undergirded by common values such as valor, public-spiritedness, and respect for others and for the law; it cannot stand unless it is populated by people who will act on motives superior to their own immediate interest. Keeping the law, respecting human life and property, loving one's family, fighting to defend national goals, helping the unfortunate, paying taxes–all these depend on the individual virtues of courage, loyalty, charity, compassion, civility, and duty" (Colson 1989:67)

From the beginning of the debate on climate change, there have been some scientists and others relying on what they understand to be science who have rejected the central thesis of the IPCC, that human behavior can change the climate. They have questioned the findings of the IPCC (and its members, including those brought out by climate gate–the hacking of emails in a climate change institute). This chapter analyzes the issues raised by the deniers, investigates the errors and hacked emails and the United Nation and IPCC's commitments to address these issues.

6.1 IPCC is a Trustworthy and Reliable Knowledge Producer

There is no doubt that climate change has been occurring. The IPCC has been providing the most reliable facts and figures to global communities and has played a significantly positive role in climate change mitigation. The scientists, who work voluntarily, are able to raise awareness regarding the critical issues surrounding climate change for a healthy Earth. As we have mentioned in the previous chapters, the IPCC has firm principles, practices and a long peer review process for the publication of its reports. The IPCC's statement regarding principles and procedures states:

> *The IPCC is a unique partnership between the scientific community and the world's governments. Its goal is to provide policy-relevant but not policy-prescriptive information on key aspects of climate change, including the physical science basis, impacts of and vulnerability to climate change in human and natural systems, options for adapting to the climate changes that cannot be avoided, and options for mitigation to avoid climate change. The IPCC relies on a combination of broad participation, rigorous oversight, and transparent, thorough adherence to carefully designed procedures to produce assessment reports that have become, over the last 20 years, the international gold standard in the scientific assessment of climate change. Any such human endeavor can never be completely error free, but IPCC assessments are as close to this goal as the international community of scientists and governments can accomplish. The members of the IPCC (who are the world's national governments), its elected leadership, and the thousands of scientists who contribute to each report are continually working to improve all aspects of future reports (*IPCC principles and procedures, 2010).

It is unnecessary to elaborate further on how well the IPCC follows the given mandates; to date, it has published five Assessment Reports one each in 1990, 1995, 2001, 2007 and 2014 respectively and is working on a sixth one, which will be published by 2022. In 2007, the IPCC was awarded the Noble Peace Prize with former US Vice President Al Gore, for its outstanding contribution on the climate change issue. This award drew a great deal of attention toward the IPCC. Our research shows that until 2007, the general public in the developing world had rarely heard about the IPCC, had no idea what its role was, or how it functioned (survey from India, Nepal, Pakistan and Bangladesh). After 2007, the IPCC received vast media coverage and people began to read its publications. At the same time, the general public, governments and scientists of all fields increased their expectations and intended roles for the committee, which it has been trying to fulfill.

6.2 More Reputation-more Expectations

There always have been two types of people who do not support the scientific outcomes—the deniers—because of their own business or identical interests. However, as Kemp (2010) notes *"Climate-change denial could have disastrous consequences, if it delays global action to cut carbon emissions. Denialism is gaining popularity because people have difficulty differentiating deniers' twisted arguments from the legitimate concerns of genuine sceptics.....Denial of the science of climate change is eroding public understanding of the issue and seems to be undermining trust in scientist......Denialism is motivated by conviction rather than evidence. It has been applied to a wide range of issues, including evolution and the link between HIV and AIDS. Deniers use strategies that invoke conspiracies, quote fake experts, denigrate genuine experts, deploy evidence selectively and create impossible expectations of what research can deliver*" (Kemp 2010:673). Similar to Kemp, there are hundreds of documents, researches, proofs, evidences based on ground truthing based on satellite data, weather pattern data, person experiences, and changes on ecosystems, which clearly indicates that the cause of climate change began to accelerate mostly from industrial era, which is direct impact of human excessive intervention on planet natural process. The scholarly domain clearly indicates that anthropogenic disturbances in nature is the major cause rapid change; however, there are also few who are trying to reject the facts due to economic self-interest [a list of relevant reference (mainstream and denial) is included in annex].

In such situations, it is always crucial for the climate change scientists need to take such denials claims and need to work more seriously. As IPCC stands for reliable knowledge producer, its (involved) scientists that the research should be fully supported with clear evidence. Even a minor mistake could draw immense public attention and can damage the image of the organization (as it happened in 2007).

It is also known that there are some people who believe the climate change has been occurring but do not accept it is a result of anthropogenic causes. The people who are in this category do not have sufficient peer reviewed papers, but there are several newspaper articles and blogs in operated by the climate change deniers. These include the conservative politicians both from the North and South, some big business firms particularly the mining and oil companies or organizations that benefit from them. "*Money and grid matter a lot,*" is one in particular that has created a great deal of opposition through political power, media, etc. The Source Watch Organization summarizes the main points of these skeptics.

"Skeptics' disputes with the IPCC and advocates of action on climate change are commonly along one or more of the following lines:

- *There is no conclusive evidence that climate change is happening;*
- *The changes in measured temperatures are part of the natural cycle;*
- *Even if the changes are human induced, the scale is not sufficiently large to make changes beyond sensible "least cost" measures; and*
- *The economic impact of making substantial cuts in greenhouse gas emissions on the scale suggested by the IPCC or other groups is too large.*
- *The dire predictions of global warming are based on computer models, but those models don't include such highly correlated data as sun spots and global temperature changes.*

> *While some skeptics challenge a particular technical point or approach in the IPCC reports - for instance, that there is global warming, but it is not due to human activity - others have a more sweeping rejection of climate change science and proposed responses in general"* (Source Watch Organization, 2010).

However, there is no firm ground to support these claims. As we show, the cases of climate change denial are not a new phenomenon in the United States or the rest of the world. Several prominent authors of the social sciences and conscious group of society have been exploring the root interest of climate change deniers, as well as how the counter movements on climate science have been organized and how they are funded to produce incorrect information to confuse the general public and influence to the politicians (Jacques, Dunlap and Freeman, 2008; Freudenburg, Gramling and Davidson, 2008; Mccright and Dunlap, 2003). Jacques, Dunlap and Freeman (2008) have analyzed the funding mechanism for the production of false science to counter the climate science; where they analyzed 141 English-language environmentally skeptical books published between 1972 and 2005, where they found 92% were published in the United States and associated with the conservative think tanks. In their own words:

> *"Environmental skepticism denies the seriousness of environmental problems, and self-professed 'skeptics' claim to be unbiased analysts combating 'junk science'... Further, we analyze conservative*

> *think tanks involved with environmental issues and find that 90 percent of them espouse environmental skepticism. We conclude that skepticism is a tactic of an elite-driven countermovement designed to combat environmentalism, and that the successful use of this tactic has contributed to the weakening of US commitment to environmental protection* (Jacques, Dunlap and Freeman, 2008: 349).

This is a serious issue, because it is falsifying the fact, which can at least create confusion for the general public and the power centric conservative politicians. Likewise, McCright and Dunlap (2003) have also conducted a similar type of research about the motive of conservative think tanks.

The work of McCright and Dunlap (2003) assert that there are ample empirical research and practical evidence that climate change is occurring because of anthropogenic activities. Keeping this fact as background, they explore the motive and politics behind the skeptics. They analyzed the role of the Influential Conservative Think Tanks Addressing Global Warming between 1990 and 1997 and how they impacted policy maker in the United States; however, they did not elaborate on how these conservative Think Tanks were funded. Their research shows that:

> "*these countermovement organizations aligned themselves with prominent American climate change skeptics known for their staunch criticism of mainstream climate research and their affiliations with the fossil fuels industry) .... the rhetoric as well as actions of the Bush Administration clearly suggest that the conservative movement's efforts in the 1990s to redefine global warming as non-problematic and thereby undermine the Kyoto Protocol are having an enduring influence. Once again, the conservative movement and especially conservative think tanks appear to have successfully affected our nation's policymaking, this time with international implications* (McCright and Dunlap 2003: 348 and370).

This clearly indicates that the dinners are funneled by the emission producer companies, and they hold a prominent part in power politics.

The notion is clear that some of the big emission producer industries, including Exxon Mobil and Koch Industries among others, are the major funding sources for the climate change deniers and to the politicians. On March 30, 2010, Greenpeace International USA published a comprehensive report entitled "Koch Industries: Secretly Funding the Climate Denial

Machine" containing firm evidence regarding how much money the industries have paid to create the skepticism on climate change. Greenpeace International is an extremist among the conservation organizations; however, the facts and figures shown in the research report supports whatever the climate science scholars have been trying to explore. The research document states that:

> "*Most Americans have never heard of Koch Industries, one of the largest private corporations in the country, because it has no Koch-branded consumer products, sells no shares on the stock market and has few of the disclosure requirements of a public company. Although Koch intentionally stays out of the public eye, it is now playing a quiet but dominant role in a high-profile national policy debate on global warming....Koch Industries has become a financial kingpin of climate science denial and clean energy opposition. This private, out-of-sight corporation is now a partner to ExxonMobil, the American Petroleum Institute and other donors that support organizations and front-groups opposing progressive clean energy and climate policy. In fact, Koch has out-spent ExxonMobil in funding these groups in recent years. From 2005 to 2008, ExxonMobil spent $8.9 million while the Koch Industries-controlled foundations contributed $24.9 million in funding to organizations of the 'climate denial machine".... The Koch Industries and its associates have quietly funneled nearly $50 million to climate-denial front groups that are working to delay policies and regulations aimed at stopping global warming is no joking matter* (Greenpeace 2010).

The Exxon Mobil case was already a well-known case that, in 1998, devised a plan to stall action on global warming. The plan was outlined in an internal memo (see in the endnote for detail). It noted that "*Victory will be achieved when uncertainties in climate science become part of the conventional wisdom*" for the "average citizens" and "the media." Further, Exxon noted that the company would recruit and train new scientists who lack a "*history of visibility in the climate debate*" and "*develop materials depicting supports of action to cut greenhouse emissions as "out of touch with reality*" (Exxon Mobil 1998). We cannot disregard the fact current skeptics have the influence from the environmentally skeptical books and journal articles written to make emissions emitters like Exxon Mobil or Koch Industry group happy.

Similarly, George Monbiot, climate science researcher, Guardian columnist and the author of the best seller book on Global Warming (as well as others) also have been identifying who the climate change deniers are and their motives, which includes people involved with the Heartland Institute; politicians; journalists; and even university faculties who do consulting for pollution producing companies. There are also several people who have been trying to identify skeptics and their motives through their blogs, news paper articles and scholarly discussions through various list serves. The former Australian Prime Minister Mr. Kevin Michael Rudd (2009) offers a very good explanation about these climate change skeptics and deniers. He states:

> *"the climate change skeptics, the climate change deniers, the opponents of climate change action are active in every country. They are a minority. They are powerful. And invariably they are driven by vested interests....Powerful enough to so far block domestic legislation in Australia, powerful enough to so far slow down the passage of legislation through the US Congress. And ultimately – by limiting the ambition of national climate change commitments – they are powerful enough to threaten a deal on global climate change both in Copenhagen and beyond"....The opponents of action on climate change fall into one of three categories. (1) First, the climate science deniers (2) Second, those that pay lip service to the science and the need to act on climate change but oppose every practicable mechanism being proposed to bring about that action (3) Third, those in each country that believe their country should wait for others to act first"* (Rudd 2009).

Mr. Rudd's explanation is appropriately applicable in ongoing debates regarding climate change. It is a good indicator that world government leaders (we exclude the skeptics) have a clear stand regarding climate change and regardless of the opposition put forth, the urgency to tackle climate change in the long run will not be diminished. United States President Barack Obama also accepts that deniers exist but will be defeated soon with the sound commitment and policies. President Obama states:

> *"The naysayers, the folks who would pretend that this is not an issue (climate change focused added), they are being marginalized. But I think it's important to understand that the closer we get, the harder the opposition will fight and the more we'll hear from those whose interest or ideology run counter to the much-needed*

> *action that we're engaged in. There are those who will suggest that moving toward clean energy will destroy our economy — when it's the system we currently have that endangers our prosperity and prevents us from creating millions of new jobs"* (Obama 2009).

Further in his speech the President acknowledges that the task is not easy, but with the necessary commitment, nothing is impossible. The role in the Copenhagen submit he played has laid some groundwork for new breakthroughs.

What we found through these sources is that these deniers are aware of climate change; however, they simply ignore it because of their business interests. In reality there are thousands of academic publications, books and journal articles with solid evidence identifying the reality of global climate change. The IPCC is the major pillar in bringing forth the evidence to produce the reliable knowledge. However, the current few errors in its report, particularly in the Working Group 2, have provided ample opportunities for climate change skeptics to raise their voices.

6.3 Something Went Wrong: Urgency to Address

In reality, the recent debates are not mounted only from the skeptics, but also from scholars who actually believe in climate change. Similarly, there are also several authors who doubt the evidence on the IPCC reports; some state the information is incomplete, and others claim the reports are wrong. The major criticism comes from the scientists themselves claiming the IPCC team does not represent the experts, is poorly representative of the developing world, the procedure of nomination is unnecessarily lengthy and unscientific, and the lack of compatibility among the scientists because of the governments' nominations (not necessarily able to capture scientific fact).

These claims are not necessarily true; however, there are valid points for which climate scientists, general public, government officials and academician are seriously concerned regarding the IPCC 2007 WG 2. In this report, they found the wrong, misinterpreted or inappropriately cited paragraphs, table and texts including the prediction of the Himalayan Glacier melting time forecast, an illustration of the biophysical situation of the Netherlands, the Amazonian forest rapid change, and the African crop face rain fed. In recent years, these cases have been highly publicized, especially by news houses such as Reutter News agency, UK, British Broadcasting Corporation, UK, Guardian News and Media Limited, The Times, Telegraph, and Daily

Mail, UK, and New York Times, Washington Post, USA and several other News and media agencies of Asia, Europe and Africa.

In this chapter we will first review the IPCC report errors identified and publicized by the news and media houses, followed by the email cases and the occurrences.

"You have been involved in several IPCC assessment reports. There has been a lot of criticism of the IPCC recently. What is your comment on that? There was one significant mistake that was made. This related to the rate of retreat of the Himalayan glaciers. But it was not an error that was carried through into the policy summary of the IPCC assessment. Yet it resonated heavily because it came at a time when there were criticisms of research on climate change, for example concerning that at the University of East Anglia, UK. The climate-change 'deniers' highlighted the issue and the IPCC was to slow to respond because it was not 'media-savvy'. It was a single mistake, one amongst millions of 'facts' that the IPCC reports. It had no effect on policy and it has now been corrected. But it is important, nevertheless, that the IPCC has tightened up on its checking procedures because a mistake like that should not have slipped through the net. I think that there is an emerging view that the IPCC needs to evolve with the demands on it. When it was set up 20 years ago there was no international pressure on it for speed of reporting. The sheer amount of information is now probably 100 times more than it was then. About 20 years ago, you could have put all the books on climate change impacts on two shelves. Now you would need an entire library. I think the IPCC will re-think its ways of working" (interview with Martin Parry- by Georg Götz-2012:37).

"During the last years, climate research has moved extremely into the focus of politics. Decisions of huge impact are made. Nicholas Stern (London School of Economics) has estimated that reaching even moderate goals for the reduction of CO2 emission will cost at least 1% of the global Gross National Product. How does the transfer from knowledge and assessments from science to politics happen? What is your experience? My personal experience is not too good. I was in fact involved in the IPCC assessments from the beginning, I participated in the first three reports (1990–2001). It started with fairly direct connections from scientists to politicians, but then there came what I call "Climatogracy": There are large groups, even at universities, that specialize in translating the scientific results of climate research for the

policy makers. And then they summarize, and they summarize the summaries. I think this happens in a time in which we anyhow have got a "consulting culture" in politics. When politicians need advice on some topic, they go to a consulting bureau and pay them a lot of money.
And is it worth the money? Are your results transferred correctly to the politicians? No, not at all. I think the IPCC is still the best we have. It has been criticized a lot, sometimes with reason. But there is no organization like that, that could work without criticism. The IPCC is still fairly unique. But all the initiatives on the more national and local scales are completely useless. We have some groups here in the Netherlands that claim the IPCC reports are too conservative about the sea level rise, a statement based on only a few publications on glaciers and Greenland. So, they say: well, let's do it our self and they established this Delta commission and so on. This is useless, rubbish. Many people are involved, it a huge machine. But the number of scientists that look at fundamental processes in the climate system is very small compared to the number of people that do something with their results. It's a pyramid upside down. Once, I talked about that to our minister and said: Why are you asking these consulting bureaus for advice about climate problems? If you are worried about the Greenland ice sheet—we are working there, we have a good track record. If you need information, phone me and within a week I will write down in 2 pages, what I think is the state of scientific knowledge. And you do not even have to pay me since you are already paying me ((interview with Hans Oerlemans- by Georg Götz-2012:6).

6.4 The Timing of the Himalayan Glacier Melt: Error Overview

The IPCC's report indicating the melting time of the Himalayan Glacier in 2035 is a critical error, drawing concern from around the globe from the people who take climate change seriously. This error also bolstered climate change skeptics' ability to question the creditability of IPCC's working modalities. In fact, the magnitude of skeptics is so strong UN Secretary General Ban Ki-Moon urged global scientists and governments to reject the attempt of skeptics (The Guardian, 2010).

> *"The UN secretary-general Ban Ki-moon today urged environment ministers to reject attempts by skeptics to undermine efforts to forge*

> *a climate change deal, stressing that global warming poses "a clear and present danger.... To maintain the momentum, I urge you to reject last-ditch attempts by climate skeptics to derail your negotiations by exaggerating shortcomings in the ... report... Tell the world that you unanimously agree that climate change is a clear and present danger... Ban said in the statement read at the start of an annual UN meeting of environmental officials from 130 countries on the Indonesian resort island of Bali"* (The Guardian, 2010).

This simple urge indicates how determined the climate change skeptics are. Furthermore, when exploring how people are reacting to this report, a simple google search using the text "the IPCC report on Himalayan Glacier meting time" will result in 35,600 links. This clearly indicates the positive (or negative for skeptics) impact the IPCC has on the climate change issue. More important, most of the major online or printed daily journals, weekly news papers and television and radio stations have aired this issue as headline news. Similarly, in the academic, conservation, international organizations and other public domains maintaining a network of people with similar objectives (e.g. American Sociological Association, Geographical Association, Mountain forum, Forestry, etc.), it has been a major topic of discussion since the Copenhagen Submit (December 7-18, 2009). Below is a small example from the American Sociological Association (ASA) Environmental Sociology Section list-serve. We posted a query on the list-serve on February 5, 2010 stating that:

> *There has been a good discussion around about the role of IPCC (insufficient role) mostly during the COP 15 and afterwards regarding the reports of the IPCC. Recently IPCC chairman R K Pachauri accepted that there was an error regarding the melting glacier. "The debate has centered on the statement made in the IPCC AR4 Working Group II report that the Himalayan glaciers are retreating faster than in any other part of the world and at the present rate of retreat could disappear by the third decade of this millennium. This has culminated with the statement from the IPCC on 20 January 2010 retracting this one statement in AR4 but reiterating that the broader conclusion of the report is unaffected"* (by mountain forum Moderator, 2010). How do you see IPCC's role and what is your opinion about the current debate?

To date, we have received 345 direct and indirect reactions. The responses are from those who have been advocating for the climate change issue, teaching about it and trying to change the public lifestyles to minimize the risk from adverse effect. In each of the responses, they have provided the theoretical grounds for their arguments and additional links of related publications. The members of the ASA have spent a great deal of time and energy giving valuable feedback. This indicates that the IPCC holds a strong position among the academician and other audiences who care for the environment. This also indicates that if such a valuable organization makes errors, it affects the society. The following are few randomly chosen anonymous examples of public concern:

How much do they value the IPCC (which is supported by about 55 of others):

> *"The misstatement about Himalayan glaciers occurred in one paragraph of the WG2 report. The WG1 (physical science) report, more carefully peer-reviewed by glacier scientists, gave a more accurate description of what was then known about the heterogeneous responses of Himalayan glaciers. A much more detailed and up-to-date overview was presented in connection with the American Geophysical Union meetings last fall, including this comment about the WG2 paragraph that has drawn so much publicity from climate-change denialists... "This was a bad error. It was a really bad paragraph..."* ***(emphasis added)****... However, contrary to the denialist spin, the AGU authors do not think this error invalidates either the IPCC process or concern about climate change, as they concluded... "Global climate change is a huge factor in this region."... So far as I know, the bad paragraph from WG2 was not echoed in the summaries or synthesis report, and was not presented as a major IPCC conclusion. Rather, it seems to have been unnoticed by most glacier experts until recently. The IPCC itself has published a note correcting the error"* (personal correspondence).

This simple response illustrates that the scientific world still strongly supports the IPCC. Similar supportive examples can be seen in Dr Ralph J. Cicerone's recent article in Science Magazine editorial, where he states:

> *"My reading of the vast scientific literature on climate change is that our understanding is undiminished by this incident; but it has*

raised concern about the standards of science and has damaged public trust in what scientists do.... It is essential that the scientific community work urgently to make standards for analyzing, reporting, providing access to, and stewardship of research data operational, while also establishing when requests for data amount to harassment or are otherwise unreasonable. A major challenge is that acceptable and optimal standards will vary among scientific disciplines because of proprietary, privacy, national security and cost limitations. Failure to make research data and related information accessible not only impedes science, it also breeds conflicts" (Cicerone, 2010: 624).

The major focus of Cicerone's (2010) article is the assurance of the scientific facts of climate change; however, this also indicates the urgency for standardization of the scientific research. This sentence "*it has raised concern about the standards of science and has damaged public trust in what scientists do*" is referring to the East Angelia University's email scandal and the recently identified errors of the IPCC 2007 report. What was the error? The IPCC WG 2 2007 report stated that:

Glaciers in the Himalaya are receding faster than in any other part of the world (see Table 10.9) and, if the present rate continues, the likelihood of them disappearing by the year 2035 and perhaps sooner is very high if the Earth keeps warming at the current rate. Its total area will likely shrink from the present 500,000 to 100,000 km2 by the year 2035. The receding and thinning of Himalayan glaciers can be attributed primarily to the global warming due to increase in anthropogenic emission of greenhouse gases."

The IPCC report further elaborates, stating that:

*"The receding and thinning of Himalayan glaciers can be attributed primarily to the global warming due to increase in anthropogenic emission of greenhouse gases. The relatively high population density near these glaciers and consequent deforestation and land-use changes have also adversely affected these glaciers. The 30.2 km long Gangotri glacier has been receding alarmingly in recent years (*Figure 10.6*). Between 1842 and 1935, the glacier was receding at an average of 7.3 m every year; the average rate of recession between 1985 and 2001 is about 23*

> *m per year (Hasnain, 2002). The current trends of glacial melts suggest that the Ganga, Indus, Brahmaputra and other rivers that criss-cross the northern Indian plain could likely become seasonal rivers in the near future as a consequence of climate change and could likely affect the economies in the region. Some other glaciers in Asia – such as glaciers shorter than 4 km length in the Tibetan Plateau – are projected to disappear and the glaciated areas located in the headwaters of the Changjiang River will likely decrease in area by more than 60% (Shen et al., 2002)" (Section 10.6.2 The Himalayan glaciers; Box TS.6. The main projected impacts for regions, Asia section D [*10.6.2*])* .

This supports the argument that it was not a typing error but was misinformation regarding such a serious issue. More important, on January 20, 2010, the IPCC on published a press release with the title "IPCC statement on the melting of Himalayan glaciers;" however, the release did not note the 2035 date was wrong (see full statement in footnote 1), but merely presented the results as being accurate. There is no argument among the scientists and academicians that climate change has been occurring and the anthropogenic activities are major causes for this; however, the problem of the IPCC's response still remains.

The case was brought to light not by the climate change deniers or skeptics, but from the people who believe the IPCC produces the factual data and provides the knowledge to the governments of UN members. Several individuals and groups have tried to explore how this mistake occurred. One detailed operational history of how the error occurred is written by Bidisha Banerjee and George Collins on February 4, 2010 at the *Yale Forum on Climate Change and the Media*, USA; and a similar but brief clarification has been published by the Climate Science Watch (in the January 19, 2010 issue). According to these two recent online publications, IPCC's incorrect citation of WWF (2005) and WWF's wrong citation of Mridula Chettri's article on Down to the Earth (of April 30, 1999) created the error. Most important, Banerjee and Collins (2010), have provided the identical table Chettri used in 1999, which the IPCC copied and pasted without citing Chettri (see Table 10.9 in IPCC Section 10.6.2:). In the words of Banerjee and Collins (2010):

> *"The IPCC's table is almost identical to Chettri's - a fact which has been largely missed in the discussion so far. The glacier names,*

the headings (except for capitalization), the dates, the figures, and even the (arbitrary) order are the same, with only one exception: a second study on Gangotri Glacier has been added. This study is credited in the IPCC text as Hasnain (2002)"(in the Yale Forum on the Climate Change and the Media 2010).

Shown below are the tables used by Chettri, 1999 and the IPCC, 2007:

Chettri 1999 (in Down to the Earth)

Glacier	Period	Retreat of Snout (metre)	Average retreat of Glacier (metre/yr)
Triloknath Glacier (Himachal Pradesh)	1969-1995	400	15.4
Pindari Glacier (Uttar Pradesh)	1845-1966	2,840	135.2
Milam Glacier (Uttar Pradesh)	1909-1984	990	13.2
Ponting Glacier (Uttar Pradesh)	1906-1957	262	5.1
Chota Shigri Glacier (Himachal Pradesh)	1986-1995	60	6.7
Bara Shigri Glacier (Himachal Pradesh)	1977-1995	650	36.1
Gangotri Glacier (Uttar Pradesh)	1977-1990	364	28.0
Zemu Glacier (Sikkim)	1977-1984	194	27.7

It has 8 rows.

IPCC 2007 (table 10.9)

Glacier	Period	Retreat of snout (metre)	Average retre of glacier (met
Triloknath Glacier (Himachal Pradesh)	1969 to 1995	400	15.4
Pindari Glacier (Utteranchal)	1845 to 1966	2,840	135.2
Milam Glacier (Utteranchal)	1909 to 1984	990	13.2
Ponting Glacier (Utteranchal)	1906 to 1957	262	5.1
Chota Shigri Glacier (Himachal Pradesh)	1986 to 1995	60	6.7
Bara Shigri Glacier (Himachal Pradesh)	1977 to 1995	650	36.1
Gangotri Glacier (Uttaranchal)	1977 to 1990	364	28.0
Gangotri Glacier (Uttaranchal)	1985 to 2001	368	23.0
Zemu Glacier (Sikkim)	1977 to 1984	194	27.7

IPCC has added 2nd last row Uttarnchal.

These two tables clearly indicate that the IPCC was not accurate in citing the resources. The IPCC has even not mentioned the source of the table.

Very interestingly, Banerjee and Collins have also provided evidence from the 911.olotila.net publication by Janet Raloff (published with the title IPCC's Himalayan Glacier 'Mistake' No Accident on 2010-01-25) that even the date 2035 was intentionally kept. This article has mentioned that:

Authors knew that "there were no solid data to support the report's claim that Himalayan glaciers – the source of drinking and irrigation water for downstream areas throughout Asia – could dry up by 2035" and "We thought that if we can highlight it, it will impact policy makers and politicians and encourage them to take some concrete action." In other words,... "last night admitted [the scary figure] was included purely to put political pressure on world leaders". It futher states in the section notes that *"WWF article [from which the 2035 date was picked up] also contained a basic error in its arithmetic. A claim that one glacier was retreating at the alarming rate of 134 metres a year should in fact have said 23 metres"* (Raloff 2010; Banerjee and Collins 2010).

However, we do not have direct contact with any IPCC authors or with those who made the appealable stories in newspapers. We do believe the IPCC has a firm role and very strong case that climate change has been occurring at an alarming rate and one of the major causes is anthropogenic activities. We believe it was unnecessary for IPCC to make incorrect claims for political reasons.

Further in the case of Himalayan Glaciers controversies, several climatologists have also tried to explore the actual facts. One example is the 2010 NASA symposium in Arizona where the issue was discussed extensively. A nineteen-person group lead by Professor Jeffrey S. Kargel of University of Arizona presented a paper with an explanation of the scenario as follows:

"Confusion about the future of Himalayan glaciers:

Two recent conjectures about Himalayan glaciers have caused much confusion. A letter submitted (by Cogley, Kargel, Kaser and Van derVeen) to the editor of Science, summarized here with some further elaboration, attempts to clear up the confusion.

- *Himalayan rates of recession are not exceptional*
- *The first "2035" is from WWF 2005, which cites a news story about an unpublished study that does not estimate a date for disappearance of Himalayan glaciers*
- *The second "2035", an apparent typographic error, is not in WWF 2005, but can be traced circumstantially to a rough estimate of the shrinkage of all extra polar glaciers (excluding those in basins of internal drainage) between the present and 2350*
- *In conflict with knowledge of glacier-climate relationships, disappearance by 2035 would require a 25-fold acceleration during 1999–2035 from the loss rate estimated for 1960–1999*
- *Poses a legitimate question about how to improve IPCC's review process. It was not a conspiracy*
- *The error does not compromise the IPCC Fourth Assessment, which for the most part was well reviewed and is highly accurate" (Kargel et.al., 2010: 67).*

These findings give the overall scenario of how the error occurred, how the IPCC should react and its current situation. In the following sections, we briefly examine the other errors of the IPCC 2007 WG 2 reports.

6.5 The Netherland 55 Percent Land is Below Sea Level: Error Overview

The Netherland case has not drawn the same global attention as the Himalayan Glaciers case, but it is also in the discussion. This case is also from the 2007 WG 2 report (Climate change 2007: Impacts, adaptation and vulnerability) and appears in the sentence:

> *"The Netherlands is an example of a country highly susceptible to both sea-level rise and river flooding because 55% of its territory is below sea level where 60% of its population lives and 65% of its Gross National Product (GNP) is produced. As in other regions, natural ecosystems in Europe are more vulnerable to climate change than managed systems such as agriculture and fisheries (Hitz and Smith, 2004)"* (from IPCC WG 2 Report section 12.2.3 Current adaptation and adaptive capacity).

On February 12, 2010, the IPCC published a clarification or press release that states: the "*The sea level statistic was used for background information only, and the updated information remains consistent with the overall conclusions*" (Reuters: OSLO; Sat Feb 13, 2010 12:09pm EST). However, Rob Kievit's (2010) article "Sea level blunder enrages Dutch minister" notes the Netherlands' government strong concern in his statement:

> *The Dutch environment minister, Jacqueline Cramer, on Wednesday demanded a thorough investigation into the 2007 report by the UN's Intergovernmental Panel on Climate Change after a Dutch magazine uncovered it incorrectly states 55 percent of the country lies below sea level. The Dutch national bureau for environmental analysis has taken responsibility for the incorrect figure cited by the IPCC (Kievit 2010).*

Similarly, NRC Handelsblad news reporter (2010) noted that:

> *The Dutch environment minister, Jacqueline Cramer, wrote a letter to the IPCC, saying she was "not amused" there were mistakes in the scientific report she bases the Dutch environmental policies on. Now she is confronted with errors in the data about her own country. "This can't happen again," the minister told reporters in The Hague on Wednesday. "The public trust in science and politics has been badly damaged" (NRC, 2010).*

In this case, even government officials claimed that because of the error in the data, the trust in science and politics was damaged. However, the IPCC used the data provided by Netherlands Environmental Assessment Agency (BPL) for their report; therefore, BPL has taken the responsibility and suggested:

> *This should have read that 55 per cent of the Netherlands is at risk of flooding; 26 per cent of the country is below sea level, and 29 per cent is susceptible to river flooding. Examples of the latter are the near flooding, in the mid-1990s, of areas along the rivers Meuse and Waal – areas that are well above sea level (BPL 2010).*

Below is where the confusion occurs. The Netherlands Governments' Ministry of Transport, Public Works and Water Management document published on February 20, 2007 states that:

> *The Netherlands is situated in a low-lying delta where the rivers Rhine, Meuse, Scheldt and Ems debouch into the North Sea. It is a very flood prone country:* ***about 60% of the country is situated below sea-level****, 9 million people live here and about 65% of the Gross Domestic Product is earned here. Because of this, Dutch flood protection policy concentrates on prevention against floods, i.e. minimizing the probability of flooding" (*Ministry of Transport, Public Works and Water Management, 2007:1).

It is generally understood that for geography, map and population information, the government data are the reliable sources. Because the government itself is stating that about 60% of the country is situated below sea level, how can IPCC scientists ignore that fact?

This case seems not to be the fault of the IPCC panel; however, the reaction added more fuel to the climate change skeptics' argument. This case suggests that even the information coming from reliable sources needs to be verified from other sources.

6.6 The Amazonian Forest Reduction: Error Overview

The Amazonian case is different from the Glaciers and the Netherlands cases. The Glaciers case resulted from a typing error and inappropriate citation and the Netherlands case was created by dependency on single source. The Amazonian forest case is related to an incorrect citation; however, there is little debate regarding actual facts and figures.

The IPCC 2007 report in section 13.4.1 Natural ecosystems (online version, 2nd paragraph) states that:

> *"Up to 40% of the Amazonian forests could react drastically to even a slight reduction in precipitation; this means that the tropical vegetation, hydrology and climate system in South America could change very rapidly to another steady state, not necessarily producing gradual changes between the current and the future situation (Rowell and Moore, 2000)". It is more probable that forests will be replaced by ecosystems that have more resistance to multiple stresses caused by temperature increase, droughts and fires, such as tropical savannas"* (section 13.4.1 Natural ecosystems: IPCC WG 2, 2007: 596).

Senior Scientist Daniel Nepstad, who has worked extensively on the Amazonian forest issue states that:

> *The Rowell and Moore review report that is cited as the basis of this IPCC statement cites an article that we published in the journal Nature in 1999 as the source for the following statement:*
>
> *"Up to 40% of the Brazilian forest is extremely sensitive to small reductions in the amount of rainfall. In the 1998 dry season, some 270,000 sq. km of forest became vulnerable to fire, due to completely depleted plant-available water stored in the upper five meters of soil. A further 360,000 sq. km of forest had only 250 mm of plant-available soil water left. [Nepstad et al. 1999]" (Rowell and Moore 2000).*

He further states:

> *"The IPCC statement on the Amazon is correct, but the citations listed in the Rowell and Moore report were incomplete. (The authors of this report interviewed several researchers, including the author of this note, and had originally cited the IPAM website where the statement was made that 30 to 40% of the forests of the Amazon were susceptible to small changes in rainfall). Our 1999 article (Nepstad et al. 1999) estimated that 630,000 km2 of forests were severely drought stressed in 1998, as Rowell and Moore correctly state, but this forest area is only 15% of the total area*

> *of forest in the Brazilian Amazon"....In sum, the IPCC statement on the Amazon was correct. The report that is cited in support of the IPCC statement (Rowell and Moore 2000) omitted some citations in support of the 40% value statement" (Nepstad 2010).*

The World Wildlife Fund (WWF) also supports the IPCC report. In its press statement, the WWF states that:

> *"IPCC has also cited its source from the joint report by the World Wildlife Fund (WWF) and the International Union for the Conservation of Nature (IUCN) entitled, "Global Review of Forest Fires," published in 2000. WWF make clear stating that, The WWF/IUCN study said: "Up to 40% of the Brazilian forest is extremely sensitive to small reductions in the amount of rainfall," but failed to include the correct citation – a 1999 report titled* "Fire in the Amazon," *by the Amazon Environmental Research Institute (IPAM). That report said: "Probably 30 to 40% of the forests of the Brazilian Amazon are sensitive to small reductions in the amount of rainfall." Absent the reference to the IPAM report, readers assumed incorrectly that the source was a 1999 Nature article cited two sentences later" (WWF 2010).*

Here, two points can be clearly contradictory of these clarifications. The Brazilian forest is under the Amazonian forest, so quoting the Brazilian forest as the Amazonian forest is not appropriate. However, to show the general trend of forest degradation, Nepstad's (2010) point is valid. There are no disagreements among the scientific communities that the Amazonian forest has been facing the climate change threat. The United Nations Environmental Program (UNEP) and Amazon Cooperation Treaty Organization (ACTO) in 2009 published a synthesis scientific report consisting of two full years of study involving more than 150 experts in the eight countries (Bolivia, Brazil, Colombia, Ecuador, Guyana, Peru, Suriname and Venezuela) that share the Amazon that also covers previously published research. The findings of this extensive research conclude that the Amazonian forest has been at risk because of climate change. The following are a few conclusions of the report:

- *Amazonia is going through a process of environmental degradation that is expressed in growing deforestation, loss of biodiversity, water pollution, deterioration of the indigenous populations and cultural values and degradation of environmental quality in urban areas. This environmental situation is the result of a set of processes and driving forces that adversely affect this complex ecosystem and its ecosystem services, which is translated into the loss of quality in the lives of the local, national and entire regional population (UNEP and ACTO 2009, Page 7).*

- *Amazonia is changing at an accelerated pace and there are profound modifications in the ecosystem environmental degradation in Amazonia is the result of internal and external factors.*
- *Climate change is a threat to Amazonia: the Amazonian region is being affected by the rise in average temperature and by the change in the accustomed pattern of precipitation. These changes affect ecosystem equilibrium and increase vulnerability of both the environment and among the human populations, especially the poorest (UNEP and ACTO 2009, Page 9).*

This extensive research supports the findings of the IPCC; however, it is important for the IPCC to extensively revisit the sources of information used in the report.

6.7 The Questions of the IPCC: The African Crop Yields Case

The African case is also related to inappropriately citations. Another case of African crop yields, The IPCC Synthesis Report states:

> *"By 2020, between 75 and 250 million of people are projected to be exposed to increased water stress due to climate change (WGII 9.4, SPM). By 2020, in some countries, yields from rain-fed agriculture could be reduced by up to 50%. Agricultural production, including access to food, in many African countries is projected to be severely compromised. This would further adversely affect food security and exacerbate malnutrition (WGII 9.4, SPM). Towards the end of the 21st century, projected sea level rise will affect low-lying coastal areas with large populations. The cost of adaptation could amount to at least 5 to 10% of GDP (WGII 9.4, SPM). By 2080, an increase of 5 to 8% of arid and semi-arid land in Africa is projected under*

> *a range of climate scenarios (high confidence)" (WGII Box TS.6, 9.4.4)* (IPCC, 2007: Synthesis Report, Page 50).

The above results are based on the study conducted by Ali Agoumi in 2003. The IPCC cited the reference in chapter 9.4 of the WG 2 report, which says: "*In other countries, additional risks that could be exacerbated by climate change include greater erosion, deficiencies in yields from rain-fed agriculture of up to 50% during the 2000-2020 period, and reductions in crop growth period" (Agoumi, 2003:5)*. This clearly indicates skeptics are voicing unnecessary criticism in the even minor cases. However, as we noted in previous sections, the IPCC must use a more serious approach toward citations and in making statements. It also appears cases like the 2020 case, are not far enough to forecast, as the documentary evidence indicates; therefore, the scientists need to reexamine whether the claims are appropriate or not and this seems lacking in the 2007 IPCC WG 2 report.

In addition to these four highlighted, incorrectly cited cases of the IPCC 2007 WG 2 report, there are other issues being debated, mostly by the skeptics. Those include "grey literature," "wrong diagram use," "one unpublished dissertation used to support the claim sea-level rise could impact people living in the Nile delta and other African coastal areas," "use of Defender of Wildlife report to show the Salmon case," "estimation of carbon-dioxide emissions from nuclear power stations," etc. However, the scientific community demonstrates solidarity with the IPCC, stating that having minor typo mistakes or a few citation errors does not provide opportunity to question the contribution of 450 lead authors, 800 contributing authors, and 2500 reviewers from more than 130 countries. The International Council for Science (ICSU) shows its concern about the current discussion on the minor errors identified in the IPCC report; however, it reflects its solidarity with the IPCC through its press release statement on the controversy around the 4th IPCC Assessment:

> *"The IPCC 4th Assessment Report* ***represents the most comprehensive international scientific assessment ever conducted****. This assessment reflects the current collective knowledge on the climate system, its evolution to date, and its anticipated future development. It is now apparent, and given the scale of the enterprise not surprising, that some errors did occur in part of the report. However, in proportion to the sheer volume of the research reviewed and analyzed, these lapses of accuracy are* ***minor*** *and they in* ***no way undermine the main conclusions****. It should be noted that the*

errors were initially revealed and made public by scientists and the misinterpretations can now be corrected accordingly. Rather than compromising the integrity and credibility of the science of climate change, this series of events is in itself a demonstration of the vigor and rigor of the scientific process".... Scientists, governments, and other societal stakeholders need to work together to ensure the quality and relevance of such assessments. We need to learn from the current controversy and make improvements where necessary. ***We should be grateful to the many thousands of scientists who give freely of their time to contribute to the IPCC and other scientific assessments.*** *And we should continue to be critical but constructively so and in ways that openly recognize the strengths and limitations of the scientific process itself"* (ICSU, 2010).

This clearly shows the scientific communities highly value the outcome of the 4th report. However, they also do not want to see such errors because they give skeptics opportunity to question results.

6.8 The Hacking of Emails in a Climate Change Institute

We have discussed the major errors of IPCC 2007 report. The rumors come mainly from the climate change skeptics, and over all there is no question of IPCC credibility. However, it is clearly seen there were minor procedural and citation errors in the report the IPCC needs to review carefully and correct as needed. The most sensitive case is the hacked emails from the Climate Research Unit (CRU) at the University of East Anglia. In the Wall Street Journal on November 23, 2010, Keith Johnson writes:

"The scientific community is buzzing over thousands of emails and documents – posted on the Internet last week after being hacked from a prominent climate-change research center – that some say raise ethical questions about a group of scientists who contend humans are responsible for global warming.... The correspondence between dozens of climate-change researchers, including many in the U.S., illustrates bitter feelings among those who believe human activities cause global warming toward rivals who argue that the link between humans and climate change remains uncertain.... In all, more than 1,000 emails and more than 2,000 other documents were stolen Thursday from the Climate Research Unit at East

Anglia University in the U.K. The identity of the hackers isn't certain, but the files were posted on a Russian file-sharing server late Thursday, and university officials confirmed over the weekend that their computer had been attacked and said the documents appeared to be genuine" (Johnson 2009).

This is clearly an illegal act of hacking emails and cannot create any question about the creditability of science. Through various UK and USA major news agencies reports, one can guess this illegal activity is from the climate change skeptics, but the hacker is not identified. However, the hacked emails and documents can be found at several locations. A public domain has been created regarding these hacked emails, from where we can download 61.9MB all documents and emails (documents include the presentations, PDF files, Word documents, funding proposal, etc). According to the web list, forty-three pages of emails are listed, totaling 1073 correspondences, (the first email date is March 7, 1996 and the last is November 12, 2009) which could be thousands of pages in actual text. Most of them are very formal correspondences regarding the issue of climate change and so on. *The first email is a personal conversation about the proposed proposal, working modality and the money, and the last one is about the succinct report.* The last email states: "Attached is a draft letter. We were keen to keep it as short, sweet and uncomplicated as possible without skipping over important details. Shorter, simpler, requests are more likely to get read and acted upon was the specific advice from international relations" (see the link). The followings are a few examples of email correspondences (those highlighted by ClimateGate; in the actual emails, names and dates are listed; however, here we have omitted the names).

The example of correspondences in hacked emails:

1. *"I've attached a cleaned-up and commented version of the matlab code that I wrote for doing the Mann and Jones (2003) composites. I did this knowing that (name) and I are likely to have to respond to more crap criticisms from the idiots in the near future, so best to clean up the code and provide to some of my close colleagues in case they want to test it, etc. Please feel free to use this code for your own internal purposes, but don't pass it along where it may get into the hands of the wrong people".*

2. *The Korttajarvi record was oriented in the reconstruction in the way that (name) said. I took a look at the original reference – the temperature proxy we looked at is x-ray density, which the author interprets to be inversely related to temperature. We had higher values as warmer in the reconstruction, so it looks to me like we got it wrong, unless we decided to reinterpret the record which I don't remember. (name), does this sound right to you?*
3. *We probably need to say more about this. Land warming since 1980 has been twice the ocean warming — and skeptics might claim that this proves that urban warming is real and important.*
4. *I've just completed (name's) Nature trick of adding in the real temps to each series for the last 20 years (ie from 1981 onwards) and from 1961 for Keith's to hide the decline.*
5. *The fact is that we can't account for the lack of warming at the moment and it is a travesty that we can't. The CERES data published in the August BAMS 09 supplements on 2008 shows there should be even more warming: but the data are surely wrong. Our observing system is inadequate.*
6. *The skeptics seem to be building up a head of steam here! ... The IPCC comes in for a lot of stick. Leave it to you to delete as appropriate! Cheers (name) PS I'm getting hassled by a couple of people to release the CRU station temperature data. Don't any of you three tell anybody that the UK has a Freedom of Information Act!*

These correspondences provide some ground skeptics may highlight considering the words used such as the clean up; the proxy; Land warming since 1980; nature trick; date wrong; and the skeptics. In addition, in 1073 emails there may be some agreements, disagreements among the scientists, which could be other issue for the skeptics. However, every person has his or her own way of discussing issues with the working colleges and these emails cannot be used as evidence to defame the IPCC.

6.9 The IPCC Acknowledges the Criticism and Takes the Steps to Correct

The United Nations family including the IPCC has taken these issues of error, email hacking and the skeptics' propaganda of "*global warming is not due to human activity*" seriously. There is no doubt among the scientists that anthropogenic activities are a major cause of climate change, which by principle all governments of the world have also accepted and agreed upon (with few exceptions). As a result, the UN (on Thursday, December 3, 2009) announced formally to investigate the East Angelia University's emails hacking case. The British government is also involved in the investigation process (The Guardian News, 2009). The Guardian also reported that "*The UEA announced yesterday that an independent review will investigate the key allegations made by climate change skeptics that a series of stolen emails showed scientists at its Climatic Research Unit (CRU) were manipulating data*". In the media response IPCC Chairman Dr. Rajendra Pachauri said "*The processes in the IPCC are so robust, so inclusive, that even if an author or two has a particular bias it is completely unlikely that bias will find its way into the IPCC report.*"

Similarly, Ed Miliband, climate change secretary, said "*We need maximum transparency including about all the data, but it's also very, very important to say one chain of emails, potentially misrepresented, does not undo the global science. The science is very clear about climate change and people should be in no doubt about that. There will be people that want to use this to try and undermine the science and we're not going to let them*" (The Guardian News, 2009). It was clear the emails were hacked by the climate change skeptics (not revealed yet from where) but it does not change scientific fact. However, it helped to bring the scientific world closer. Examples are the ICSU's statement, WWF's statement, IUCN's statement and the open letter from the 25 US renowned scientists. The National Academy of Sciences, University of California, Berkeley Professor Inez Fung; University of Washington Professor Edward Miles; and Scripps Institute of Oceanography Professor and Nobel Laureate Mario Molina (among others) wrote an open letter to the UN Congress stating:

> "*The content of the stolen emails has no impact whatsoever on our overall understanding that human activity is driving dangerous levels of global warming.... Even without including analyses from the UK research center from which the emails were stolen, the body of evidence that underlies our understanding of human-caused global warming remains robust....*" According Environment News Service (December 5, 2009) the open letter "*cites an October 21, 2009 letter to Congress from 18 U.S. scientific organizations, including the American Association for the Advancement of Science,*

> *the American Geophysical Union and the American Meteorology Society, stressing that conclusions that human activities are the primary cause of global warming are based on "multiple independent lines of evidence".... the Natural Environment Research Council and the Royal Society confirms the scientific consensus, saying, "Climate scientists from the UK and across the world are in overwhelming agreement about the evidence of climate change, driven by the human input of greenhouse gases into the atmosphere"* (Environment News Service, 2009).

These statements of IPCC officials and the scientists of the UK and United States show there is no doubt on consensus "that human activity is driving dangerous levels of global warming."

Similarly, in March 10, 2010, the UN headquarters Secretary-General Ban Ki-moon with the IPCC Chair Dr Rajendra Pachauri jointly issued a press release about the formation of Independent Review of the IPCC Assessment Process as a result of the errors of the IPCC WG 2 assessment reports of 2007. In the press release the Secretary General stated:

> Many of the world's leading scientists contributed to this landmark synthesis of what we know about climate change. The IPCC's work has been used by policymakers around the world as the most authoritative, comprehensive source for assessing climate risk. The IPCC's conclusions are clear. The *earth's climate systems are warming above and beyond natural variability*. Human activities are contributing significantly to that warming through the emission of greenhouse gases. Let me be clear: the threat posed by climate change is real. Nothing that has been alleged or revealed in the media recently alters the fundamental scientific consensus on climate change. Nor does it diminish the unique importance of the IPCC's work. Regrettably, there were a very small number of errors in the Fourth Assessment Report. Remember: this is a 3,000 pages synthesis of complex scientific data. I have seen no credible evidence that challenges the main conclusions of that report. The scientific basis for climate action remains as strong as ever. Indeed, evidence collected since the 2007 report suggests climate change is accelerating. The need for action is all the more urgent. We need to act based on the best possible science. We need to ensure full transparency, accuracy and objectivity, and minimize the potential for any errors going forward. *That is why I have initiated, in*

> *tandem with the Chair of the IPCC, a comprehensive, independent review of the IPCC's procedures and processes. This review will be conducted by the InterAcademy Council (IAC), an international scientific organization. It will be conducted completely independently of the United Nations (bold and italics our emphasis)* (United Nations News Center, 2010).

This is a firm step of the UN to ensure the world that the IPCC is the public body for the climate change knowledge. The UN ensures for the transparency of the science-base findings on climate change. The main points of the Terms of Reference to the InterAcademy Council (IAC) include:

Review IPCC procedures for preparing reports including:

- *Data quality assurance and data quality control;*
- *Guidelines for the types of literature appropriate for inclusion in IPCC assessments, with special attention to the use of non peer-reviewed literature;*
- *Procedures for expert and governmental review of IPCC material;*
- *Handling of the full range of scientific views; and*
- ***Procedures for correcting errors identified after approval, adoption and acceptance of a report*** *(our emphasis: TOR page 2).*

Schedule of the Independent Review: Because the organizational work for the Fifth Assessment Report of the IPCC has already begun, it is urgent that the IAC submits its report at the latest by 31 August 2010, to allow for the submission of a document for consideration at the 32nd Session of the IPCC in October 2010 (TOR page 3).

This assignment to the IAC comes at a very appropriate time, because the IPCC is in the process of forming an assessment panel for the fifth report which is supposed to come out by 2014. The UN has also charged the IAC with analyzing the overall IPCC process; analyzing appropriate communication strategies and the interaction of the IPCC with media; methodology of the report preparation; recommendations for amendments to the IPCC procedures; recommendations concerning strengthening the IPCC process, and so on. It clearly indicates that the IPCC is moving further with a firm step and will try to convince the world that it has been successfully fulfilling its mandate and will continue in coming days. The IPCC chairs and co-chairs show the strong commitment though their comments after assigning the independent scientific body to evaluate its stand. Here is what they say:

"I am very grateful to the Secretary General's unwavering support, not only in jointly requesting the IAC to undertake this review, but for his steadfast support of the IPCC and climate change science," (Pachauri 2010).

"We expect the recommendations from the IAC's review to inform how the IPCC prepares its fifth major assessment of global climate change, due to be published in 2013-2014. Meanwhile, the conclusions from the IPCC's 2007 report remain entirely valid: The climate is changing due to human activity, and the effects are already being felt around the globe. If anything, more recent data indicate that the IPCC's 2007 assessment underestimated the degree to which human activity is changing our climate" (Dr. Christopher Field, Co-chair of IPCC Working Group 2, 2010).

Conclusion

In this chapter we briefly examined the IPCC's errors most highlighted by the climate change skeptics. The analysis shows that the Himalayan Glaciers' melting timing is an error of the IPCC, which to some degree damaged the IPCC frame. The citation seems to be a major issue the IPCC needs to focus more attention on. It also needs to prepare concrete guidelines and policies regarding citations of advocating organizations' reports. The emails and document hacking case does not give more room to climate change skeptics. However, it also sent a message that skeptics are well equipped and scattered everywhere; therefore, the confidence and control mechanisms need to be out of reach from such skeptics.

The former Australian Prime Minister Mr. Kevin Michael Rudd (2009), appropriately states:

> *"Put more simply: these climate change skeptics around the world would be laughable if they were not so politically powerful – particularly in the ranks of conservative parties.... Attempts by politicians in this country and others to present what is an overwhelming global scientific consensus as little more than an unfolding debate, with two sides evenly represented in a legitimate scientific argument, are nothing short of intellectually dishonest. They are a political attempt to subvert what is now a longstanding scientific consensus, an attempt to twist the agreed science in the direction of a predetermined political agenda to kill climate change action" (Rudd 2009).*

Similarly, President Barak Obama also claims there is strong urgency to address the climate change issue and that we need to think, work and apply the green technology to reduce the climate change impact. He states:

> *"From China to India, from Japan to Germany, nations everywhere are racing to develop new ways to producing and use energy. The nation that wins this competition will be the nation that leads the global economy. I am convinced of that. And I want America to be that nation.... There are going to be those who make cynical claims that contradict* ***the overwhelming scientific evidence when it comes to climate change, claims whose only purpose is to defeat or delay the change that we know is necessary"*** *(President Obama* 2009).

As former Australian Prime Minister says, the skeptics are powerful and strong enough to influence the policy makers; therefore, the IPCC needs to convince all its volunteer authors that everything they are producing is based on facts and figures and a more sincere look is needed to maintain the dignity of the Nobel Prize winner organizations. President Obama (a Nobel Prize Winner of 2009), also assures that there is no doubt climate change is one pressing issue of current times. These statements clearly indicate that the scientific community need not worry about the skeptics, but they need to work vigorously on the basis of science and need to produce the real knowledge. In some of the cases we mentioned previously, the scientists sometimes put more emphasis on showing the urgency of addressing the issues, which is not the task of scientist. That should be left to politicians and policymakers. Most important, the UN's and IPCC's serious concern, investigation (through The InterAcademy Council) and action to address the report errors and email scandal are very appropriate steps. It will boost the creditability of the IPCC and minimize the questions raised by the skeptics.

Endnotes and References

Colson, C. (1989), Against the Night, Servant Publications, Ann Arbor, MI

Appendix A to the Principles Governing IPCC Work provides a detailed Procedures for the Preparation, Review, Acceptance, Adoption, Approval and Publication of IPCC Reports, Adopted at the Fifteenth Session (San Jose, 15- 18 April 1999) amended at the Twentieth Session (Paris, 19-21 February 2003), Twenty-first Session (Vienna, 3 and 6-7 November 2003), and Twenty-Ninth Session (Geneva, 31 August – 4

September 2008 and in February 2010): http://www.ipcc.ch/pdf/ipcc-principles/ipcc-principles-appendix-a.pdf (Accessed on 03/01/2010).

Statement on IPCC principles and procedures Press release on 2 February 2010, IPCC Secretariat, c/o WMO 7 bis, Avenue de la Paix C.P: 2300 CH-1211 Geneva 2 Switzerland. http://www.ipcc.ch/pdf/press/ipcc-statement-principles-procedures-02-2010.pdf (accessed on 03/10/2019)

Source Watch Organization (2010), Global warming skeptics http://www.sourcewatch.org/index.php?title=Global_warming_skeptics Source Watch is part of the Center for Media & Democracy. The Center for Media and Democracy is an independent, non-profit, non-partisan, public interest organization (accessed on 03/18/2019).

Freudenburg,William R.; Gramling, Robert Davidson, Debra J. (2008), Scientific Certainty Argumentation Methods (SCAMs): Science and the Politics of Doubt, Sociological Inquiry, Vol. 78, No. 1, February 2008, 2–38 Mccright, Aaron M. and Dunlap, Riley E. (2003), Defeating Kyoto: The Conservative Movement's Impact on U.S. Climate Change Policy, Social Problems, Vol. 50, No. 3, pages 348–373. ISSN: 0037-7791; online ISSN: 1533-8533 (In their words "Since the early 1980s a robust international consensus about the reality and seriousness of climate change has emerged, as evidenced by several comprehensive reports from the National Academy of Sciences (National Research Council 1983, 2001), Intergovernmental Panel on Climate Change (1990, 1995, 2001), and World Climate Program (1985) (Mccright and Dunlap 2003 Page 348). Jacques, Peter J., Dunlap, Riley E. and Freeman, Mark (2008) The organization of denial: Conservative think tanks and environmental skepticism', Environmental Politics, 17:3,349-385.

Greenpeace (2010) Koch Industries: Secretly Funding the Climate Denial Machine; Executive Summary, Greenpeace, USA (March 30, 2010). http://www.greenpeace.org/usa/campaigns/global-warming-and-energy/polluterwatch/koch-industries (accessed on 03/30/2010)

Environmental Defense Fund (2007), Global Warming Skeptics: A Primer Guess who's funding the global warming doubt shops? Posted: 19-Dec-2006; Updated: 28-Aug-2007; http://www.edf.org/article.cfm?ContentID =4870 Environmental Defense Fund and Exxon Mobil (1998) Internal Memo see at http://www.edf.org/documents/3860_GlobalClimateScience PlanMemo.pdf (accessed on 03/30/2019)

Monbiot, George, The Guardian UK (2010), Monbiot's royal flush: Top 10 climate change deniers (accessed on 30/03/2019) http://www.guardian.co.uk/environment/georgemonbiot/2009/mar/06/climate-change-deniers-top-10 Monbiot, George (2006), "Pundits who contest climate change should tell us who is paying them: Covert lobbying, in the UK as well as the US, has severely set back efforts to combat the world's biggest problem", Guardian (UK), September 26, 2006. Toynbee, Polly (2006) "The climate-change deniers have now gone nuclear: When the rightwing tradition of bad science comes onside, it's time to look seriously at other energy technologies", The Guardian, July 18, 2006. Thacker, Paul D. (2006), "Climate skeptics in Europe? Mostly missing in action", Society of Environment Journalists, Summer 2006 (accesses 30/03/2018).

Vance, Erik (2010), 4 Favorite Climate Change Deniers, categories: Climate Change Adaption and Mitigation, Climate Skepticism Published February 11, 2010 @ 11:04AM PT, Environment change organization, http://environment.change.org/blog/view/4_favorite_climate_change_deniers (accesses 30/03/2018).

Former Prime Minister of Australia Mr. Kevin Michael Rudd (2009), Lecture on Climate change and Australian Position, The Lowy Institute for International Policy, Sydney, Australia on November 6, 2009. http://www.pm.gov.au/node/6305 (accessed on 03/06/2019).

US President Barack Obama at MIT on October 23, 2009 http://climateprogress.org/2009/10/23/obama-at-mit-clean-energy-jobs/ (accessed on 03/06/2019).

The Guardian (2010), Reject skeptics' attempts to derail global climate deal, UN chief urges Associated Press Wednesday 24 February 2010 10.20 GMT). http://www.guardian.co.uk/environment/2010/feb/24/ban-ki-moon-un-reject-sceptics (accessed on 03/03/2010)

We used the search engine on 30/03/2019; however, this number could be very high.

Times of India (2010), Pachauri: Only one error in a 1000-page report; TNN, Jan 21, 2019, 01.31am IST, http://timesofindia.indiatimes.com/india/Pachauri-Only-one-error-in-a-1000-page-report/articleshow/5482454.cms and Nature: International Weekly Journal of Science (2010), Glacier estimate is on thin ice: IPCC may modify its Himalayan melting forecasts, Published online 19 January 2010 |Nature 463 , 276-277 (2010) http://www.nature.com/news/2010/100120/full/463276a.html (Accesses on 2/4/2019)

Cicerone, Ralph J. (2010), Ensuring Integrity in Science, Editorial, Science 5 February 2010: Vol. 327. no. 5966, p. 624 DOI: 10.1126/science.1187612. http://www.sciencemag.org/cgi/content/full/327/5966/624 (accessed on 03/03/2019) Dr. Ralph J. Cicerone is president of the U.S. National Academy of Sciences, USA.

http://www.ipcc.ch/publications_and_data/ar4/wg2/en/ch10s10-6-2.html (accessed 2/28/19: up to this date the online report was not corrected, however, IPCC on 20 January 2010, published a press release entitled "IPCC statement on the melting of Himalayan glaciers; which states "The Synthesis Report, the concluding document of the Fourth Assessment Report of the Intergovernmental Panel on Climate Change (page 49) stated: "Climate change is expected to exacerbate current stresses on water resources from population growth and economic and land-use change, including urbanization. On a regional scale, mountain snow pack, glaciers and small ice caps play a crucial role in freshwater availability. Widespread mass losses from glaciers and reductions in snow cover over recent decades are projected to accelerate throughout the 21st century, reducing water availability, hydropower potential, and changing seasonality of flows in regions supplied by melt water from major mountain ranges (e.g. Hindu-Kush, Himalaya, Andes), where more than one-sixth of the world population currently lives." This conclusion is robust, appropriate, and entirely consistent with the underlying science and the broader IPCC assessment. It has, however, recently come to our attention that a paragraph in the 938-page Working Group II contribution to the underlying assessment refers to poorly substantiated estimates of rate of recession and date for the disappearance of Himalayan glaciers. In drafting the paragraph in question, the clear and well-established standards of evidence, required by the IPCC procedures, were not applied properly. The Chair, Vice-Chairs, and Co-chairs of the IPCC regret the poor application of well-established IPCC procedures in this instance. This episode demonstrates that the quality of the assessment depends on absolute adherence to the IPCC standards, including thorough review of "the quality and validity of each source before incorporating results from the source into an IPCC Report". We reaffirm our strong commitment to ensuring this level of performance". http://www.ipcc.ch/pdf/presentations/himalaya-statement-20january2010.pdf

Banerjee, Bidisha and Collins, George (2010), in February 4, 2010 with the title of Analysis: Undoing 'The Curse' of a Chain of Errors, Anatomy of IPCC's Mistake on Himalayan Glaciers and Year 2035, The Yale Forum on Climate Change and The Media, USA: http://www.yaleclimatemediaforum.org/2010/02/anatomy-of-ipccs-himalayan-glacier-year-2035-mess/ The Yale Forum on Climate Change & The Media is an online publication and forum to foster dialogue on climate change among scientists, journalists, policymakers, and the public. The Yale Forum is an initiative of the Yale Project on Climate Change, directed by Dr. Anthony Leiserowitz of the Yale School of Forestry and Environmental Studies. http://www.yaleclimatemediaforum.org/aboutus/ (accessed on 3/01/2019).

Climate Science Watch (January 19, 2010) IPCC slips on the ice with statement about Himalayan glaciers, http://www.climatesciencewatch.org/index.php/csw/details/ipcc_slips_on_the_ice/ (accessed on 3/01/2019).

ibid

10.6.2 The Himalayan glaciers: Table 10.9. Record of retreat of some glaciers in the Himalaya at http://www.ipcc.ch/publications_and_data/ar4/wg2/en/ch10s10-6-2.html (accessed on 03/01/2019).

http://911.olotila.net/uutiset/2010/02/04/ipccs-himalayan-glacier-mistake-no-accident/ (accessed on 30/10/2019).

ibid

Kargel, Jeffrey S., Richard Armstrong, Yves Arnaud, Etienne Berthier, Michael P. Bishop, Tobias Bolch, Andy Bush, Graham Cogley, Koji Fujita, Roberto Furfaro, Alan Gillespie, Umesh Haritashya, Georg Kaser, SiriJodhaSingh Khalsa, Greg Leonard, Bruce Molnia, Adina Racoviteanu, Bruce Raup, and CornelisVan derVeen (2010), Satellite-era glacier changes in High Asia Complex and shifting Himalayan glacier changes point to complex and shifting climate driving processes, Background support presentation for NASA "Black Carbon and Aerosols" press conference associated with Fall AGU, Dec. 14, 2009. Updated and expanded Feb. 17, 2010. http://web.hwr.arizona.edu/~gleonard/2009Dec-FallAGU-Soot-PressConference-Backgrounder-Kargel.pdf (accessed on 30/04/2019)

IPCC 2007 section 12.2.3 Current adaptation and adaptive capacity: http://www.ipcc.ch/publications_and_data/ar4/wg2/en/ch12s12-2-3.html (accessed on 2/28/19)

Ireland, Louise (2010), OSLO (Reuters) U.N. climate panel admits Dutch sea level flaw, Sat Feb 13, 2010 12:09pm, http://www.reuters.com/article/idUSTRE61C1V420100213 (accessed on 2/28/10)

Kievit, Rob (2010), Sea level blunder enrages Dutch minister; A United Nations report wrongly claimed that more than half of the Netherlands is currently below sea level Published on: 4 February 2010 - 9:24am, The Radio Netherlands. http://www.rnw.nl/english/article/sea-level-blunder-enrages-dutch-minister (accessed on 03/03/2010) Radio Netherlands Worldwide provides news, information and culture via radio, television and the internet to millions of people throughout the world. We do this from a Dutch-European perspective, in ten languages. http://www.rnw.nl/english/info/what-rnws-mission

NRC Handelsblad (2010), IPCC climate report error #3: "the Netherlands is 55% below sea level" Thu, 04 Feb 2010 15:27 EST and Signs of the Times, or SOTT.net http://www.nrc.nl/international/article2476086.ece/New_mistake_found_in_UN_climate_report http://www.sott.net/articles/show/202500-IPCC-climate-report-error-3-the-Netherlands-is-55-below-sea-level- (accessed on 03/03/2019)

The Netherlands Environmental Assessment Agency -PBL (2010), Netherlands in IPCC report: Correction wording flood risks for the Netherlands in IPCC report, http://www.pbl.nl/en/dossiers/Climatechange/content/correction-wording-flood-risks.html (accessed on 03/03/2019).

Government of the Northlands, the Ministry of Transport, Public Works and Water Management: Flood maps in the Netherlands (20 February 2007), http://www.safecoast.org/editor/databank/File/folder%20engels%20def%201%20febr07.pdf (accessed on 03/05/2019)

IPCC (2007), section 13.4 Summary of expected key future impacts and vulnerabilities 13.4.1 Natural ecosystems http://www.ipcc.ch/publications_and_data/ar4/wg2/en/ch13s13-4.html#13-4-1 (accessed on 2/28/10)

Nepstad, Daniel (2010), IPCC's (AR4) statement on Amazon forest susceptibility to rainfall reduction (February 2010), The Woods Hole Research Center, USA. http://www.whrc.org/resources/online_publications/essays/2010-02-Nepstad_Amazon.htm (accessed on 03/04/2019) Daniel Nepstad is a Senior Scientist, Woods Hole Research Center and Coordinator of Research, Instituto de Pesquisa Ambiental da Amazonia.

Statement from WWF Regarding the IPCC and the Strength of Our Science (For Release: Feb 10, 2010) http://www.worldwildlife.org/who/media/press/2010/WWFPresitem15346.html (accessed on 03/04/2019)

UNEP and ACTO (2009), Environment Outlook in Amazonia: GEO Amazonia, Published by the United Nations Environment Programme (UNEP) and the Amazon Cooperation Treaty Organization (ACTO) in collaboration with the Research Center of the Universidad del Pacífico (CIUP). http://www.unep.org/pdf/GEOAMAZONIA.pdf (accessed on 03/04/2019)

IPCC (2007), Climate Change 2007: Synthesis Report, Synthesis Report; 3.3.2 Impacts on regions; Africa http://www.ipcc.ch/publications_and_data/ar4/wg2/en/ch13s13-4.html (accessed on 2/28/19)

Agoumi, Ali (2003), Vulnerability of North African Countries to Climatic Changes, Adaptation and Implementation Strategies for Climate Change, International Institute for Sustainable Development, Canada (funded project by USAID) http://www.cckn.net/pdf/north_africa.pdf (accessed on 03/06/2019)

The International Council for Science (2010), Press Release on Statement on the controversy around the 4th IPCC Assessment (23 February, 2010), ICSU, France. http://www.icsu.org/Gestion/img/ICSU_DOC_DOWNLOAD/3031_DD_FILE_IPCCstatementICSUfin.pdf (accessed on 03/08/2019).

Johnson, Keith (2009), Climate Emails Stoke Debate, Scientists' Leaked Correspondence Illustrates Bitter Feud over Global Warming, November 23, 2009, printed in The Wall Street Journal, page A3, http://online.wsj.com/article/SB125883405294859215.html (accessed on 03/06/2019).

ClimateGate: Climate center's server hacked revealing documents and emails Written by Tony Hake, Climate Change Examiner | 20 November 2009 http://www.climatechangefraud.com/climate-reports/5630-climategate-climate-centers-server-hacked-revealing-documents-and-emails which leads to two options, http://rapidshare.com/files/309710046/FOI2009.zip and http://www.filedropper.com/foi2009. Another link is http://eastangliaemails.com/index.php (this does not say who own this site) http://wattsupwiththat.com/2009/11/19/breaking-news-story-hadley-cru-has-apparently-been-hacked-hundreds-of-files-released/ (accessed on 03/06/2019)

East Anglia Confirmed Emails from the Climate Research Unit – Searchable: On 20 November 2009, emails and other documents, apparently originating from with the Climate Research Unit (CRU) at the University of East Anglia. http://eastangliaemails.com/index.php (accessed on 03/05/2019)

Ibid (to number 32)
The Guardian News and Media Limited (2009), United Nations panel to examine evidence in leaked climate email case: Independent review will analyze hacked email exchange between UEA scientists at centre of climate controversy: reported on Friday, December 4, 2009 11.17 GMT, http://www.guardian.co.uk/environment/2009/dec/04/un-panel-uae-hacked-climate-email (accessed on 03/18/2019)
Ibid
Environment News Service (ENS) (2009), The Case of the Stolen Climate Emails http://www.ens-newswire.com/ens/dec2009/2009-12-05-01.asp (accessed on 03/18/2019).
United Nations News Center (2010), Secretary-General Ban Ki-moon: UNHQ 10 March 2010: Remarks to the press on the Intergovernmental Panel on Climate Change (IPCC); formation of Independent Review of the IPCC Assessment Process http://www.un.org/apps/news/infocus/sgspeeches/statments_full.asp?statID=745 (accessed on 03/20/2019)
Terms of Reference to the InterAcademy Council (IAC) (2010), by United Nations http://genevalunch.com/files/2010/03/tor-independent-review-10032010.pdf (accessed on 03/20/2019)
IPCC (2010), Press Release: Scientific Academy to Conduct Independent Review of the Intergovernmental Panel on Climate Change's Processes and Procedures at Request of United Nations and IPCC; http://www.ipcc.ch/pdf/press/pr-1003210-UN.pdf (accessed on 03/20/2019)
Ibid (to number 5)
Ibid (number 6)
The InterAcademy Council (2010), The 18-member InterAcademy Council Board is composed of presidents of 15 academies of science and equivalent organizations representing Argentina, Australia, Brazil, China, France, Germany, India, Indonesia, Japan, South Africa, Turkey, the United Kingdom, and the United States. It also includes the African Academy of Sciences and the Academy of Sciences for the Developing World (TWAS) as well as representatives of the InterAcademy Panel (IAP) of scientific academies, the International Council of Academies of Engineering and Technological Sciences (CAETS), and the InterAcademy Medical Panel (IAMP) of medical academies. The IAC Secretariat is hosted by the Royal Netherlands Academy of Arts and Sciences in Amsterdam. http://www.interacademycouncil.net/Object.File/Master/12/737/IAC%20Co-Chairs%202009-2013.pdf

http://www.interacademycouncil.net/CMS/3239.aspx (accessed on 03/20/2010).
Götz, Georg (2012), Global Change, Interviews with Leading Climate Scientists, (Springer Briefs in Earth System Sciences, DOI: 10.1007/978-3-642-23444-6_1), Springer Heidelberg Dordrecht London New York
Kemp, Jeremy (2010), Sceptics and deniers of climate change not to be confused, Nature- Opinion- CorrespondencelVol 464l1
Collomb, Jean-Daniel (2014), The Ideology of Climate Change Denial in the United States, European journal of American studies [En ligne], 9-1 ldocument 5, mis en ligne le 01 avril 2014, consulté le 30 avril 2019. URL: http://journals.openedition.org/ejas/10305; DOI: 10.4000/ejas.10305

Annex – A List of Relevant Reference (Mainstream and Denial)

Allin, Craig W. (2008), The Politics of Wilderness Preservation, Fairbanks, AK: The University of Alaska Press.
Armey, Dick and Matt Kibbe (2010), Give Us Liberty: A Tea Party Manifesto, New York, Harper Collins.
Barnejee, Neela (2012), "Mitt Romney Worked to Combat Climate Change as Governor," The Los Angeles Times, (June 13, 2012), http://articles .latimes.com/2012/jun/13/nation/la-na-romney-energy-20120613
Boccia, Romina, Jack Spencer, and Robert Gordon Jr. (2013), "Environmental Conservation Based on Individual Liberty and Economic Freedom." The Heritage Foundation. Backgrounder no 2758: 1–8. http://www.heritage.org/research/reports/2013/01/environmental-conservation-based-on-individual-liberty-andeconomic-freedom
Borowski, Julie (2012), "Earth Day Special: Private Property Protects the Environment," Freedom Works. http://www.freedomworks.org/blog/jborowski/private-property-protects-environment
Brownstein, Ronald (2007), The Second Civil War: How Extreme Partisanship Has Paralyzed Washington and Divided America. New York: Penguin
Callicott, J. Baird (2009), "From the Land Ethic to the Earth Ethic: Aldo Leopold and the Gaia Hypothesis." In Gaia in Turmoil: Climate Change, Biodepletion, and Earth Ethics in an Age of Crisis, edited

by Eileen Crist, H. Bruce Rinker, and Bill McKibben, 177-194. Cambridge, MA: The MIT Press

Callicott, J. Baird (2010), "5 Questions." In Sustainability Ethics, edited by Ryne Raffaelle, Wade Robinson, and Evan Selinger, 57-70. Copenhagen: Automatic Press/VIP

Collomb, Jean-Daniel (2014), The Ideology of Climate Change Denial in the United States, European journal of American studies [En ligne], 9-1 |document 5, mis en ligne le 01 avril 2014, consulté le 30 avril 2019. URL: http://journals.openedition.org/ejas/10305; DOI: 10.4000/ejas.10305

Cornwall Alliance (2010), "An Evangelical Declaration on Global Warming." http://www.cornwallalliance.org/articles/read/an-evangelical-declarat ion-onglobal-warming/

Courtillot, Vincent (2004), State of Fear. New York: Harper Collins

Crichton, Michael (2005), "The Role of Science in Environmental Policy-Making," US Senate Committee on Environment and Public Works Hearing Statements, (October 28, 2005), http://epw.senate.gov/heari ng_statements.cfm?id=246766

Crichton, Michael (2009), State of Fear, New York: Harper Collins.

Dunlop, Norton (2006), "Federalism and Free Markets: The Right Environmental Agenda," The Heritage Foundation, http://www.heritage.org/a bout/speeches/federalism-and-free-markets-the-rightenvironmental-a genda

Fleischer, Ari (2001), "Press Briefing," (May 7, 2001) http://georgewbushw hitehouse.archives.gov/news/briefings/20010507.html

Geman, Ben (2012), "House GOP Leaders Pledge to Oppose Climate Tax," The Hill, (November 15, 2012), http://thehill.com/blogs/e2-wire/e2-w ire/268289-house-gop-leaderspledge-to-oppose-climate-tax

Gerson, Michael (2012), "Climate Change and the Culture War," The Washington Post, (January 17, 2012), http://www.washingtonpost.com/opi nions/climate-and-theculture-war/2012/01/16/gIQA6qH63P_story.h tml

Goklany, Indur (2008), "What to Do About Climate Change." Cato Institute Policy Analysis 609: 1-28. http://www.cato.org/sites/cato.org/files/p ubs/pdf/pa-609.pdf

Götz, Georg (2012), Global Change, Interviews with Leading Climate Scientists, (Springer Briefs in Earth System Sciences, DOI: 10.1007/978-3-642-23444-6_1), Springer Heidelberg Dordrecht London New York

Gore, Al (2007), The Assault on Reason, London: Bloomsbury, Kindle edition.

Gore, Al (2011), "Climate of Denial," Rolling Stone, http://www.rollingstone.com/politics/news/climate-of-denial-20110622

Gottlieb, Robert (1993), Forcing the Spring: The Transformation of the American Environmental Movement. Washington, DC: Island Press.

Green, Kenneth P. (2010), "Not Going Away: America's Energy Security, Jobs and Climate Challenges." Statement before the House Select Committee on Energy Independence and Global Warming, http://www.aei.org/speech/energy-and-the-environment/climate-change/not-going-away-americas-energysecurity-jobs-and-climate-challenges/

Groves, Steven (2009), "National Sovereignty May Melt at Climate Conference." The Heritage Foundation. Commentary http://www.heritage.org/research/commentary/2009/12/national-sovereignty-may-melt-at-climate-conference.

Groves, Steven (2009), "The 'Kyoto II' Climate Change Treaty: Implications for American Sovereignty," The Heritage Foundation, Copenhagen Consequences, Analysis of the 2009 Copenhagen UN Climate Change Conference n°5:3, http://www.heritage.org/research/reports/2009/11/the-kyoto-ii-climate-change-treatyimplications-for-american-sovereignty.

Groves, Steven (2009),"National Sovereignty May Melt at Climate Conference," The Heritage Foundation, Commentary (December 4, 2009), http://www.heritage.org/research/commentary/2009/12/national-sovereignty-may-melt-at-climate-conference

Groves, Steven (2010), "Why Does Sovereignty Matter to America?," The Heritage Foundation, Understanding America, 7-8, http://www.heritage.org/research/reports/2010/12/why-does-sovereignty-matter-to-america

Grunwald, Michael (2012), The New New Deal. New York: Simon & Schuster.

Hoggan, James (2009), Climate Cover-Up: The Crusade to Deny Global Warming, Vancouver: Greystone Books.

Inhofe, James (2011), "Energy Tax Prevention Act: The Only End to Cap and Trade," Human Events, (March 28, 2011), http://www.humanevents.com/2011/03/28/energy-taxprevention-act-the-only-end-to-cap-and-trade/

Isaac, Rael Jean (2012), Roosters of the Apocalypse, Chicago: The Heartland Institute.

Jacques, Peter J., Riley E. Dunlap, and Mark Freeman (2008), "The Organization of Denial: Conservative Think Tanks and Environmental Scepticism." Environmental Politics 17.3: 349-385.

James, Sallie (2009), "A Harsh Climate for Trade: How Climate Change Proposals Threaten Global Commerce." Trade Policy Analysis 41:1-20. http://www.cato.org/sites/cato.org/files/pubs/pdf/tpa-041.pdf

Johnson, Jason Scott (2008), "A Looming Policy Disaster." Regulation 31.3:38-44. http://www.cato.org/sites/cato.org/files/serials/files/regulation/2008/9/v31n3-1.pdf

Kemp, Jeremy (2010), Sceptics and deniers of climate change not to be confused, Nature- Opinion- Correspondence|Vol 464| 1

Krauthammer, Charles (2008), "Carbon Chastity," The Washington Post, (May 30, 2008), http://articles.washingtonpost.com/2008-05-30/opinions/36813249_1_socialismcarbon-chastity-co2into

Loris, Nicolas and Brett D. Schaefer (2013), "Climate Change: How the United States Should Lead," The Heritage Foundation, Issue Brief no 3841 www.heritage.org/research/reports/2013/01/climate-change-how-the-us-should-lead.

Mann, Michael E. (2012), The Hockey Stick and the Climate Wars: Dispatches from the Front Lines. New York: Columbia University Press.

McKibben, Bill (2005), "Climate of Denial," Mother Jones, http://www.motherjones.com/politics/2005/05/climate-denial

Michaels, Patrick J. (2009), "Global Warming and Climate Change," In Cato Handbook for Policymakers, 7th ed., edited by David Boaz, 474-488. Washington, DC: Cato Institute.

Milloy, Steve (2009), Green Hell: How Environmentalists Plan to Control your Life and What You Can Do to Stop Them. Washington, DC: Regnery Publishing Inc.

Mooney, Chris (2006), The Republican War on Science, New York: Basic Books.

Morgan, Derrick (2012), "A Carbon Tax Would Harm US Competitiveness and Low-Income Americans Without Helping the Environment," The Heritage Foundation, Backgrounder no 2720:4, http://www.heritage.org/research/reports/2012/08/a-carbontax-would-harm-us-competitiveness-and-low-income-americans-without-helping-theenvironment

Oreskes, Naomi (2010), "Beyond the Ivory Tower: The Scientific Consensus on Climate Change," Science 306 (5702):1686.

Oreskes, Naomi and Erik M. Conway (2012), Merchants of Doubt: How a Handful of Scientists Obscured the Truth on Issues from Tobacco Smoke to Global Warming, London: Bloomsbury.
Pinchot, Gifford (1987), Breaking New Ground, Washington, DC: Island Press.
Pooley, Eric (2010), The Climate Wars: True Believers, Power Brokers, and the Fight to Save the Earth. New York: Hyperion.
Posner, Eric A., and Cass R. Sumstein (2008), "Global Warming and Social Justice." Regulation 31.1:14-20. http://www.cato.org/sites/cato.org/files/serials/files/regulation/2008/2/v31n1-3.pdf
Richards, Jay Wesley (2009), "The Economy Hits Home: Energy and the Environment" (Washington: The Heritage Foundation, Undated), 14, http://thf_media.s3.amazonaws.com/2009/pdf/EconHitsHome_Environment.pdf
Romney, Mitt (2010), No Apology: Believe in America, New York: Saint Martin's Griffin.
Scissors, Derek (2009), "China Will Follow the United States: A Climate Change Fable." The Heritage Foundation. WebMemo no 2327 www.heritage.org/research/reports/2009/03/china-will-follow-the-us-a-climate-changefable.
Stimson, James A (2009), Tides of Consent: How Public Opinion Shapes American Politics. New York: Cambridge University Press.
Taylor, Jerry (2009), "Environmental Policy." In Cato Handbook for Policy-makers, 7th ed., edited by David Boaz, 463-474. Washington, D.C.: Cato Institute.
Will, George F (2010), "Global Warming Advocates Ignore the Boulders." The Washington Post, (February 21, 2010), http://newstrust.net/stories/853779/toolbar
Wing, Nick (2010), "John Shimkus, GOP Rep. Who Denies Climate Change on Religious Grounds, Could Lead House Environmental Policy." The Huffington Post, (November 13, 2010). http://www.huffingtonpost.com/2010/11/13/john-shimkusclimate-change_n_782664.html
Winner, Langdom (1986), The Whale and the Reactor: A Search for Limits in an Age of High Technology. Chicago, IL: Chicago University Press.
Zernike, Kate (2010), Boiling Mad: Inside Tea Party America, New York: Times Books.

7

The Composition of the IPCC

One of the strongest arguments to support the IPCC is that its composition guarantees a scientifically-sound consensus. This chapter analyzes the composition (by country, type of institution, profession) of the people who worked on the third and fourth appraisals. It shows the extent to which the composition guarantees a scientifically respectable consensus.

7.1 The Intergovernmental Organization: The Ipcc as Knowledge Producer

Managing of any organization (small or large, national or international, nongovernmental, governmental or intergovernmental), needs a framework and management procedure. The rules and procedures differ and depend on the goals of the organization. In the current context, study of international organizations not only covers nation-states, international regimes and security alliances, but also covers the international form of organizations that focus on non-state actors. In this context, the role of IGOs is not solely centered in implementation of political agendas, but also focuses on the social, cultural and economic power dynamism, and most importantly, in the context of the IPCC, scientific knowledge formation and dissemination to the member nations.

Theoretically, organizational research broadly examines (1) producing units and what factors determine organizational effectiveness or productivity and (2) sets of individuals whose well-being is affected by the terms of organizational membership and whose motivation to continue that membership depends on their assessment of its comparative contribution to their well-being (Kahn and Zald1990:3). "*The idea of an international organization is the outcome of an attempt to bring order into international relations by establishing lasting bonds across frontiers between governments or social groups wishing to defend their common interests within the context of permanent bodies, distinct from national institutions, having their own individual characteristics, capable of expressing their own will and whose role it is to perform certain functions of international importance*" (Gerbet, 1977:7, as cited in Archer, 1992:36). This notion can be applied to investigating the role of international organizations because they follow more complex formalities than domestic formal organizations. The roles of organizations depend on the motives behind why, how and for what purpose organizations were formed. In the case of the IPCC, it is based on a formal instrument of agreement between the WMO and UNEP member countries who are the major stakeholders of the climate change assessment process and knowledge users. As we discussed in the previous chapters, the IPCC was formed on the recommendations of world scientists through various meetings as a result of many conferences where they expressed the global impact of climate change and recommended that, because the climate change is a global problem, it needed to be assessed with global participation and a globalization process.

The process of globalization has intensified the global interconnections of world societies. This process has made it routine to discuss social life in a global frame rather than a national or local one. This trend of globalization has also helped the standardization of organizational structures, organizational quality maintenance and the regulation of organization management practices. Thus, standardized homogeneity can be seen in the IPCC assessment processes which directly involve the government and, through the government's channel, other stakeholders (Drori, Mayer, and Hwang, 2006).

To illustrate this issue, Rischard (2001) has summarized the environmental challenges posed through globalization in three major headings: (1) "Sharing our planet: Issues involving and global commons;" such as global warming, biodiversity loss, deforestation and desertification, water deficit, fisheries depletion, maritime security and pollution. He has also identified the major issues to be considered urgently in section (2) "Sharing our humanity: Issues requiring a global commitment;" such as the fight against poverty, peace keeping and conflict prevention, education for all, global infectious diseases, digital divide, natural disaster prevention and mitigation; and in section (3) "Sharing our rulebook: Issues needing a global regulatory approach;" reinventing taxation for the 21st century, biotechnology rules, global financial architecture, illegal drugs, trade, investment and competition rules, intellectual property rights, e-commerce rules, and international labor and migration rules (Rischard, 2002: 6, as cited by Held 2004: 12). INGOs have been highlighting those issues and creating a high-pressure environment through globalized connectedness and the mass media (mostly focused on the global commons). This notion of connectedness and networking is a positive aspect of globalization that IGOs have been utilizing to create an international regime in climate change. This principle of globalization equally applies in the global environment conservation movement and scientific knowledge construction through the IPCC assessment processes.

7.2 The IPCC Scientists and Reviewers: The Public Face

> *"The IPCC is an intergovernmental body, and it is open to all member countries of UN and WMO. Governments are involved in the IPCC work as they can participate in the review process and in the IPCC plenary sessions, where main decisions about the IPCC work program are taken and reports are accepted, adopted and approved. The IPCC Bureau and Chairperson are also elected in the plenary sessions"......... The IPCC is a scientific body. It reviews and assesses the most recent scientific, technical and socio-economic information produced worldwide relevant to the understanding of climate change. It does not conduct any research nor does it monitor climate related data or parameters. Thousands of scientists from all over the world contribute to the work of the IPCC on a voluntary basis. Review is an essential part of the IPCC process, to ensure an objective and complete assessment of current information. Differing viewpoints existing within the scientific community are reflected in the IPCC reports* (IPCC 2010).

IPCC experts come through the nomination from the United Nation and the WMO member countries. However, the IPCC also can invite to the international, intergovernmental or nongovernmental organizations to contribute to the work of the IPCC Working Groups and Task Forces. The member countries governments should be informed in advance of invitations extended to experts from their countries and they may nominate additional experts (Principle 9 of the Principles Governing IPCC Work: 2010). The IPCC secretariat sends an invitation letter to the focal point of each country if exist otherwise to the foreign ministry to send the nomination to work as lead authors, authors or for reviewer, with the full qualifications and expertise of the nominated person. In general, each government has an IPCC focal point that coordinates IPCC related activities in the country (IPCC 2010).

Chart 1: The composition of the IPCC

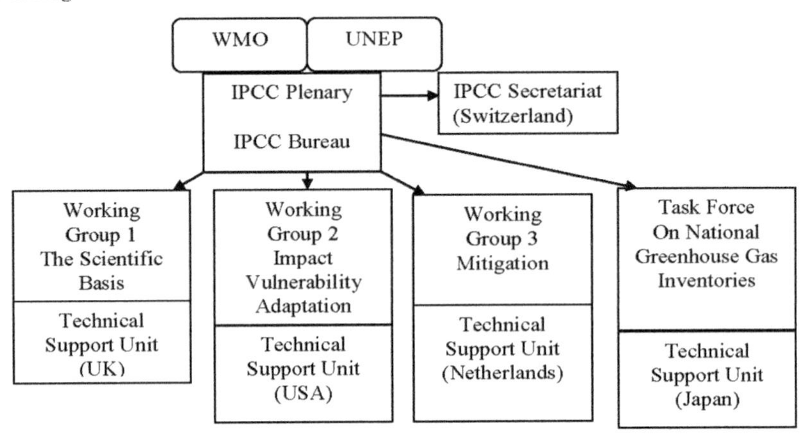

Source: IPCC 2010

IPCC is responsible to conduct climate change assessment tasks allotted by the WMO Executive Council and UNEP Governing Council resolutions and decisions as well as on actions in support of the UN Framework Convention on Climate Change (UNFCC) process. The chart 1, illustrate the composition of the IPCC processes. The IPCC plenary makes the major decisions in the plenary meetings.

> The plenary "*meets approximately once a year at the plenary level of government representatives. The sessions are attended by hundreds of officials and experts from relevant ministries, agencies and research institutions from member countries and from participating organizations. All major decisions are taken by the Panel in plenary session such as on IPCC's principles, procedures and structure, mandate of working groups and task forces, work plan and budget. The Panel decides whether to prepare a new report, its scope and outline, and it accepts reports. It elects also the IPCC Chair and the Bureau*" (IPCC 2004:4).

The IPCC Bureau, the IPCC Working Group Bureaux and the Bureaux of any Task Forces of the IPCC (by principle) reflect balanced geographic representation with due consideration for scientific and technical requirements.

> *Members of the IPCC Bureau are normally elected for the duration of the preparation of an IPCC Assessment Report (5-6 years). They*

should be experts in the field of climate change and all regions should be represented in the IPCC Bureau. The Bureau is chaired by the Chair of the IPCC and is composed of the Co-Chairs of the three IPCC Working Groups and the Task Force Bureau on National greenhouse Gas Inventories, IPCC Vice-Chairs and Vice-Chairs of the Working Groups. Presently the IPCC Bureau has 30 members.

The Working Groups and any Task Forces constituted by the IPCC shall have clearly defined and approved mandates and work plans as established by the Panel and shall be open-ended (IPCC 2010).

- *IPCC Working Groups and any Task Force of the IPCC have clearly defined mandates and work plans agreed by the Panel and are led by two Co-chairs. IPCC Working Groups at the plenary level of government representatives agree on the scope of a report prepared by the respective working group, provide guidance on the selection of authors and later accept the content of the report and approve its Summary for Policymakers. The reports are prepared by teams of authors (see later), but governments are invited to provide comments during government reviews.*
- *Working Group I assesses the scientific aspects of the climate system and climate change; Working Group II assesses the scientific, technical, environmental, economic and social aspects of the vulnerability (sensitivity and adaptability) to climate change of, and the negative and positive consequences (impacts) for, ecological systems, socio-economic sectors and human health, with an emphasis on regional, sectoral and cross-sectoral issues; Working Group III assesses the scientific, technical, environmental, economic and social aspects of the mitigation of climate change.*
- *The Task Force is mandated to carry out work on inventory-related methodologies and practices. The Bureau of the Task Force is elected by the Panel, but only the Co-chairs of the Task Force Bureau are members of the IPCC Bureau. Reports prepared by the Task Force are accepted by the Panel.*
- *The activities of each IPCC Working Group and of the Task Force on Inventories are supported by Technical Support Units (TSU). The government of the developed country Co-Chair assumes the primary responsibility for funding the TSU, which is normally hosted by a research institute in that country. The Technical Support Units are presently located in Boulder (United States of America), Exeter (United*

Kingdom), Bilthoven (the Netherlands) and Hayama (Japan) (IPCC 2004).

IPCC Assessment Preparation Process

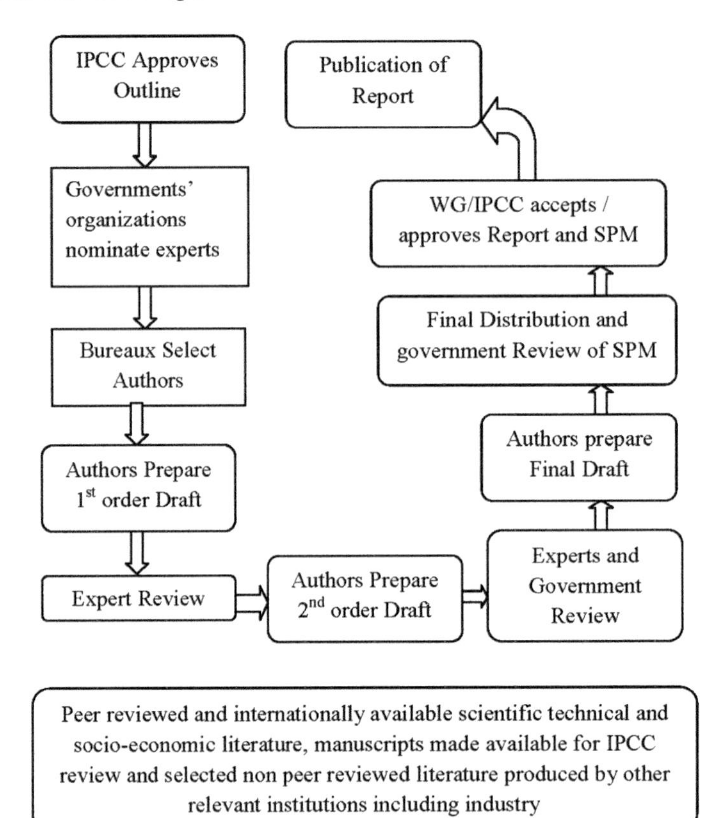

Source: IPCC 2010

IPCC Assessment Preparation Process is well defined and strait forward. It prepares at regular intervals comprehensive assessment reports of scientific, technical and socio-economic information relevant for the understanding of human induced climate change, potential impacts of climate change and options for mitigation and adaptation and also produces the special reports covering aviation, regional impacts of climate change, technology transfer, emissions scenarios, land use, land use change and forestry, carbon dioxide capture and storage and on the relationship between safeguarding the ozone layer and the global climate system including The reports' summaries for policymakers; methodology reports, which needs to meet the UNFCC criteria; technical papers as needed and produces the supporting material on the basis of its meetings and workshops (IPCC 2010).

The IPCC is composed of a much larger group of persons. In order to prepare the drafts, there is a complex system of lead authors, coordinating lead authors, contributing authors, expert reviewers, review editors and government focal points. As set out in the procedures found in Appendix A of the Principles (4.2.1), the compilation of lists is a very open-ended process:

> *At the request of Working Group Bureau Co-Chairs through their respective Working Group Bureau, and the IPCC Secretariat, governments, and participating organizations and the Working Group Bureaux should identify appropriate experts for each area in the Report who can act as potential Coordinating Lead Authors, Lead Authors, Contributing Authors, expert reviewers or Review Editors. To facilitate the identification of experts and later review by governments, governments should also designate their respective Focal Points. IPCC Bureau Members and Members of the Task Force Bureau should contribute where necessary to identifying appropriate Coordinating Lead Authors, Lead Authors, Contributing Authors, expert reviewers, and Review Editors in cooperation with the Government Focal Points within their region to ensure an appropriate representation of experts from developing and developed countries and countries with economies in transition. These should be assembled into lists available to all IPCC Members and maintained by the IPCC Secretariat* (Appendix A to the Principles Governing IPCC Work, 4.2.1).

The process is intended to be inclusive, but with some accountability, and if it is successful, the participants will clearly represent the range of scientific opinion and knowledge. The full composition of the IPCC has not been analyzed before. For this chapter, we analyzed the formal lists of participants for 2001 and 2007 in terms of country of nationality, type of organization for which they worked and profession.

7.3 The Expert Participation in Terms of Place of Employment

As we discussed, the IPCC is formally intergovernmental, but the participants can come from a variety of sources. While all must be accepted by the governments who oversee the Panel, slightly less than half of the participants work for the government. Most of the rest come from universities. The only change over time has been a slight decline in the proportion coming from government and a corresponding increase in personnel from universities (Figure 1). The Figure 1 shows that in the third assessment (TAR) the Government officials' participation was about the half of the total participants having 48.2 percent, which decreased 41.4 percent (by 6.8 points). Interestingly in AR4, governments nominated 43.2 percent of university faculty, which was 37 percent in TAR. In contrast the scholars' participation from the nongovernmental organization (NGOs) dropped from 8.2 percentages to 6.6 percentages, which leaves the ground for question because the role of NGOs and international nongovernmental organization (INGOs) is growing in the international arena (Ahsan, DelValls, and Blasco 2009, United Nations 2010). Similarly, private sectors participation remains the same 4.6 percent in both TAR and AR4 and participation of international organization increases from 2 percentages in TAR and reaches to 4.2 percent, which seems still negligible.

As NGOs, the role of INGOs for environment conservation and action programs to address the climate change issue has been raised exponentially. For example, the program to address the climate change mitigation process was not dominant in main framework of INGOs until the Rio Conference of 1992.

The agenda 21, chapter 27states that:

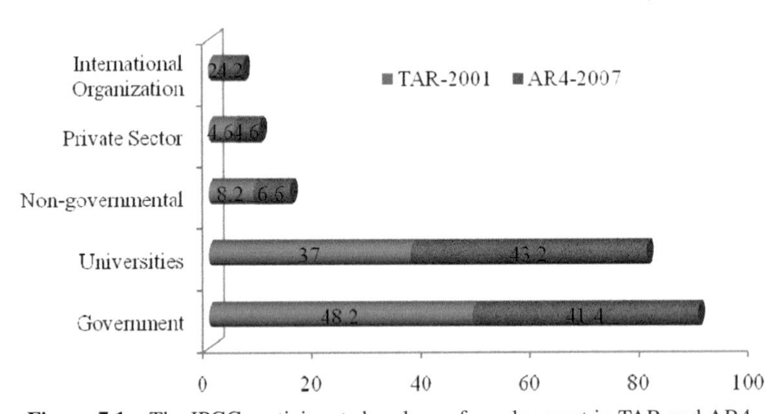

Figure 7.1 The IPCC participants by place of employment in TAR and AR4.

> *Non-governmental organizations play a vital role in the shaping and implementation of participatory democracy. Their credibility lies in the responsible and constructive role they play in society. Formal and informal organizations, as well as grass-roots movements, should be recognized as partners in the implementation of Agenda 21. The nature of the independent role played by non-governmental organizations within a society calls for real participation; therefore, independence is a major attribute of non-governmental organizations and is the precondition of real participation* (United Nation 1992).

As the outcome of Rio 1992, adopted noted agenda 21, the leader of the NGOs/ INGOs and civil society organization made a firm commitment to have collaborations where they signed the agreement with the commitment for the future. In the following articles they noted that:

> *For civil society, especially, the positive side is that, after the 1992 UNCED process, it will now be impossible for governments or international institutions to decide on our future without hearing our voices. On the basis of our new awareness and autonomy, we will fight for the democratization of states, international organizations and the United Nations (UN) itself. We will fight for the active participation of citizens in the various decision-making mechanisms and in control over their policies* (Rio-NGOs declaration 1992, Article 10).

And in article 18 they stated:

> *To speak of environment and development is to speak of life as a whole. To try to address this whole over the past several days, we have broken out a number of separate issues: climate, biodiversity, forests, savannahs, deserts and semi-arid areas, fresh water and oceans, toxic and nuclear waste, energy, fisheries, human settlements, industrial working conditions, land reform, sustainable agriculture, new technologies, communication, poverty, urban and rural violence, racism, militarism, population issues, indigenous issues, children and adolescents, women, foreign debt, international trade, transnational corporations, General Agreement on Tariffs and Trade (GATT), International Monetary Fund (IMF), World Bank, global decision-making mechanisms and environmental education*(Rio-NGOs declaration 1992, Article 18).

These are very important agreement from the non-governmental society of the world. However, the stake of nongovernmental sector in the IPCC assessment process was not well represented in TAR and AR4. In the articles of The NGO Alternative Treaties (1992), the global NGO forum noted that "they are important part of the world environment and development" but seems they are not strong enough to influence the IPCC focal points or the government agency to nominate to participate in the process (until AR4). Currently, the IPCC Bureau is constituted by 31 members, which provides fair amount of involvement of developing world, but has failed to capture the stake of global NGOs and its commitments. However, the current IPCC chair Dr. Rajendra K. Pachauri, himself comes from the NGO sector, following all three vice-chairs Dr. Ogunlade Davidson (Sierra Leone); Dr. Jean-Pascal van Ypersele (Belgium) and Dr. Hoesung Lee (Republic of Korea) all three from the Universities respectively. Similarly, the co-chairs (9); vice-chairs (17) and in the Bureau of the Task Force on National Greenhouse Gas Inventories- co-chairs (2) and other members also dominantly come from the universities (14), with few exceptions. Though as we noticed the participation of universities faculties increased by 6.8 points from TAR to AR4, there is a strong hold of academia in the IPCC mechanism.

Similarly, in terms of composition by working group and host organizations, the distribution pattern reveals the slightly different scenario (figure 2).

The Figure 2 shows that in the working group 1 of 2001 (TAR), the authors and reviewers participation from the governments' or governmental agencies', were 51.6 percent, in working group 2 (WR2), 53.4 percent and in working group 3 (WR3), 25 percent; however in 2007 (AR4) that combination sharply dropped in WR2 to 30.8 percent (from 53.4 percent by 22.6 points). In contrast the participation from the universities increased from 34.7 percent to 50.8 percent (increased by 16.1 points). In the case of nongovernmental organizations (NGOs) participation, in TAR-WR1 it was 8.4 percent which dropped by 4.2 points in the AR4, however remained increased from TAR to AR4 in WR2, but decreased in the WR3 case. Involvement of private sector in WR1 and WR2 in TAR and AR4 was lees then 2 percent in WR1 and 4.2 percent in TAR and 3.2percent in AR4, however in WR3 case it increased from 12.7 percent to 15.2 percent. The involvement of international organization (IO) in the process was 6.4 percent in TAR-WR3 which dropped to 5.9 percent in AR4; however, the IO's participation WR2 increased from 1.7 percent TAR to 7.6 percent in AR4. These situation shows that still the IPCC is predominated by the government officials and universities professor.

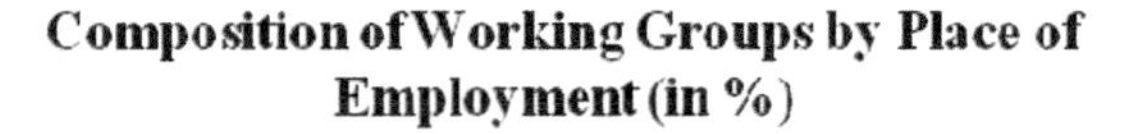

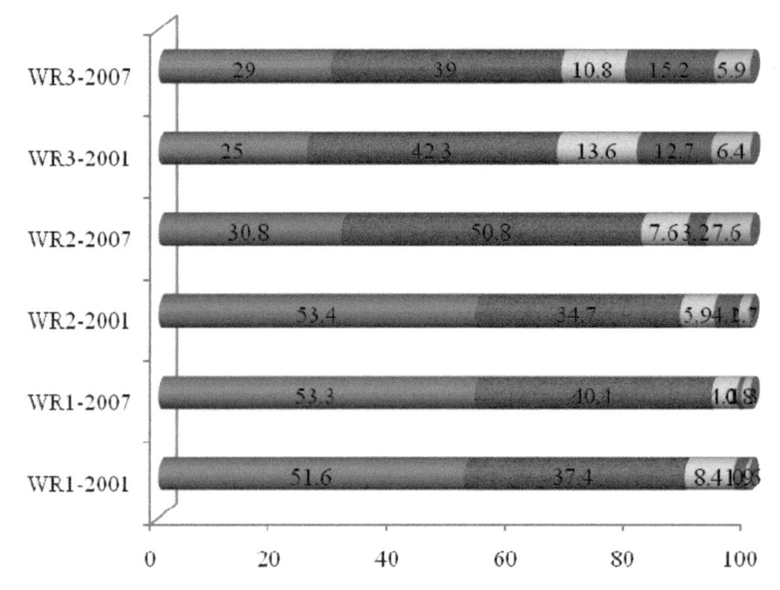

Figure 7.2 The IPCC participant's composition by working groups and place of employment in TAR and AR4.

7.4 Working Groups Composition by Country and Host Organization in TAR and AR4

Fundamentally, there is no significant change in composition TAR and AR4 processes. The total government participation in 2001 was 46.5 percent which dropped to 44.4 percent in 2007. Likewise, the universities participation remains all most the same 39.3 in 2001 and 39.8 in 2007 and participation from other agencies including NGOs/ INGOs and private sector changed from 14.2 percent in 2001 to 19 percent in 2007 respectively (Figure 2). The following Figure 2 illustrates the overall scenarios of the participation in two years in relation to host organization.

The above Figure 3 shows that the participation from the African nations is dominated by the university's professors in both by 51 percent and 48.4

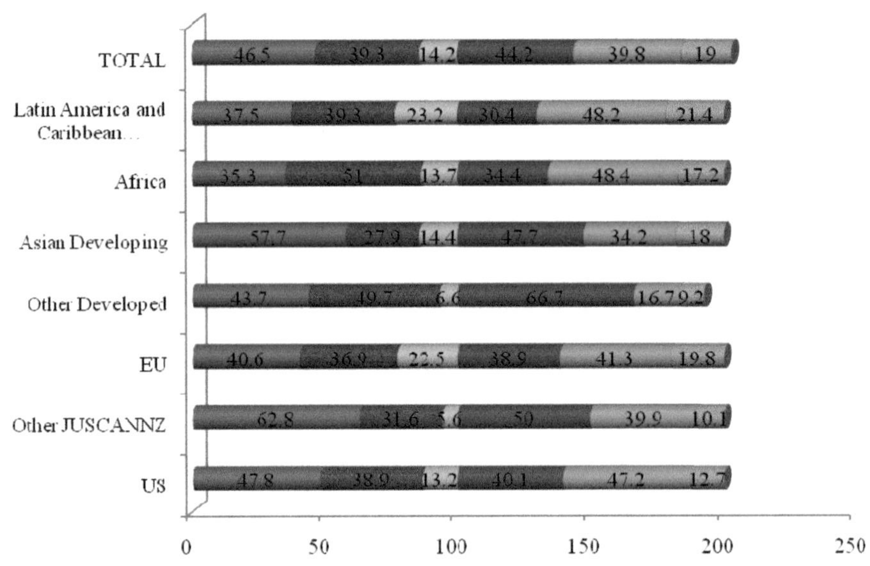

Figure 7.3 Participation in TAR and AR4 by countries and host organization; example from working group 1.

percent in TAR and AR4 respectively; the Latin America and Caribbean Developing nations also show the same trend, where, 39.3 percent participants were from the universities in TAR, which reached to 48.2 in AR4. However, the other JUSCANNZ: Japan, Canada, Australia, Norway and New Zealand have the highest government's participants in both TAR 62.8 percent and AR4 50 percent respectively; followed by the Asian Developing countries by 57.7 percent in TAR and 47.7 percent in AR4. Likewise, other Developed (includes the Russian Federation and other those not joined the Group of 77 as well as South Korea, Mexico and Switzerland) had the highest participation of 66.7 percent in the AR4, which was only 43.7 percent in TAR. There is no drastic change in participation from other sectors in both TAR and AR4. The overall scenarios of participation show the minimal involvement from NGOs, INGOs and private sector which is categorized as other in the Figure 2.

7.5 Repetition of Participation in TAR and AR4

We noticed that all the authors and reviewers, whichever is the host organization respected governments need to approve them. It is likely that the person who involves in IPCC assessment process he or she holds more experience and needs less training for the new report. Therefore, there is some overlap on the participants in between the two assessments TAR and AR4. We found that in average 33.05 percent WR1, 27.05 percent in WR2 and 34.65 percent in WR3 authors and reviewers were repeated in these two assessments (Figure 4).

The Figure 4 shows that repetition of scientist in each working group are close in WR3 and WR1, however WR2 repetition percent increased by 11.1 point from TAR to AR4.

The similar attempt of analyzing scientist's repetition from TAR to AR4 was presented at the IPCC twenty-second session held in Delhi, India by the working group three in 2004, New Delhi. According to those data, out of 159 authors 63 were from TAR and 96 from Non-TAR. The following Figure 5 shows the situation. This number was increased in review, therefore does not challenge to the Figure 4, but support to illustrate that repetition in first term to next is common phenomena in the IPCC assessment processes (Figure 5). Working group 3, in New Delhi session, also presented the participation by gender (Figure 6). Out of total 159 authors 87 percent or 139 people were male and only 20 were female authors. This shows that the female participation in WR3 for AR4 was minimal.

The working group three also presented its authors composition terms of geographical representation and gender (in New Delhi 22nd session). The report shows that 50 percent authors were from North America (19) and

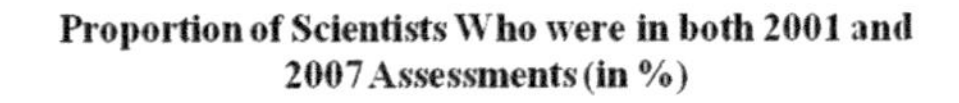

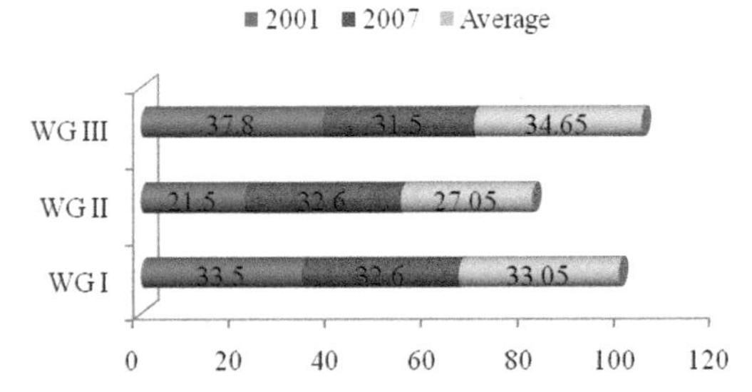

Figure 7.4 Repetition of scientist in TAR and AR4.

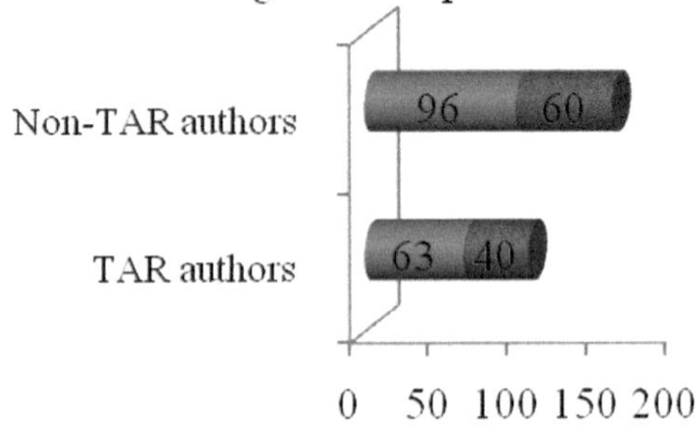

Figure 7.5 Involvement in TAR and AR4 (WR3).
Source: IPCC 2004

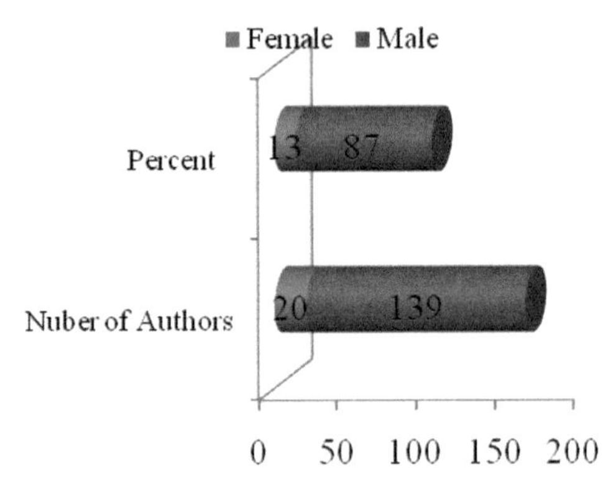

Figure 7.6 Gender distribution (WR3).
Source: IPCC 2004

Europe (31) respectively; 15 percent from Asian Developing countries; and 11 percent from Africa, Japan/Korea/Oceania and Latin America/Caribbean (all equally). The following Figure 7 provides the outline. Similarly, in combination of all industrialized, economy in transition and developing countries (Figure 8), industrialized nation representation was 55 percent, followed by

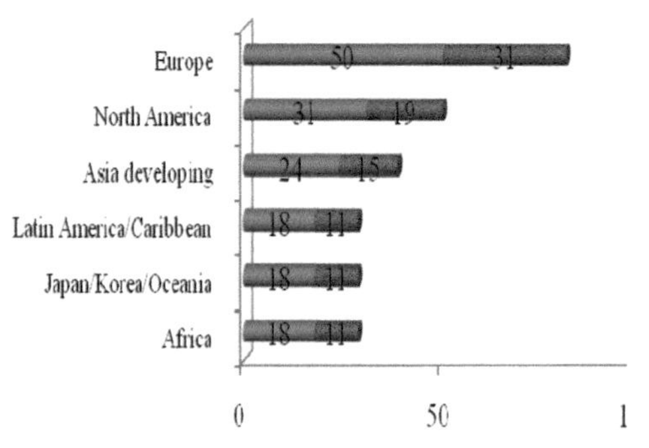

Figure 7.7 Geographical representation in WR3 for AR4 by country group.
Source: IPCC 2004

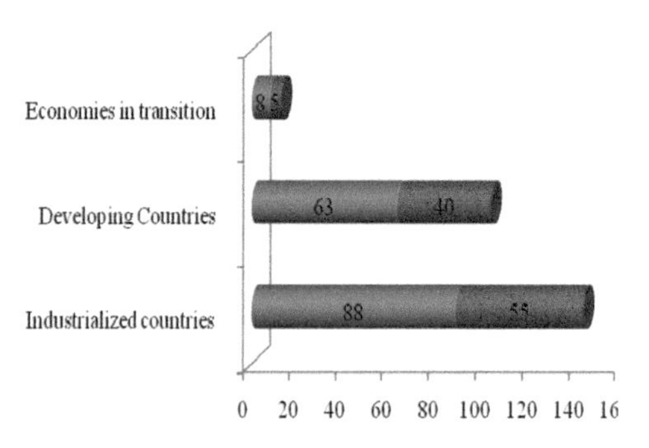

Figure 7.8 Representation by industrialized and developing country group (WR3).
Source: IPCC 2004

the 40 percent developing and 5 percent from the countries of economy in transition.

As we discussed above the IPCC has the clear mandate and principles in governing IPCC work (IPCC 2010). In addition to the principles and procedure mandated by the WMO and UNEP, the IPCC, provides a guideline to all authors and reviewers (to both expert scientists and to the government), through the scoping meetings, which is the first step of Assessment process.

The Summary Description of the IPCC Process, released on December 2009 states that:

> *"The assessment process begins with a Scoping Meeting attended by scientific experts whose task it is to outline the report..... The IPCC Bureau selects the experts attending the Scoping Meeting after an open call for nominations to Governments and IPCC observer organizations....... A call for nominations of authors for the report is sent out to Governments and observer organizations. The individual Working Group Bureaus select the authors for respective contributions to the Assessment Report, ensuring that the composition of the author teams reflects a range of views, scientific expertise and geographical diversity....... Over the ensuing 3-4 years, writing teams collaborate on drafting chapters – including four Lead Author meetings, plus chapter meetings as necessary.................. The author teams prepare two externally reviewed drafts and then a final version of their respective chapters. All chapters undergo a rigorous writing and open review process to ensure consideration of all relevant scientific information from established journals with robust peer review processes or from other sources which have undergone robust and independent peer review"* (IPCC 2009).

We have seen that, over the time, the importance of the assessments to governments in helping drive the negotiation process has clearly increased. The Fifth Assessment is expected by 2014 and initial scoping discussions have already been held. Many of these took place at a special expert group meeting in Venice in July 2009. The Figure 9 shows the composition of the participants by country group and type of organization. The largest group of participants was composed of academics with university affiliations, reflecting the trend noted above. There were, however, differences. The Group of 77 participants were more heavily from governments, as were the participants from other developed countries. The United States had the lowest relative participation of governmental experts. This suggests that different country groups had different constituencies for the assessment. In addition, there were 17 representatives from different international organizations.

The Figure 9 shows the growing participation scenarios in the IPCC scoping meetings. The IPCC scoping meeting from 13–17 July 2009, held in Venice, Italy was participated by the 180 countries, where 20 percent participating members were alone from the United States, which was dominated

Participation in July 2009 Expert Meeting on Scoping for the Fifth Assessment (in%)

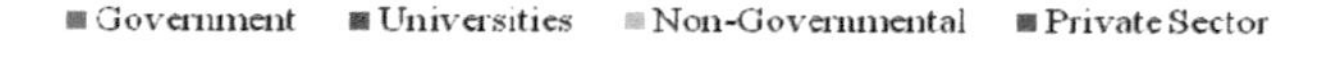

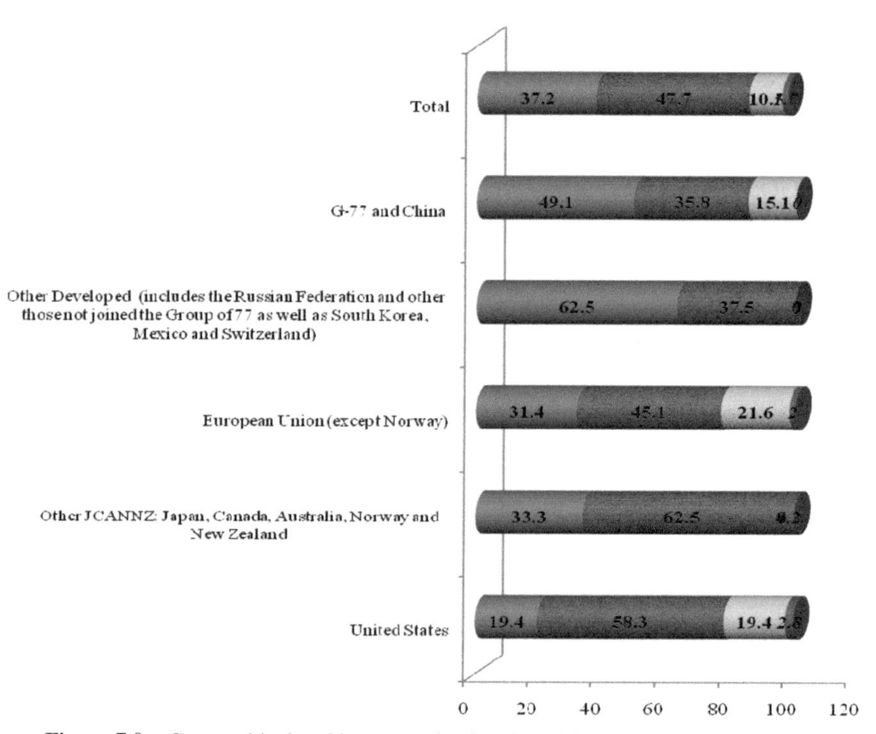

Figure 7.9 Geographical and host organizational participation in special meeting.

by the universities faculties by 58.3 percent, followed by the government officials and NGOs leaders by 19.4 percent and 2.8 percent from the private sector. Likewise, "Other JCANNZ: Japan, Canada, Australia, Norway and New Zealand" have 62.5 percent participants from the universities and 33.3 percent from the government agencies, with "0" percent participation from NGOs. In contrast, the participation from the Other Developed (includes the Russian Federation and other those not joined the Group of 77 as well as South Korea, Mexico and Switzerland) government agencies in highest level by 62.5 percent, followed by 37.5 percent universities faculties and "0" participation form NGOs and private sectors. Likewise, G-77 (130 member countries) and China's which have the largest number of participation (53 person a 29.4 percent of total participants) where 49.1 percent from the government agencies, 35.8 percent from the universities and 15.1 percent from the NGOs sector. Overall the situation shows that the IPCC expert body is dominated by the universities, followed by the government agencies. However, from "Other JCANNZ: Japan, Canada, Australia, Norway and New Zealand" and Other Developed (includes the Russian Federation and other those not joined the Group of 77 as well as South Korea, Mexico and Switzerland) participation from NGOs sector was "0" percent overall their presence was 10.5 and private sectors participation 1.7 percent. This indicates that NGOs involvement seems in growing trend in the IPCC processes.

Conclusion

The organizational performances depend on how organization is governed. The case of the IPCC is clear and straightforward e.g. it is under the United Nations major agencies who deal the global environment and climate change the UNEP and WMO, which has plenary (IPCC decision making body) followed by the bureau and three working groups one task force and the technical support units for all four. The IPCC secretariat's role is to maintain the relationships with UNEP and WMO and with the member government, including facilitation. The secretariat requests the governments to nominate the experts from their countries; however, the plenary selects the scientists among the government nominee, they can be from government agencies, universities, NOGs or INGOs and private agencies or individual. The overall scenarios show that still majority of the participants (including authors and reviewers) are from government agencies; however, in excluding the reviewers the majority of authors come from the universities. The involvement NGOs and INGOs and private sector is very low in the previous year's assessment, including female.

The majority of the lead and other authors are from the industrialized nation, which seems changing SAR to TAR and AR4. The IPCC is preparing for the fifth assessment and is selecting the authors and reviewers form the government nominee. The Fifth Assessment will build on the work of the previous four. The various participants in the working groups will be selected during 2010, but the scoping suggests that while many will be new, many will also carry over from previous assessments. This process of renewal coupled with consistency will ensure that the Fifth Assessment will be as credible as its predecessors

"Our description of the Knowing Organization provides a unified view of the principal ways in which an organization can make use of information strategically. By attending to and making sense of signals from its environment, the organization is able to adapt and thrive. By mobilizing the knowledge and expertise of its members, the organization is constantly learning and innovating. By designing action and decision routines based on what its members know and believe, the organization is able to choose and commit itself to courses of action" (Choo 1996: page 340).

Endnotes and References

Kahn, R. L. and M. N. Zald (1990), Organizations and Nation-States: New Perspectives on Conflict and Cooperation, San Francisco: Jossey-Bass.

Archer, Clive (1992), International Organizations (2nd Edition), Routledge, USA. Gerbet, Pierre (1977), Rise and Development, International Social Science Journal, Vol. 29, no. 1: 7-26.

Drori, Gili S., John W. Meyer and Hokyu Hwang (2006), Globalization and Organization: World Society and Organizational Change. Oxford: Oxford University Press.

There is also vast literature on globalization and climate change issues. Climate change issues are different than environmental problems because of its global character, unequal distribution of cause and effects, fundamental relationship with energy consumption and high degree of uncertainty regarding effects for different regions (Addink, Arts and Mol 2003:75).

Held, D. (2004), Globalization: The dangers and the answers. Available online: www.opendemocracy.net/globalization-vision reflections/article 1918.jsp (accessed on 03/23/2010) Rischard, Jean-François (2002), High Noon: 20 Global Problems, 20 Years to Solve Them (New York: Basic Books). Rischard, Jean-François (2001), High Noon: We Need New Approaches to Global Problem-Solving, Fast, Journal of International Economic Law 2001 4(3):507–525.

IPCC (2010), Organizational (home page) http://www.ipcc.ch/organization/organization.htm (accessed on 30/21/2010)

IPCC (2010), Guidance document for IPCC Government Focal Points on the nomination of potential Lead Authors, Coordinating Lead Authors and Review Editors, IPCC, Geneva. http://www.ipcc.ch/pdf/ar5/ar5-fp-guidance.pdf http://www.ipcc.ch/pdf/ar5/Web%20version%20-%20

ORG%20Nomination%20of%20Authors%20for%20the%20AR5.pdf (letter of invitation) http://www.ersilia.org/canvi_climatic/documents/IPPC/IPCC_Whoiswho.pdf (accessed on 03/21/2010) IPCC (2010), Expert participation in IPCC assessment process: Participation in the work of the IPCC is open to all UNEP and WMO Member countries; Invitations to participate in the sessions of the Panel and its Working Groups, Task Forces and IPCC workshops shall be extended to Governments and other bodies by the Chairman of the IPCC; Experts from WMO/UNEP Member countries or international, intergovernmental or nongovernmental organizations may be invited in their own right to contribute to the work of the IPCC Working Groups and Task Forces. Governments should be informed in advance of invitations extended to experts from their countries and they may nominate additional experts.

IPCC (2010), principles governing IPCC work http://www.ipcc.ch/pdf/ipcc-principles/ipcc-principles.pdf http://www.ipcc.ch/pdf/ipcc-principles/ipcc-principles-appendix-a.pdf (accessed on 03/21/2010)

IPCC (2004), Membership WHO is WHO in the IPCC, http://www.ersilia.org/canvi_climatic/documents/IPPC/IPCC_Whoiswho.pdf (accessed on 03/21/2010)

IPCC (2010), Assessment Preparation Process http://www1.ipcc.ch/ipccreports/index.htm. (accessed on 03/21/2010)

IPCC (2010), Publications and Data: Reports, IPCC home page http://www.ipcc.ch/publications_and_data/publications_and_data.htm (accessed on 03/21/2010).

To obtain the list of participants we used the secondary documentary sources published by the IPCC itself, mostly online. We compared and compiled the list of IPCC experts from the 2001 and 2007 publications' appendixes where list of experts is listed and put them into an Excel spreadsheet for coding and analysis. Coding was done in each case by two coders using standard inter-coder reliability checks. When information was missing, we used on-line search engines. In addition to the secondary data, we also collected field data (through interviews) from India, Nepal, Bangladesh, and Pakistan, to get to know whether environmental experts of South Asia were actually aware of the contribution of IPCC's role.

Ahsan, D. A., DelValls, T.A. and Blasco, J. (2009), The Relationship of National and International Environmental NGOs in Bangladesh and Their Role in Wetland Conservation, Int. J. Environ. Res., 3(1):23-34, Winter 2009. UN (2010), The United Nations and Civil Society;

http://www.un.org/en/civilsociety/index.shtml "Our times demand a new definition of leadership - global leadership. They demand a new constellation of international cooperation - governments, civil society and the private sector, working together for a collective global good." Secretary-General Ban Ki-moon Speech at World Economic Forum Davos, Switzerland (29 January 2009) (assessed on 301/21/2010)

UNFCC (1992) Agenda 21: chapter 27, Strengthening The Role Of Non-governmental Organizations: Partners For Sustainable Development, basis for action listed in 1: http://habitat.igc.org/agenda21/a21-27.htm (accessed on 03/21/2010).

The NGO Alternative Treaties (1992) Rio De Janeiro Declaration, From the Global Forum at Rio de Janeiro, June 1-15, 1992; http://habitat.igc.or g/treaties/at-02.htm (accessed on 03/21/2010).

IPCC (2004), Progress of Working Group III towards the IPCC fourth assessment report (AR4) (Submitted by the Co-chairs of Working Group III) Intergovernmental Panel on Climate Change IPCC-XXII/Doc. 11 (3.XI.2004) twenty-second session Agenda item: 3.1 New Delhi, 9-11 November 2004 http://www.ipcc.ch/meetings//session22/doc11.pdf (accessed on 03/21/2010).

Ibid (15)

IPCC (2010), PRINCIPLES GOVERNING IPCC WORK: Approved at the Fourteenth Session (Vienna, 1-3 October 1998) on 1 October 1998, amended at the 21st Session (Vienna, 3 and 6-7 November 2003) and at the 25th Session (Mauritius, 26-28 April 2006) http://www.ipcc.ch/pdf/ipcc-principles/ipcc-principles.pdf (accessed on 03/22/2010)

IPCC (2009), Summary Description of the IPCC Process; December 2009; Working Group II Contribution to the IPCC Fourth Assessment Report, Climate Change 2007: Impacts, Adaptation, and Vulnerability; http://www.ipcc-wg2.gov/WGIIsummary.pdf (accessed on 03/21/2010).

Choo, C W (1996), The Knowing Organization: How Organizations Use Information to Construct Meaning, Create Knowledge and Make Decisions, International Journal of Information Management, Vol. 16, No. 5, pp. 329-340. http://www.sciencedirect.com/science?_ob=MImg&_imagekey=B6VB4-3VWT0YX-1-1&_cdi=5916&_user=783137&_pii=0268401296000205&_orig=search&_coverDate=10%2F31%2F1996&_sk=999839994&view=c&wchp=dGLbVzb-zSkzS&md5=93b78166a1d8c7fa6fe44cae8dfd56c8&ie=/sdarticle.pdf (accessed on 03/21/2010).

8

Science and Regime Creation

The chapter will examine how international regime creation requires a consensus on the facts that is often the most difficult part of an agreement. It will compare climate change with other regimes such as law of the sea, international trade and international crime.

8.1 Regime Theory and International Organizations

As international relations theories are closely interlinked with the IOs, IOs actually create regimes[1]. "Regime is sets of implicit or explicit principles, norms, rules and decision-making procedures around which actors' expectations converge in a given area of international relations" (Krasner 1983; Krasner's approach to international regimes defines them as: (1) principles and values; (2) norms; (3) rules; and (4) enforcement mechanisms; as cited by Cogburn 2003, 136). Keohane and Nye (1983) state that regimes are sets of governing arrangements [that include] networks of rules, norms, and procedures that regularize behavior and control its effects. Viotti and Kauppi (1999) clarify, stating that regimes are sets of rules which may have international and nongovernmental organizations associated with them (Viotti and Kauppi 1999:215). International regimes are structures designed to foster international co-operation among participants' countries. Every country needs help to solve transnational problems. Examples to be seen

[1]"Over these several decades scholars around the world have documented the emergence and efficacy of International regimes in a wide variety of issue areas within the world-system, including:(1) international shipping; (2) international air transport; (3) international post; (4) international atomic energy and weapons; (5) international environmental issues; (6) the global "commons" (e.g., the high seas and outer space); and (7) even for commodities (e.g. diamonds) (Zacher & Sutton, 1997). However, one of the oldest and most successful international regimes has been the international telecommunications regime (Cowhey, 1990) (as cited by Cogburn 2003, page 136).

in the current environment are global terrorism and the fight against the transnational drug problem, and the fight to minimize HIV and AIDS, which cannot be resolved by the single state. International organizations can create powerful tools to solve a particular problem, which helps to increase their power, access, and authority through collaborative efforts, mutual agreements and policy formation. This situation creates a favorable environment for formulating new regimes, where solutions can be contemplated. This is one aspect of international regimes. The second aspect focuses on implementation of formulated policy through institutionalization of rules and international agreements that devise and control solutions to the initial problem(s) (Krasner 1983, Young 1989 and Cogburn 2003).

Power-based realist theorists give emphasis to the role of anarchy and the impact of the relative distribution of capabilities. In this type of realist regime, the guiding ideal is hegemonic stability; the argument is that regimes are established and maintained when a state holds a preponderance of power resources, as the United States did after the Second World War and still continues to do. Another type of regime is interest-based, which makes the claim that international regimes can play a role in helping states to realize their common interests. Similarly, knowledge-base theorists argue that state interests are not given but created. Knowledge can be shared by decision makers through the influence of transnational epistemological communities. A knowledge-base regime can be formed through social construction (Viotti and Kaupppi 1999). This knowledge of regime can equally be applied to the study of international organizations from an organizational sociological approach to examine the effectiveness and contribution to social wellbeing.

Regime studies use similar epistemology to International Relations. The major approaches of regime studies are liberal and neo-liberal, realist, (classical realism, defensive realism), neo-realist and Marxist, neo-Marxist (Western or Hegelian Marxism, such as neo-Gramscian theory), and post-modernist (Cogburn 2003). However, there are several other approaches in use embedded within these epistemologies, such as balance of power (Paul, Wirtz, and Fortmann 2004), behavioralism (Viotti and Kauppi 1987), complex interdependence (Keohane and Nye 1977), constitutive theory and constructivism (Wendt 1992), cosmopolitanism (Cheah and B. Robbins 1998), dependency, feminism, game theory (Viotti and Kauppi 1987), globalization (Dash 1998), globalism developed through Marxist and dependency theory (Viotti and Kauppi 1987, Cogburn 2003), conflict prevention (Samson and Charrier 1997), hegemony (Gramsci 1971; DuBoff 2003; Chomsky 2003), imperialism (Morgenthau 1948), inter-governmentalism (Moravcsik 1993), normative theory (Viotti and Kauppi 1987) and so on. These approaches are equally important and commonly used approaches of the organizational sociologist. The only difference is that political scientists examine these approaches as political power dynamics. They put the state first and social dynamism comes later. On the other side for the sociologist is that society or social environment is given first priority. Both regime and international relations theories follow the same basic theories and principles, equally support the importance of international organizations, and are equally influenced by development theory and globalization. Except for Ness and Brechin, sociologists have mostly not examined how international relations and the regime concept can be applied to study social problems. Ness and Brechin (1988:258) state that "international relations can be enhanced if we pay greater attention to how modern international organizations emerged and what they do in action - in short, if we pay greater attention to organizational performance". They argue that technology is the prime factor in organizational change. I see a clear link between technology organization and international regime as proposed by Ness and Brechin. Technology produced both world consensus and conflict (in the case of the Condom distribution) however external technology has greatly benefited the organization (Ness and Brechin 1988:258). This account can be related to contemporary regime formation through technology. Another example is the expansion of information technology to regime generated by the McDonalds fast food corporation. However, the incremental formation of such regimes will not necessarily solve associated political or social problems.

Social problems have been increasing on a global scale (HIVAID, inequality, transnational migration, terrorism, environmental problems), and these problems themselves create certain types of regime and format their own networks and relationships. In this context, sociologists can inform international relations theorists and regime theorists about how knowledge forms in local contexts (social collaboration) and how this knowledge can be transferred to the international context.

Economic integration, however defined, may be based on political motives and frequently begets political consequences' (Haas, 1958: 12).

8.2 Regime as Process

'...integration is the process whereby political actors in several distinct national settings are persuaded to shift their loyalties, expectations and political activities toward a new centre, whose institutions possess or demand jurisdiction over the pre-existing national states' (Haas, 1958: 16)

'As the process of integration proceeds, it is assumed that values will undergo change, that interests will be redefined in terms of regional rather than purely national orientation and that the erstwhile set of separate national group values will gradually be superseded by a new and geographically larger set of beliefs' (Haas, 1958: 13).

'successful intergovernmental regime designed to manage economic interdependence through negotiated policy co-ordination' (Moravcsik, 1993: 474).

'the substantive and institutional development of which may be explained through the ... analysis of national preference formation and intergovernmental strategic interaction' (Moravcsik, 1993: 481).

'The decision to join all but the most minimalist of regime involves some sacrifice of national autonomy, which increases the political risk to each Member State ....[I]n the inter-governmentalist view, the unique institutional structure of the EC is acceptable to national governments only insofar as it strengthens, rather than weakens, their control over domestic affairs, permitting them to attain goals otherwise unachievable' (Moravcsik, 1993: 507).

Constructivists, functionalists, and institutionalists argue that global politics is increasingly organized around regimes and institutions that foster cooperation by providing information and organizational structure, promoting norms and common belief systems, and reducing transaction costs (Hass 1964 as cited by Boehmer; Gartzke and Nordstrom 2004, Page 3)

There has been a growing trend of global environmental conference, convention and treatise which ultimately create regime. The following figure of from Meyer *et.al* (1997) provides the growing trend of agreement to address the global environmental problem.

Figure counts of international environmental activities, 1870-1990: Adopted from Meyer; Frank; Hironaka; Schofer and Tuma (1997: Page 625)

The environmental regime has been creating through international treaties, conventions and agreements (the detail can be found in chapter three of this book). Among them the followings are the major including Ramsar Convention, Stockholm Declaration of the United Nations Conference on the Human Environment, The Rio Declaration on Environment and Development, Convention on Biological Diversity, Convention on Long-Range Trans-boundary Pollution, Convention on the Prevention of Marine Pollution by Dumping of Wastes and Other Matter, Convention on International Trade

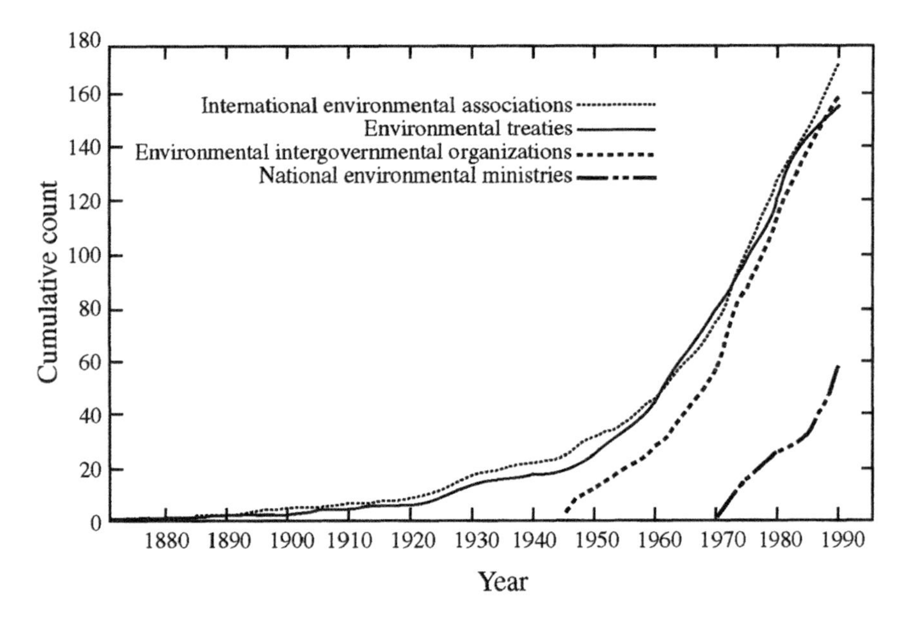

in Endangered Species (CITES), Basel Convention on the Control of Transboundary Movements of Hazardous Wastes and Their Disposal, Convention Concerning the Protection of the World Cultural & Natural Heritage, United Nations Convention to Combat Desertification, United Nations Convention on Law of the Sea (UNCLOS), United Nations Framework Convention on Climate Change and the Kyoto Protocol on Global Warming and many others), and capture the notion of international multicultural and socio-economic politics. Many researchers have examined the successes and failures of international treaties and agreements. These researches accept the role and responsibilities of nation, international organizations, civil societies, NGOs and advocacy group. The hegemonic power relationship is still in force within the current neo-liberal world.

However, the major step in fostering environmental regime begins from the first UN Conference on the Human Environment, Stockholm (1972), followed by the second Earth Summit in Rio de Janeiro 1992, where 172, 108 people participated including head of the states, business personnel and other experts. In the first time about 2,400 representatives of non-governmental organizations (NGOs) participated in Rio summit. Summit produced agenda 21 declaration on environment and development, the statement of forest principles, the United Nations framework convention on climate change and the United Nations convention on biological diversity. Since Rio summit global concern on environment management and policy reform became common agenda to the entire world. Most of the states in the world started focus and monitor on patterns of production (i.e. toxic components, gasoline, and poisonous waste), investigation on alternatives for the fossil fuels (which is major cause for global climate change), alternatives for the public transportation (to reduce air pollution and smog) and water resource management. Blueprint provides a comprehensive structure for the modernization of national / transnational environment protection and environment reform which includes the framework for sustainability and offers the links between economic growth using science and technology to solve the environmental problems with the application of multi-driven approaches.

The world conferences based on environmental reforms have been broadly focusing on the natural resource management, searching options to reduce the environment impact due to economic activities with the application of new technology.

Several studies suggest that the increase of environmental concern is a global phenomenon (Brechin and Kempton. 1994, 1997, Dunlap and Mertig 1995, 1996, 1997, Inglehart 1995, 1996, Abramson 1997, Brechin 1999, Escobar 2001). The parameters and hypotheses Frank, Hironaka and Schofer (2000b) have presented are common because they represent both national and international concern of the current world. Their finding which shows the growing concern for environment conservation such as cumulative increase of (1) protected area (2) environmental organizations (3) state memberships (4) environment impact laws (5) ministries and agencies. Globally environment concern is also growing as political agenda (i.e. US Democratic Party, Germany, the Netherlands, France, India, Nepal, Slovakia, Ukraine and Latvia Green Party etc.) and has global connection and networking. International organization such as World Conservation Union has states, non-governmental organization and individual members. Insofar, environmental conservation has global connection through various organizational networks (i.e. The Global Climate Network (GCN); World Nature Organization (WNO); United Nations Environment Programme (UNEP); International Union for Conservation of Nature (IUCN); Intergovernmental Panel on Climate Change (IPCC); Global Environment Facility (GEF); Earth System Governance Project (ESGP) (Pariona 2017).

In general term, environment protection is institutionalizing, blueprints are becoming common and nations are more receptive to address the environment problem and the diffusion of such concern is accelerating. The United Nations Conference on Environment and Development (UNCED), also known as the Rio de Janeiro Earth Summit, the Rio Summit, the Rio Conference, and the Earth Summit (Portuguese: ECO92), a major United Nations conference (held in Rio de Janeiro from 3 to 14 June in 1992) paved a new direction on environmental regime.

> The Rio Summit adopted Agenda 21, a major nonbinding policy document, the implementation of which is overseen by the UN Commission on Sustainable Development. Principle 12 of the Rio Declaration advances three key elements:
>
> Environmental measures dealing with transboundary or global problems should be based on international agreements
>
> Unilateral action to deal with such problems should be avoided and
>
> Environmental measures should not be arbitrarily or unjustifiably discriminatory or a disguised restriction on trade (Rio Declaration on Environment and Development, A/CONF.151/26 (Vol. 1), 12 August 1992. Principle 12 of the Rio Declaration reads: "States should cooperate to promote a supportive and open international economic system that would lead to economic growth and sustainable development in all countries, to better address the problems of environmental degradation. Trade policy measures for environmental purposes should not constitute a means of arbitrary or unjustifiable discrimination or a disguised restriction on international trade. Unilateral actions to deal with environmental challenges outside the jurisdiction of the importing country should be avoided. Environmental measures addressing transboundary or global environmental problems should, as far as possible, be based on an international consensus) (page 272).

Rio Summit (1992), was a major event on addressing the overall environmental challenges including climate change, which played the crucial role in bringing all nations together to minimize the impact of broader spectrum. It boasted climate regime and diffused not only at the governmental arena. The nongovernmental sectors also played important role, they protest, appeared, showed the severity of environmental crisis and also exhibited the willingness to contribute to over come this problem. There was minimum presence of academic scholars (in my personal experience), however, they got the subject matter of research, the contents for theorizing the notions of climate change science, sustainability and developmental agendas connecting within the societal, environmental and economic domain. Rio Summit gave a new direction and created a global agenda of sustainable development, including climate change.

Another milestone of climate change regime formation / boasting and spreading can be seen through formation of The Conference of Parties, known as COP, United Nations Framework Convention on Climate Change. "The COP is the supreme decision-making body of the Convention. All States that are Parties to the Convention are represented at the COP, at which they review the implementation of the Convention and any other legal instruments that the COP adopts and take decisions necessary to promote the effective implementation of the Convention, including institutional and administrative arrangements". ... "A key task for the COP is to review the national communications and emission inventories submitted by Parties. Based on this information, the COP assesses the effects of the measures taken by Parties and the progress made in achieving the ultimate objective of the Convention.

The COP meets every year, unless the Parties decide otherwise. The first COP meeting was held in Berlin, Germany in March 1995. The COP meets in Bonn, the seat of the secretariat, unless a Party offers to host the session. Just as the COP Presidency rotates among the five recognized UN regions - that is, Africa, Asia, Latin America and the Caribbean, Central and Eastern Europe and Western Europe and Others – there is a tendency for the venue of the COP to also shift among these groups" (UNFCCC 2020). COP administers through The Bureau, which comprised of a President, seven vice-presidents, the chairs of the subsidiary bodies established by Articles 9 and 10 of the Convention, and a rapporteur. "Bureau is responsible for advising the President and taking decisions with regard to the overall management of the intergovernmental process. The Bureau has overall responsibility for questions of process. Bureau members often consult with their regional groups on issues. The Bureau is not a forum for political negotiations" ..... "The Bureau is mainly responsible for questions of process management. It assists the President in the performance of his or her duties by providing advice and by helping with various tasks (e.g. members undertake consultations on behalf of the President). The Bureau is responsible for examining the credentials of Parties, reviewing the list of IGOs and NGOs seeking accreditation, and submitting a report thereon to the Conference" (UNFCCC 2020). "The Bureau of the COP, CMP, and CMA consists of 11 members comprising: The President, Vice-Presidents, the Chairs of the subsidiary bodies and the Rapporteur. Each of the five regional groups is represented by two Bureau members, with one member from the Small Island Developing States (SIDS). Members are elected for an initial term of one year and may serve for not more than two consecutive terms of one year. They remain in office until their successors are elected. It is customary to invite a representative of the host country of the next COP to attend meetings of the Bureau if that Party is not already represented on the Bureau" (UNFCCC 2020). This can be seen as whole mechanism of validating the conference- which is a direct process of bolstering the climate change regime.

As commonly known throughout its history from 1995 to date (2019), governments or the conference of parties held meeting, close door plus open doors and each conferences thousands of climate concerned stakeholder gather organize the main and side events. The COP conference is a platform for all which particularly provides the venue to express concerns of governments, general public, NGOs, INGOs, social / environment activists, academician and so on. Basically, COP conferences can be taken as regime penetration, in both way top to bottom- in way that governments present the policy- bottom up- because public sector raises its voice, whether it heard or not and media plays prime role to raise the awareness towards climate change regime formation. The table shows the COP timeline and its output.

Table 8.1 COP timeline and output.

Year	Location	Conference/Outcome
1995	Berlin, Germany	COP 1- first meeting-paved the way forward
1996	Geneva, Switzerland	COP 2- Accepted the scientific findings on climate change proffered by the Intergovernmental Panel on Climate Change (IPCC) in its second assessment (1995); Rejected uniform "harmonized policies" in favor of flexibility; Called for "legally binding mid-term targets".
1997	Kyoto, Japan	COP 3- one of the most important conference, which adopted the Kyoto Protocol-The Kyoto Protocol was adopted on 11 December 1997. Owing to a complex ratification process, it entered into force on 16 February 2005. Currently, there are 192 Parties to the Kyoto Protocol (https://unfccc.int/kyoto_protocol). the Kyoto Protocol operationalizes the United Nations Framework Convention on Climate Change by committing industrialized countries to limit and reduce greenhouse gases (GHG) emissions in accordance with agreed individual targets. The Convention itself only asks those countries to adopt policies and measures on mitigation and to report periodically. The Kyoto Protocol is based on the principles and provisions of the Convention

(Continued)

Table 8.1 Continued

		and follows its annex-based structure. It only binds developed countries, and places a heavier burden on them under the principle of "common but differentiated responsibility and respective capabilities", because it recognizes that they are largely responsible for the current high levels of GHG emissions in the atmosphere (https://unfccc.int/kyoto_protocol).
1998	Buenos Aires, Argentina	COP 4- Determined to strengthen the implementation of the United Nations Framework Convention on Climate Change and prepare for the future entry into force of the Kyoto Protocol to the Convention, and to maintain political momentum towards these aims .... Preparations for the first session of the Conference of the Parties serving as the meeting of the Parties to the Kyoto Protocol, including work on the elements of the Protocol related to compliance and on policies and measures for the mitigation of climate change (decision 8/CP.4); (https://unfccc.int/resource/docs/cop4/16a01.pdf)
1999	Bonn, Germany	COP 5- Review of the implementation of commitments and of other provisions of the Convention. Preparations for the first session of the Conference of the Parties meetings as the Meeting of the Parties to the Kyoto Protocol (decision 8/CP.4). Proposal by the President. Decision 1/CP.5. Implementation of the Buenos Aires Plan of Action (https://unfccc.int/process-and-meetings/conferences/past-conferences/bonn-climate-change-conference-october-1999/cop-5/cop-5-documents-UNFCCC 2020). Cooperation with the IPCC
2000	The Hague, Netherlands	COP 6- Input to the Third United Nations Conference on the Least Developed Countries; Second compilation and synthesis of initial national communications from Parties not included in Annex I to the Convention (https://unfccc.int/process-and-meetings/conferences/past-conferences/the-hague-climate-conference-november-2000/decisions-the-hague-climate-change-conference-november-2000)
2001	Bonn, Germany	COP 6- it resumed again in Germany- (President of the United States and had rejected the Kyoto Protocol in March 2001) however, COP made further important decision without USA- which adopted - flexibility mechanisms; Carbon sinks; Compliance and financial mechanism (UNFCCC 2020).
2001	Marrakech, Morocco	COP 7- Operational rules for international emissions trading among parties to the Protocol and for the CDM and joint implementation; A compliance regime that outlined consequences for failure to meet emissions targets but deferred to the parties to the Protocol, once it came into force,

(Continued)

Table 8.1 Continued

		the decision on whether those consequences would be legally binding; Accounting procedures for the flexibility mechanisms; A decision to consider at COP 8 how to achieve a review of the adequacy of commitments that might lead to discussions on future commitments by developing countries....... The date of the World Summit on Sustainable Development (August–September 2002) was put forward as a target to bring the Kyoto Protocol into force. The World Summit on Sustainable Development (WSSD) was to be held in Johannesburg, South Africa (UNFCCC 2020).
2002	New Delhi, India	COP 8- Parties Agree that, in order to respond to the challenges, we face now, and, in the future, climate change and its adverse effects should be considered within the framework of sustainable development through enhanced international cooperation. (https://unfccc.int/cop8/latest/delhidecl_infprop.pdf)
2003	Milan, Italy	COP 9- The 2003 United Nations Climate Change Conference took place between 1–12 December 2003 in Milan, Italy. The conference included the 9th Conference of the Parties (COP9) to the United Nations Framework Convention on Climate Change (UNFCCC). The parties agreed to use the Adaptation Fund established at COP7 in 2001 primarily in supporting developing countries better adapt to climate change. The fund would also be used for capacity-building through technology transfer. At the conference, the parties also agreed to review the first national reports submitted by 110 non-Annex I countries (https://caps-conference.eu/2003-united-nations-climate-change-conference/).
2004	Buenos Aires, Argentina	COP 10- The parties discussed the progress made since the first United Nations Climate Change Conference ten years ago and its future challenges, with special emphasis on climate change mitigation and adaptation. To promote developing countries better adapt to climate change, the Buenos Aires Plan of Action[1] was adopted. The parties also began discussing the post-Kyoto mechanism, on how to allocate emission reduction obligation following 2012, when the first commitment period ends (https://unfccc.int/cop4/resource/docs/cop4/16a01.pdf).
2005	Montreal, Canada	COP 11/CMP 1- Decides that the eligibility to participate in the mechanisms by a Party included in Annex I shall be dependent on its compliance with methodological and reporting requirements under Article 5, paragraphs 1 and 2, and Article 7, paragraphs 1 and 4, of the Kyoto Protocol.

(Continued)

Table 8.1 Continued

		Oversight of this provision will be provided by the enforcement branch of the Compliance Committee, in accordance with the procedures and mechanisms relating to compliance as contained in decision 24/CP.7, assuming approval of such procedures and mechanisms by the Conference of the Parties serving as the meeting of the Parties to the Kyoto Protocol in decision form in addition to any amendment entailing legally binding consequences, noting that it is the prerogative of the Conference of the Parties serving as the meeting of the Parties to the Kyoto Protocol to decide on the legal form of the procedures and mechanisms relating to compliance; Decides that certified emission reductions, emission reduction units and assigned amount units under Articles 6, 12 and 17, as well as removal units resulting from activities under Article 3, paragraphs 3 and 4, may be used to meet commitments under Article 3, paragraph 1, of the Parties included in Annex I, and can be added as provided for in Article 3, paragraphs 10, 11 and 12, of the Kyoto Protocol and in conformity with the provisions contained in decision 13/CMP.1, and that emission reduction units, assigned amount units and removal units can be subtracted as provided for in Article 3, paragraphs 10 and 11, and in conformity with the provisions contained in decision 13/CMP.1, without altering the quantified emission limitation and reduction commitments inscribed in Annex B to the Kyoto Protocol. (*It was one of the largest intergovernmental conferences on climate change ever. The event marked the entry into force of the Kyoto Protocol. Hosting more than 10,000 delegates, it was one of Canada's largest international events ever and the largest gathering in Montreal since Expo 67. The Montreal Action Plan was an agreement to "extend the life of the Kyoto Protocol beyond its 2012 expiration date and negotiate deeper cuts in greenhouse-gas emissions". Canada's environment minister at the time, Stéphane Dion, said the agreement provides a "map for the future"*) (https://unfccc.int/resource/docs/2005/cmp1/eng/08a01.pdf; http://www.universalrights.net/news/display.php?id=1665)
2006	Nairobi, Kenya	COP 12/CMP 2- The conference also included, from 6 to 14 November, the twenty-fifth session of the Subsidiary Body for Scientific and Technological Advice (SBSTA 25), the twenty-fifth session of the Subsidiary Body for Implementation (SBI 25), and the second session of the Ad Hoc Working Group on Further Commitments for Annex I

(Continued)

Table 8.1 Continued

		Parties under the Kyoto Protocol (AWG 2) including an in-session workshop. Further guidance to an entity entrusted with the operation of the financial mechanism of the Convention, for the operation of the Special Climate Change Fund (https://unfccc.int/process-and-meetings/conferences/past-conferences/nairobi-climate-change-conference-november-2006/decisions-nairobi-climate-change-conference-november-2006). BBC reporter Richard Black coined the phrase "climate tourists" to describe some delegates who attended "to see Africa, take snaps of the wildlife, the poor, dying African children and women" (Black 2006).
2007	Bali, Indonesia	COP 13/CMP 3- he Bali Climate Change Conference brought together more than 10,000 participants, including representatives of over 180 countries together with observers from intergovernmental and non-governmental organizations and the media. Governments adopted the Bali Road Map, a set of decisions that represented the various tracks that were seen as key to reaching a global climate deal. The Bali Road Map includes the Bali Action Plan, which launched a "new, comprehensive process to enable the full, effective and sustained implementation of the Convention through long-term cooperative action, now, up to and beyond 2012", with the aim of reaching an agreed outcome and adopting a decision at COP15 in Copenhagen. Governments divided the plan into five main categories: shared vision, mitigation, adaptation, technology and financing. Please click here for the full text of the Bali Action Plan. Other elements in the Bali Road Map included: A decision on deforestation and forest management; A decision on technology for developing countries; The establishment of the Adaptation Fund Board; The review of the financial mechanism, going beyond the existing Global Environmental Facility (https://unfccc.int/process-and-meetings/conferences/past-conferences/bali-climate-change-conference-december-2007/bali-climate-change-conference-december-2007-0).
2008	Poznañ, Poland	COP 14/CMP 4- The Poznan Climate Change Conference drew 9,250 participants, including almost 4,000 government officials, 4,500 representatives of UN bodies and agencies, intergovernmental organizations and nongovernmental organizations, and more than 800 accredited media representatives. The meeting produced a number of useful results: It launched the Adaptation Fund under the Kyoto Protocol. The Fund was to be filled by a 2% levy on projects

(Continued)

Table 8.1 Continued

		under the Clean Development Mechanism. Parties agreed that the Adaptation Fund Board should have legal capacity to grant direct access to developing countries. It saw Parties endorse an intensified negotiating schedule for 2009. It identified divergences of views on key issues related to increasing the level of available funding for adaptation and improvements to the Clean Development Mechanism, which required resolution in the coming year (https://unfccc.int/process-and-meetings/conferences/past-conferences/poznan-climate-change-conference-december-2008/poznan-climate-change-conference-december-2008-0).
2009	Copenhagen, Denmark	COP 15/CMP 5- The Conference was an exceptional event that attracted unprecedented participation and resulted in: Attendance by 120 Heads of State and Government, Â raising climate discussions to a new level. Record numbers of participants including 10,500 delegates, 13,500 observers and coverage by more than 3,000 media representatives. Intensive negotiations characterized by over 1,000 official, informal and group meetings among Parties. Observers discussed climate change in more than 400 meetings and media attended over 300 press conferences. A vibrant program of over 200 side events. Over 220 exhibits from Parties, UN, IGOs and civil society. A total of 23 decisions adopted by the COP and the CMP. Governments engaged at the highest political level, and the outcome of that engagement was reflected in the Copenhagen Accord. While much attention has focused on the Accord, the Conference in Copenhagen also made good progress in a number of areas including improvements to the clean development mechanism, amending Annex I to the Convention to add Malta, guidance on REDD+, and draft decisions on adaptation, technology, and capacity-building. However, the Bali Roadmap negotiations could not be concluded and negotiations will continue in 2010 (https://unfccc.int/process/conferences/pastconferences/copenhagen-climate-change-conference-december-2009/statements-and-resources/outcome-of-the-copenhagen-conference).
2010	Cancún, Mexico	COP 16/CMP 6- The Cancun Climate Change Conference drew almost 12,000 participants, including 5,200 government officials, 5,400 representatives of UN bodies and agencies, intergovernmental organizations and nongovernmental organizations, and 1,270 accredited members of the media. The meeting produced the basis for the most comprehensive and far-reaching international response to climate change the world had ever seen to reduce carbon emissions and build a

(Continued)

Table 8.1 Continued

		system which made all countries accountable to each other for those reductions. Among the highlights, Parties agreed: to commit to a maximum temperature rise of 2 degrees Celsius above pre-industrial levels, and to consider lowering that maximum to 1.5 degrees in the near future. to make fully operational by 2012 a technology mechanism to boost the innovation, development and spread of new climate-friendly technologies; to establish a Green Climate Fund to provide financing to projects, programs, policies and other activities in developing countries via thematic funding windows; on the Cancun Adaptation Framework, which included setting up an Adaptation Committee to promote the implementation of stronger, cohesive action on adaptation. On the mitigation front, developed countries submitted economy-wide emission reduction targets and agreed on strengthened reporting frequency and standards and to develop low-carbon national plans and strategies. Developing countries submitted nationally appropriate mitigation actions (NAMAs), to be implemented subject to financial and technical support. Work continued on shaping the form and functions of a registry for NAMAs to enable the matching of such actions with finance and technology. Developing countries were also encouraged to develop low-carbon national plans and strategies. Work also progressed on reducing emissions from deforestation and forest degradation (REDD), boosting capacity-building in developing countries, and how to deal with any consequences of response measures to action on climate change. Governments also agreed to include carbon capture and storage (CCS) in the projects under the Clean Development Mechanism (CDM), subject to technical and safety standards (https://unfccc.int/process-and-meetings/conferences/past-conferences/cancun-climate-change-conference-november-2010/cancun-climate-change-conference-november-2010-0)
2011	Durban, South Africa	COP 17/CMP 7- COP delivered a breakthrough on the international community's response to climate change. In the second largest meeting of its kind, the negotiations advanced, in a balanced fashion, the implementation of the Convention and the Kyoto Protocol, the Bali Action Plan, and the Cancun Agreements. The outcomes included a decision by Parties to adopt a universal legal agreement on climate change as soon as possible, and no later than 2015. The President of COP17/CMP7 Maite Nkoana-Mashabane said: "What we have achieved in Durban will play a central role in saving tomorrow, today

(Continued)

Table 8.1 Continued

		(https://unfccc.int/process-and-meetings/conferences/past-conferences/durban-climate-change-conference-november-2011/durban-climate-change-conference-november-2011).
2012	Doha, Qatar	COP 18/CMP 8- Approaches to address loss and damage associated with climate change impacts in developing countries that are particularly vulnerable to the adverse effects of climate change to enhance adaptive capacity.....Promoting gender balance and improving the participation of women in UNFCCC negotiations and in the representation of Parties in bodies established pursuant to the Convention or the Kyoto Protocol (https://unfccc.int/process-and-meetings/conferences/past-conferences/doha-climate-change-conference-november-2012/cop-18/cop-18-decisions).
2013	Warsaw, Poland	COP 19/CMP 9- Key decisions adopted at this conference include decisions on further advancing the Durban Platform, the Green Climate Fund and Long-Term Finance, the Warsaw Framework for REDD Plus, the Warsaw International Mechanism for Loss and Damage and other decisions (https://unfccc.int/process-and-meetings/conferences/past-conferences/warsaw-climate-change-conference-november-2013/warsaw-climate-change-conference-november-2013-0).
2014	Lima, Peru	COP 20/CMP 10- The conference delegates held negotiations towards a global climate agreement (https://unfccc.int/process-and-meetings/conferences/past-conferences/lima-climate-change-conference-december-2014/cop-20).
2015	Paris, France	COP 21/CMP 11-main outcome Paris agreement- The Paris Agreement builds upon the Convention and for the first time brings all nations into a common cause to undertake ambitious efforts to combat climate change and adapt to its effects, with enhanced support to assist developing countries to do so. As such, it charts a new course in the global climate effort. *The Paris Agreement* central aim is to strengthen the global response to the threat of climate change by keeping a global temperature rise this century well below 2 degrees Celsius above pre-industrial levels and to pursue efforts to limit the temperature increase even further to 1.5 degrees Celsius. Additionally, the agreement aims to strengthen the ability of countries to deal with the impacts of climate change. To reach these ambitious goals, appropriate financial flows, a new technology framework and an enhanced capacity building framework will be put in place, thus supporting action by developing countries and the most vulnerable countries, in line with their own national

(Continued)

Table 8.1 Continued

		objectives. The Agreement also provides for enhanced transparency of action and support through a more robust transparency framework (https://unfccc.int/process-and-meetings/the-paris-agreement/the-paris-agreement; (please see annexes) https://unfccc.int/process-and-meetings/the-paris-agreement/what-is-the-paris-agreement 2020).
2017	Bonn, Germany	COP 23/CMP 13/CMA 1-2- main purpose was to meet to advance the aims and ambitions of the Paris Agreement and achieve progress on its implementation guidelines. The COP23 was organized following an innovative concept of "one conference, two zones". During the two weeks of the conference, a vast area of the city of Bonn became the Climate Campus that was organized in two zones: the "Bula Zone" and the "Bonn Zone". This approach focused on a close integration of the zones to ensure that negotiations, events and exhibits were integrated into one conference (https://unfccc.int/process-and-meetings/conferences/un-climate-change-conference-november-2017/about/un-climate-change-conference-november-2017).
2018	Katowice, Poland	COP 24/CMP 14/CMA 1-3- The conference agreed on rules to implement the Paris Agreement, which will come into force in 2020, that is to say the rulebook on how governments will measure, and report on their emissions-cutting efforts (https://unfccc.int/process-and-meetings/conferences/katowice-climate-change-conference-december-2018/katowice-climate-change-conference-december-2018).
2019	Madrid, Spain	COP 25/CMP 15/CMA 2- The conference was designed to take the next crucial steps in the UN climate change process. Following agreement on the implementation guidelines of the Paris Agreement at COP 24 in Poland last year, a key objective was to complete several matters with respect to the full operationalization of the Paris Climate Change Agreement. The conference furthermore served to build ambition ahead of 2020, the year in which countries have committed to submit new and updated national climate action plans. Crucial climate action work was taken forward in areas including finance, the transparency of climate action, forests and agriculture, technology, capacity building, loss and damage, indigenous peoples, cities, oceans and gender (https://unfccc.int/about-the-un-climate-change-conference-december-2019).

Sources: (https://unfccc.int/ 2020) (please see annexes for Paris Agreement and IPCC response)

Conclusion

Each of COP conferences focuses on how to combat with the global environmental challenges (Climate Change, Air Pollution; Global Warming; Overpopulation; Natural Resource Depletion; Waste Disposal; Loss of Biodiversity; Deforestation; Ocean Acidification; Ozone Layer Depletion; Acid Rain; Soil and Water Pollution; Urban Sprawl; Public Health Issues; Genetic Engineering), with major focus on climate change and its impacts. Each of these conferences' governments- as it is government mechanism; however, thousands of general public, representatives of NGOs, CBOs, INGOS, and media personal participate, advocate, advice, argue, agree, disagree show the pathway of how these challenges can be mitigated. Because the objective of all concern stakeholders who gather in such conferences is to show the worrisome, they have hold in their soul and how much they are worried. These international conferences / events create the various positive spectrums in regime creation, formulation and theorization. COP mechanism can be considered one of the major pillars to support and activate climate change regime. IPCC is one of the active intergovernmental body, which has been playing critical role on climate change regime expansion. "The IPCC is an intergovernmental body sponsored by UNEP and WMO. The main decision-making body is the "Panel" which meets at regular intervals in plenary sessions at the level of government representatives of all 194 IPCC member countries. Its role as defined in the "Principles Governing IPCC Work" is "to assess on a comprehensive, objective, open and transparent basis the scientific, technical and socio-economic information relevant to understanding the scientific basis of risk of human-induced climate change, its potential impacts and options for adaptation and mitigation. IPCC reports should be neutral with respect to policy, although they may need to deal objectively with scientific, technical and socio-economic factors relevant to the application of particular policies. Review is an essential part of the IPCC process. Since the IPCC is an intergovernmental body, review of IPCC documents should involve both peer review by experts and review by governments" (IPCC 2010). As noted on Yearbook of International Co-operation on Environment and Development (2002/2003), the roles of mega conferences are: Setting global agendas; Facilitating 'joined-up' thinking; Endorsing common principles: Providing global leadership; Building institutional capacity; and Legitimizing global governance through inclusivity (Syfang and Jordan 2002). Regimes "are more specialized arrangements that pertain to well-defined activities, resources, or geographical areas and often involve only some subset of the members of international society", (Young 1989). International organizations like IPCC create or validate the climate change knowledge, formulate principles, prepare global policies, rules and implement penetrating mechanism for the national government, through feasible decision-making procedures for the given time frame, which can be considered as functions of regime. Regime changes according to demand of time, regime creation continues as new challenges immerge or resolved.

Annex-1 The Paris Agreement

*"**Long-term temperature goal** (Art. 2) – The Paris Agreement, in seeking to strengthen the global response to climate change, reaffirms the goal of limiting global temperature increase to well below 2 degrees Celsius, while pursuing efforts to limit the increase to 1.5 degrees. **Global peaking and 'climate neutrality'** (Art. 4) – To achieve this temperature goal, Parties aim to reach global peaking of greenhouse gas emissions (GHGs) as soon as possible, recognizing peaking will take longer for developing country Parties, so as to achieve a balance between anthropogenic emissions by sources and removals by sinks of GHGs in the second half of the century. **Mitigation** (Art. 4) – It establishes binding commitments by all Parties to prepare, communicate and maintain a nationally determined contribution (NDC) and to pursue domestic measures to achieve them. **Sinks and reservoirs** (Art.5) – It also encourage Parties to conserve and enhance, as appropriate, sinks and reservoirs of GHGs as referred to in Article 4, paragraph 1(d) of the Convention, including forests. **Voluntary cooperation/Market- and non-market-based approaches** (Art. 6) – It recognizes the possibility of voluntary cooperation among Parties to allow for higher ambition and sets out principles – including environmental integrity, transparency and robust accounting – for any cooperation that involves internationally transferal of mitigation outcomes. **Adaptation** (Art. 7) – It establishes a global goal on adaptation – of enhancing adaptive capacity, strengthening resilience and reducing vulnerability to climate change in the context of the temperature goal of the Agreement. It aims to significantly strengthen national adaptation efforts, including through support and international cooperation. It recognizes that adaptation is a global challenge faced by all. All Parties should engage in adaptation, including by formulating and implementing National Adaptation Plans, and should submit and periodically update an adaptation communication describing their priorities, needs, plans and actions. The adaptation efforts of developing countries should be recognized; **Loss and damage** (Art.*

8) – It recognize the importance of averting, minimizing and addressing loss and damage associated with the adverse effects of climate change, including extreme weather events and slow onset events, and the role of sustainable development in reducing the risk of loss and damage. Parties are to enhance understanding, action and support, including through the Warsaw International Mechanism, on a cooperative and facilitative basis with respect to loss and damage associated with the adverse effects of climate change. **Finance, technology and capacity-building support** *(Art. 9, 10 and 11) – It reaffirms the obligations of developed countries to support the efforts of developing country Parties to build clean, climate-resilient futures, while for the first time encouraging voluntary contributions by other Parties.* **Climate change education, training, public awareness, public participation and public access to information** *(Art. 12) is also to be enhanced under the Agreement.* **Transparency** *(Art13),* **implementation and compliance** *(Art. 15) – It relies on a robust transparency and accounting system to provide clarity on action and support by Parties, with flexibility for their differing capabilities of Parties.* **Global Stocktake** *(Art. 14) – A "global stocktake", to take place in 2023 and every 5 years thereafter, will assess collective progress toward achieving the purpose of the Agreement in a comprehensive and facilitative manner (*https://unfccc.int/process-and-meetings/the-paris-agreement/what-is-the-paris-agreement)

Annex 2

An IPCC Special Report on the impacts of global warming of 1.5°C above pre-industrial levels and related global greenhouse gas emission pathways, in the context of strengthening the global response to the threat of climate change, sustainable development, and efforts to eradicate poverty

"Understanding Global Warming of 1.5°C Human activities are estimated to have caused approximately 1.0°C of global warming above pre-industrial levels, with a likely range of 0.8°C to 1.2°C. Global warming is likely to reach 1.5°C between 2030 and 2052 if it continues to increase at the current rate. (high confidence)

Warming from anthropogenic emissions from the pre-industrial period to the present will persist for centuries to millennia and will continue to cause further long-term changes in the climate system, such as sea level rise, with associated impacts (high confidence), but these emissions alone are unlikely to cause global warming of 1.5°C (medium confidence).

Climate-related risks for natural and human systems are higher for global warming of 1.5°C than at present, but lower than at 2°C (high confidence). These risks depend on the magnitude and rate of warming, geographic location, levels of development and vulnerability, and on the choices and implementation of adaptation and mitigation options (high confidence).

Projected Climate Change, Potential Impacts and Associated Risks

Climate models project robust differences in regional climate characteristics between present-day and global warming of 1.5°C, and between 1.5°C and 2°C. These differences include increases in: mean temperature in most land and ocean regions (high confidence), hot extremes in most inhabited regions (high confidence), heavy precipitation in several regions (medium confidence), and the probability of drought and precipitation deficits in some regions (medium confidence).

By 2100, global mean sea level rise is projected to be around 0.1 metre lower with global warming of 1.5°C compared to 2°C (medium confidence). Sea level will continue to rise well beyond 2100 (high confidence), and the magnitude and rate of this rise depend on future emission pathways. A slower rate of sea level rise enables greater opportunities for adaptation in the human and ecological systems of small islands, low-lying coastal areas and deltas (medium confidence).

On land, impacts on biodiversity and ecosystems, including species loss and extinction, are projected to be lower at 1.5°C of global warming compared to 2°C. Limiting global warming to 1.5°C compared to 2°C is projected to lower the impacts on terrestrial, freshwater and coastal ecosystems and to retain more of their services to humans (high confidence).

Limiting global warming to 1.5°C compared to 2°C is projected to reduce increases in ocean temperature as well as associated increases in ocean acidity and decreases in ocean oxygen levels (high confidence). Consequently, limiting global warming to 1.5°C is projected to reduce risks to marine biodiversity, fisheries, and ecosystems, and their functions and services to humans, as illustrated by recent changes to Arctic sea ice and warm-water coral reef ecosystems (high confidence).

Climate-related risks to health, livelihoods, food security, water supply, human security, and economic growth are projected to increase with global warming of 1.5°C and increase further with 2°C. Most adaptation needs will be lower for global warming of 1.5°C compared to 2°C (high confidence). There are a wide range of adaptation options that can reduce the risks of climate change (high confidence). There are limits to adaptation and adaptive capacity for some human and natural systems at global warming of 1.5°C, with associated losses (medium confidence). The number and availability of adaptation options vary by sector (medium confidence).

Emission Pathways and System Transitions Consistent with 1.5°C Global Warming

In model pathways with no or limited overshoot of 1.5°C, global net anthropogenic CO_2 emissions decline by about 45% from 2010 levels by 2030 (40–60% interquartile range), reaching net zero around 2050 (2045–2055 interquartile range).

For limiting global warming to below 2°C CO_2 emissions are projected to decline by about 25% by 2030 in most pathways (10–30% interquartile range) and reach net zero around 2070 (2065–2080 interquartile range). Non-CO_2 emissions in pathways that limit global warming to 1.5°C show deep reductions that are similar to those in pathways limiting warming to 2°C. (high confidence)

Pathways limiting global warming to 1.5°C with no or limited overshoot would require rapid and far-reaching transitions in energy, land, urban and infrastructure (including transport and buildings), and industrial systems (high confidence). These systems transitions are unprecedented in terms of scale, but not necessarily in terms of speed, and imply deep emissions reductions in all sectors, a wide portfolio of mitigation options and a significant upscaling of investments in those options (medium confidence).

All pathways that limit global warming to 1.5°C with limited or no overshoot project the use of carbon dioxide removal (CDR) on the order of 100–1000 $GtCO_2$ over the 21st century. CDR would be used to compensate for residual emissions and, in most cases, achieve net negative emissions to return global warming to 1.5°C following a peak (high confidence). CDR deployment of several hundreds of $GtCO_2$ is subject to multiple feasibility and sustainability constraints (high confidence).

Significant near-term emissions reductions and measures to lower energy and land demand can limit CDR deployment to a few hundred $GtCO_2$ without reliance on bioenergy with carbon capture and storage (BECCS) (high confidence).

Strengthening the Global Response in the Context of Sustainable Development and Efforts to Eradicate Poverty

Estimates of the global emissions outcome of current nationally stated mitigation ambitions as submitted under the Paris Agreement would lead to global greenhouse gas emissions in 2030 of 52–58 $GtCO_2$ eq yr-1 (medium confidence). Pathways reflecting these ambitions would not limit global warming to 1.5°C, even if supplemented by very challenging increases in *the scale and ambition of emissions reductions after 2030 (high confidence). Avoiding overshoot and reliance on future large-scale deployment of carbon dioxide removal (CDR) can only be achieved if global* CO_2 *emissions start to decline well before 2030 (high confidence).*

The avoided climate change impacts on sustainable development, eradication of poverty and reducing inequalities would be greater if global warming were limited to 1.5°C rather than 2°C, if mitigation and adaptation synergies are maximized while trade-offs are minimized (high confidence).

Adaptation options specific to national contexts, if carefully selected together with enabling conditions, will have benefits for sustainable development and poverty reduction with global warming of 1.5°C, although trade-offs are possible (high confidence).

Mitigation options consistent with 1.5°C pathways are associated with multiple synergies and trade-offs across the Sustainable Development Goals (SDGs). While the total number of possible synergies exceeds the number of trade-offs, their net effect will depend on the pace and magnitude of changes, the composition of the mitigation portfolio and the management of the transition. (high confidence).

Limiting the risks from global warming of 1.5°C in the context of sustainable development and poverty eradication implies system transitions that can be enabled by an increase of adaptation and mitigation investments, policy instruments, the acceleration of technological innovation and behaviour changes (high confidence).

Sustainable development supports, and often enables, the fundamental societal and systems transitions and transformations that help limit global warming to 1.5°C. Such changes facilitate the pursuit of climate-resilient development pathways that achieve ambitious mitigation and adaptation in conjunction with poverty eradication and efforts to reduce inequalities (high confidence).

Strengthening the capacities for climate action of national and subnational authorities, civil society, the private sector, indigenous peoples and local communities can support the implementation of ambitious actions implied by limiting global warming to 1.5°C (high confidence). International cooperation can provide an enabling environment for this to be achieved in all countries and for all people, in the context of sustainable development. International cooperation is a critical enabler for developing countries and vulnerable regions (high confidence)" (https://www.ipcc.ch/site/assets/uploads/sites/2/2019/06/SR15_Headline-statements.pdf).

References

Abramson, Paul R. (1997), "Postmaterialism and Environmentalism: A Comment on a Analysis and a Reappraisal." *Social Science Quarterly* 78: 21–23.

Arturo Escobar (1998) Whose Knowledge, Whose nature? Biodiversity, Conservation, and the Political Ecology of Social Movements, Journal of Political Ecology Vol.5 1998

Black, Richard (November 18, 2006), "Climate talks a tricky business". BBC News. Archived from the original on June 18, 2010. Retrieved June 19, 2010. http://news.bbc.co.uk/2/hi/science/nature/6161998.stm

Boehmer, Charles; Gartzke, Erik and Nordstrom, Timothy (2004), Do Intergovernmental Organizations Promote Peace? *World Politics* 57 (October 2004), 1–38

Brechin, Steven R. (1999), "Objective Problems, Subjective Values and Global Environmentalism: Evaluating the Postmaterialism Argument and Challenging a New Explanation." *Social Science Quarterly* 80: 793–806.

Brechin, Steven R., and Willett Kempton. (1994), "Global Environmentalism: A Challenge to the Postmaterialism Thesis?" *Social Science Quarterly* 75: 245–69.

Brechin, Steven R., and Willett Kempton. (1997), "Beyond Postmaterialist Values: National Versus Individual Explanations of Global Environmentalism." *Social Science Quarterly* 78: 16–20.

Breitmeier, Helmut, Oran R. Young, and Michael Zurn. (2007), Analyzing International Environmental Regimes. Cambridge, MA: The MIT Press.

Buttel, Fredrick H. (2000), "World Society, the Nation-State, and Environmental Protection." *American Sociological Review* Vol. 65, No. 1:117-121.

COP (1999), Outcomes of the Fourth Conference of the Parties to the Convention on Climate Buenos Aires, 2 – 13 November 1998- https://unfccc.int/process/bodies/supreme- bodies/conference-of-the-parties-cop

Cram, L. (1996), 'Theories of European Integration' in J.J. Richardson (ed) *European Union: Power and Policymaking* London: Routledge.

Deutsch, K. (1966), *Nationalism and Social Communication* Cambridge, MA: MIT Press.

Diekmann, Andreas, and Axel Franzen. (1999), "The Wealth of Nations and Environmental Concern." *Environment and Behavior* 31: 540–49.
Dunlap, Riley E., and Angela G. Mertig. (1995), "Global Concern for the Environment: Is Affluence a Prerequisite?" *Journal of Social Issues* 51: 122–37.
Dunlap, Riley E., and Angela G. Mertig. (1996), "Weltweites Umweltbewusstsein. Eine Herausforderung für die sozialwissenschaftliche Theorie." *Kölner Zeitschrift für Soziologie und Sozialpsychologie* 36: 193–218.
Dunlap, Riley E., and Angela G. Mertig. (1997), "Global Environmental Concern: An Anomaly for Postmaterialism." *Social Science Quarterly* 78: 24–29.
Dunlap, Riley E., George H. Gallup, and Alec M. Gallup. (1993), "Of Global Concern: Results of the Health of the Planet Survey." *Environment* 35: 7–15, 33–39.
Escobar Arturo (1988), Power and Visibility: Development and the Invention and Management of the Third World *Cultural Anthropology*, Vol. 3, No. 4. 428-443.
Escobar Arturo (2001), Culture sits in places: reflections on globalism, and subaltern strategies of localization, Political Geography 20 (2001) 139–174
Frank, David John, Ann Hironaka and Evan Schofer. (2000a), "Environmentalism as a Global Institution: Reply to Buttel." *American Sociological Review* Vol. 65, No. 1:122-
Frank, David John, Ann Hironaka and Evan Schofer. (2000b.), "The Nation-State and the Natural Environment over the Twentieth Century." *American Sociological Review* Vol. 65 No. 1:96-117.
Frank, David John. (1997), "Science, Nature, and the Globalization of the Environment, 1870–1990." *Social Forces* 76: 409–37.
Haas, E.B, (1958), *The Uniting of Europe* Stanford University Press.
Haas, E. B. (1964), *Beyond the Nation-State: Functionalism and International Organization,* Stanford University Press
Haggard, Stephan, and Beth A. Simmons. (1987), Theories of international regimes. International Organization 41, no. 3: 491-517.
Hasenclever, Andreas, Peter Mayer, and Volker Rittberger. (1997), Theories of International Regimes. New York: Cambridge University Press.
Hoffmann, S. (1966), 'Obstinate or Obsolete? The Fate of the Nation State and the Case of Western Europe' in *Daedalus* 95, 892-908.

Inglehart, Ronald. (1995), "Public Support for the Environmental Protection: Objective Problems and Subjective Values in 43 Societies." *PS: Political Science & Politics* 28: 57–72.

Inglehart, Ronald. (1997), *Modernization and Postmodernization: Cultural, Economic, and Political Change in 43 Societies*.

IPCC (2010), The Role of IPCC and its role, IPCC, Geneva https://www.ipcc.ch/site/assets/uploads/2018/04/role_ipcc_key_elements_assessment_process_04022010.pdf (https://www.ipcc.ch/2010/02/04/the-role-of-the-ipcc-and-key-elements-of-the-ipcc-assessment-process/)

James, Paul; Palen, Ronen (2007), Globalization and Economy, Vol. 3: Global Economic Regimes and Institutions. London: Sage Publications.

Jay Lehr and Diane Carol Bast (2001), Blueprint for New-Era Environmentalism, The Heartland Institute *Environment News,* April 2001 http://www.heartland.org/Article.cfm?artId=1138

Kenneth A. Oye, (1986), "Explaining Cooperation Under Anarchy: Hypotheses and Strategies" In Kenneth A. Oye (ed.) Cooperation Under Anarchy. Princeton, NJ: Princeton University Press. pp. 1–24.

Keohane, Robert O. and David G.Victor. (2010), "The Regime Complex for Climate Change" Discussion Paper 2010-33, Cambridge, Mass.: Harvard Project on International Climate Agreements, January 2010. Robert Axelrod 1984. The Evolution of Cooperation. New York: Basic Books.

Keohane, Robert O. and Lisa L. Martin. 1995. "The Promise of Institutionalist Theory." International Security 20/1(Summer):39–51.

Kidd, Quentin, and Aie-Rie Lee. 1997. "Postmaterialist Values and the Environment: A Critique and Reappraisal." *Social Science Quarterly* 78: 1–15.

Krasner, Stephen D. (ed). 1983. International Regimes. Ithaca, NY: Cornell University Press.

Krasner, Stephen D. 1982. "Structural Causes and Regime Consequences: Regimes as Intervening Variables." International Organization 36/2 (Spring). Reprinted in Stephen D. Krasner, ed., International Regimes, Ithaca, NY: Cornell University Press, 1983.

Liberman, Peter. 1996. "Trading with the Enemy: Security and Relative Economic Gains." International Security 21/1(Summer), 147–165.

Matthews, John C. 1996. "Current gains and Future Outcomes: When Cumulative Relative Gains Matter." International Security 21/1(Summer), 112–146.

Mearsheimer, John. 1994/5. “The False Promise of International Institutions.” International Security 19/3 (Winter): 5–49.

Mitrany, D. (1930), ‘Pan-Europa - A Hope or a Danger?’ in *Political Quarterly* 1, 4.

Mitrany, D. (1943), *A Working Peace System* Chicago: Quadrangle.

Moravcsik, A. (1993), ‘Preferences and Power in the EC: A Liberal Intergovernmentalism Approach’ in *Journal of Common Market Studies* 31: 4, 473-524.

Moravcsik, A. (1995), ‘Liberal Intergovernmentalism and Integration: A Rejoinder’ in *Journal of Common Market Studies* 33: 4, 611-628.

Müller, B. (February 2008), “Bali 2007: On the road again!” (PDF). Oxford Energy and Environment Comment. Retrieved 2009-05-20.

Müller, B. (February 2008), “Bali 2007: On the road again!” (PDF). Oxford Energy and Environment Comment. Retrieved 2009-05-20.

Pariona, Ameber (2017), “Major International Organizations Fighting Environmental Destruction.” World Atlas, Apr. 25, 2017, worldatlas.com/articles/major-international-environmental-organizations-operating-in-the-world-today.html.

Puchala, D. (1970), ‘International Transactions and Regional Integration’ in *International Organization* 24, 732-763.

Rentz, Henning (1999), Outcomes of the Fourth Conference of the Parties to the Convention on Climate, Energy & Environment: 10:2:157–190

Rittberger, Volker. Mayer, Peter (1993), Regime theory and international relations. Clarendon Press. ISBN 1280813563.

Schmitter, P.C. (1996), ‘Examining the Present Euro-Polity with the Help of Past Theories’ in G. Marks et al (eds) *Governance in the European Union* London: Sage.

Schofer and Hironaka (2005) “World Society, the Nation-State, and Environmental Protection.” American Sociological Review, 65 (1): 117-121.

Seyfang, G., and Jordan, A., (2002), ‘The Johannesburg Summit and Sustainable Development: How Effective Are Environmental Conferences?’, in Olav Schram Stokke and Øystein B. Thommessen (eds.), Yearbook of International Co-operation on Environment and Development 2002/2003 (London: Earthscan Publications), 19–39. pp. 23

Shah, Anup (2014), “COP14–Poznañ Climate Conference”. Retrieved 7 April 2014.

Snidal, Duncan. 1986. “The Game Theory of International Politics.” In Kenneth A. Oye (ed) Cooperation Under Anarchy. Princeton, NJ: Princeton University Press. pp. 25–57.

Sushant, Chandra and Aviral, Dhirendra (2009), Multilareral Environment Agreements Versus
the World Trade Organization System; American Journal of Economics and Business Administration 1 (4): 270-277.
UNFCCC (2020), Conference of the Parties (COP), United Nations Framework Convention on Climate Change, Germany. https://unfccc.int/process/bodies/supreme-bodies/conference-of-the-parties-cop
UNFCCC (2020), Conference of the Parties (COP), United Nations Framework Convention on Climate Change, Bonn, Germany.
Weeks, Edythe E., Doctoral Dissertation, "The Politics of Space Law in a Post-Cold War Era: Understanding Regime Change", Northern Arizona University, Department of Politics and International Affairs, 2006.
Wincott, D. (1995), 'Institutional Interaction and European Integration: Towards an 'Everyday' Critique of Liberal Intergovernmentalism' *Journal of Common Market Studies* 33: 4, 597–609.

9

The Future of the IPCC

In previous chapters, I discussed the history of climate change science, the role of international organizations, the origin of the IPCC's foundation and its role, procedures, climate change denier's stand due to IPCC's own weaknesses (and also errors in the assessment report of 2007), the unbalanced composition of the IPCC including its major stand as knowledge producing intergovernmental organization. This chapter presents the future of the IPCC on the basis of its strength and weaknesses through organizational sociological perspectives (Dunlap and Brulle 2015).

9.1 The Unpredictable Future

As such future cannot be predicted. Sometime, fictional predictions seem accurate; however, these accuracies have nothing to do with realities. They are just consequences or random occurrences, or those predictions are out lairs and have no significance in the reality or fictional predictions has no impact on reality or validity test.However, by evaluating current trends of floating market, current demand structures, production pattern, relied stakeholders of the product, there is a tiny room of possibility to predict the future direction or future demand of the organization; though, it applies only, if the current trend supply and demand remained unchanged, production's quality improves or at least it remains in the same pattern of demand and supply; and also the consumers behavior and the reliability of the product remains nearly unchanged. In the normal case, insofar, there is no mechanism to forecast that, the demand will be continued or will be remain stable or will have steady growth or will in deceleration. Any time unprecedented event/challenge/problem can happen, and the pathways of any production organizations can be broken or disrupted. One of the good examples of such U-turn situation is Covid 19, epidemics. Covid completely changed or changing not only the economic, social and cultural pattern of the world, it created such a phobia that, world became almost standstill (Bhandari 2020). So, future prediction of any organization is just an assumption or conditional hypothesis which can be true or false depending on unseen future. In the case of knowledge producing organization, factual published knowledge will not be disqualified, therefore, there will no U-Turn. Deceleration could happen; however, the creditability of the product can save organization until foreseeable future.

"Any invention, any discovery, which consists essentially in the elaboration of a radically new concept cannot be predicted, for a necessary part of the prediction is the present elaboration of the very concept whose discovery or invention was to take place only in the future. The notion of the prediction of radical conceptual innovation is itself conceptually incoherent" (MacIntyre 1985: 93, as Cited in Tsoukas and Shepherd 2004:2).

In general term- organizations future depends on its mechanism of its talent management, organizational structure, workplace, its networks and how it creates its own value. In the case of the IPCC [*On the basis of current recognitions; reliability and validity; and its stand as knowledge base, governments backed- structured and governments administrated global organization; the products of IPCC the assessments, and special reports*] will have more demandable in the future. Another important points to note are, the climate change challenged has no quick solutions, without excessive investments, which seems not possible due to changed global circumstances, needs very hard active policy and program implementation to minimize the challenges, however, to bring equilibrium condition of global environment is nearly impossible. At this point, we have not even fully understood the degree of climate change impacts on earth's ecosystems, there is no immediate solution the climate change challenges. We need to understand the complexity of climate change, from micro to macro level and need to be able to analyze the impact scenarios. Once, we have such sequential analysis of climate change impact, actual path analysis is possible. In that case, of course, more, innovative ideas, technology, infrastructures, institutional architectures are needed. The knowledge of impact is in high demand and will remain, because of the complexity of impacts as well as even the lack of knowledge and tools to examine the damaged or being damaged or future damage of environment "due to continuation of anthropogenic intervention in the nature" is not fully understood. This scenario indicates that, until the problems are there, the intend of search will remain on demand. The efforts to fill the knowledge will be continued and the new knowledge will continuously be produced. In such cases, the tasks of knowledge producing organization like IPCC will increase exponentially. In general, the IPCC remain the demanded organization- it has good future. However, the epidemic like Covid 19, or anything like this can altered the demand; however, in long run again, the demand of knowledge producing organization in any field will remain on high demand (Bhandari 2020).

9.2 The Current Direction of Knowledge Production

The directions of current trend of knowledge production is more towards the techno centric, which is fostering to individualistic society. Social, and economic strata are being created due to individualistic mode of society. Social, economic, cultural, gender, nationalities, religious inequalities are increasing in almost in every continent. The competitions among individuals to individuals, communities to communities- at local to global level is increasing. There social, cultural and religious cohesions are no more able to drive the society (Bhandari 2019). The blame culture is also increasing. The increasing distance between north and south, the competition among developed large economy countries, among developing countries and within and among the all countries is creating problems to foster the climate science regime. The causes of major global tension (to combat with climate crisis, humanitarian crisis of middle east, the gender, social and economic inequalities) one way or another, are the or byproduct of hegemonic power (Gramsci 1920; Robinson 1972; Bates 1975; Hoare 1977; Salamini 1984; Adamson 1985; Scott 1997; Kendie 2006). The main cause of middle east crisis is not created within but created by unnecessary interventions of external powers. Those power are not allowing middle eastern countries to remain in peace and stand by themselves. The middle east continuously suffering as a demonstration ground of wars weapons. The fossil fuel depended economic wars are spreading. Insofar, the environmental impacts of wars are even not in the agenda of scientific research. These unexplored field of studies will increase the future demand of Knowledge producing organizations like IPCC.

In sum, the future the IPCC depends on how it fulfills core vision, mission and objectives and how it completes its role. The role of the IPCC is "*to assess on a comprehensive, objective, open and transparent basis the scientific, technical and socio-economic information relevant to understanding the scientific basis of risk of human-induced climate change, its potential impacts and options for adaptation and mitigation. Review by experts and governments is an essential part of the IPCC process. The Panel does not conduct new research, monitor climate-related data or recommend policies. It is open to all member countries of WMO and UNEP*" (IPCC Brochure 2007). The task is very important, therefore, producing reliable, trustworthy knowledge has been major challenge and will be more challenging journey for the IPCC, since climate change impacts are becoming more serious day by day.

No question on summary of IPCC Assessment reports: Assurance of climate change is occurring mainly due to anthropogenic cause; it is reliable and valid.

The Intergovernmental Panel on Climate Change (IPCC) Assessment Reports (2007, 2014) provides serious picture of global climate change and its impacts. It reports that warming of the climate system is unequivocal, and that delay in reducing emissions significantly constrains opportunities to achieve lower stabilization levels and increases the risk of more severe climate change impacts. Distinguishing that deep cut in global emissions requires achieving the ultimate objective of the United Nations Framework Convention on Climate Change (UNFCCC) and emphasizing the urgency to address climate change. However, managing climate change will be a greatest challenge for the national and international public sector for the several decades. As largely accepted notion, climate change adaptation and mitigation require international institutions to undertake functions on a scale without illustration. It requires decision-making by governments that challenge their ability to create policies and programs and agree on new obligations, which alters with the exiting regime of the governance particularly in the developing world. In effect, it means creating a climate change management regime and making it effective. Unprecedented climate change has exacerbated the existing the social and environmental problems. This has resulted in the mass exodus of both elites and agricultural workers. Eventually, climate change has become not just an environmental issue, but it has led to severe socio-economic and political ramifications (Ide et al. 2014; Kniveton et al. 2008; Keels 2017). Increasing vulnerability due to climate change has been a regular phenomenon lately and a natural disaster has been increased remarkably in the climate prone nations. The countries with the most risks are characterized by high levels of poverty, dense populations, and exposure to climate-related events. Each year natural disasters such as flood, drought, cyclone, and land slide take bring catastrophic damages taking toll of thousands of lives, houses, and large swathes of farmland (Clark 2002; Brooks, and Adger, 2005; Yohe et al. 2007). The annual monsoon rains forced reservoirs to release massive volumes of water into already burgeoning rivers, bursting banks and submerging villages and crops in low-lying areas in climate prone countries. However, due to the lack of technology and economic resources coupled with poor institutional arrangements, uncontrolled population growth and ever-increasing poverty have constrained them from taking effective actions to minimize the impact of the climate change induced risk.

As IPCC mandate is not to conduct own research but assess the climate change scenarios on the basis of published materials from the scientific communities, international or national research institutions.... "The IPCC does not plan or carry out research, and this separation between research and assessment is essential if the IPCC is to be an objective assessor. The mandate of the IPCC is to evaluate information that must be independently documented, primarily as peer-reviewed literature. The planning and coordination of international research are best carried out by organizations such as the World Climate Research Programme, the International Geosphere-Biosphere Programme, and the International Human Dimensions Programme. These bodies often consider IPCC assessments and help provide the means for the scientific community to produce related research" (Solomon and Manning (2008). Here it is noteworthy to state that the IPCC reports are all secondhand or the illustrations and analysis of scientific papers, books and reports, newspapers, or varieties of publications, however, the products, "the assessment reports" are IPCC's production. Therefore, IPCC has the responsibilities of collecting, analyzing, synthesizing factual knowledge on climate change.

9.3 Time Matters

The climate change issue was not emerged over a year but is the products of centuries exploitations and experiments on nature on the name of industrialization, modernization or development. As it is the product of long anthropogenic disturbances on natural ecosystem of the planet, the impacts of such disturbances have not been fully understood yet. Insofar, we have thousands of research, books, articles, monographs, reports and on the name of climate change challenges thousands of meetings, workshops, conferences being held specially since 1972 Stockholm conference. As noted in the previous chapters, the formation of IPCC was outcome of public awareness. WMO and UNEP took the initiatives as an output of hundreds of workshops of scientists' concern of devastating impact of climate change. On the name of climate change challenges, thousands of organized groups and individuals are advocating; this is good symbol; however, still we know very littles what we already lost as an impact of climate change and what actually going on in the global climatic patterns. So, there is a still lack of knowledge about the current and foreseeable impact of climate change. We only know partially big challenges (Pollution; Global Warming; Deforestation; Sea Leven Rise, Ozone layer depletion; Loss of biodiversity; Overpopulation etc.) but still unknown about the disturbances of harmonious relationships within the mega ecosystems and sub-ecosystems within. We have seen weather pattern change within our lifetime- however, we do not know the altered life cycles of flora and fauna. In sum, there is a knowledge gap- "what we know and what we do not know" and demand of climate change science. In this context the role of powerful (because it is governmental panel, established with the purpose to explore, examine and assess the existing knowledge on climate science and prepare the factual assessment report based on scientific data), the horizon of IPCC increases automatically. It also indicates that, the future of this governmental organization depends on how it approaches to the produced knowledge by the climate change concerned scientists.

As such, except the climate change skeptics/deniers, or temporal selfish motive hardliners, all concerned stakeholders, who believe in "anthropogenic disturbances are one of the major causes of climate change" accept the conclusion of the IPCC. *"It also estimates cumulative CO_2 emissions since pre-industrial times and provides a CO_2 budget for future emissions to limit warming to less than 2°C. About half of this maximum amount was already emitted by 2011. ...From 1880 to 2012, the average global temperature increased by 0.85°C. Oceans have warmed, the amounts of snow and ice have diminished, and the sea level has risen. From 1901 to 2010, the global average sea level rose by 19 cm as oceans expanded due to warming and ice melted. The sea ice extent in the Arctic has shrunk in every successive decade since 1979, with 1.07 × 106 km^2 of ice loss per decade. Given current concentrations and ongoing emissions of greenhouse gases, it is likely that by the end of this century global mean temperature will continue to rise above the pre-industrial level. The world's oceans will warm, and ice melt will continue. Average sea level rise is predicted to be 24–30 cm by 2065 and 40–63 cm by 2100 relative to the reference period of 1986–2005. Most aspects of climate change will persist for many centuries, even if emissions are stopped"* (IPCC Fifth assessment report 2014- United Nations 2020[1]). The acceptance of assessment reports, validates the importance of IPCC as trustworthy organization of climate change knowledge producer. Another example of its strength is as governments organization, it can advices to governments (all UN member nations are its first stakeholders), and also

[1]United Nations (2020)- https://www.un.org/en/sections/issues-depth/climate-change/ (retrieved 3/20/20)

proposes the measures to overcome the challenges, warns governments and show the future direction. The 2018 special report can be considered as warning call as well the directives to the governments and concerned stakeholders. The report set a new target "Global Warming of 1.5°C". "*In October 2018 the IPCC issued a* special report *on the impacts of global warming of 1.5°C, finding that limiting global warming to 1.5°C would require rapid, far-reaching and unprecedented changes in all aspects of society. With clear benefits to people and natural ecosystems, the report found that limiting global warming to 1.5°C compared to 2°C could go hand in hand with ensuring a more sustainable and equitable society. While previous estimates focused on estimating the damage if average temperatures were to rise by 2°C, this report shows that many of the adverse impacts of climate change will come at the 1.5°C mark*" (IPCC 2018[2]; United Nations 2020).

One of the major **unchanged claims of IPCC** is "Anthropogenic disturbance to the natural ecosystem is the major cause of climate change [(skeptics reject this) from its first assessment reports to the 2018 especial report- Global Warming of 1.5°C)]. IPCC 2018 special report states that "*Human activities are estimated to have caused approximately 1.0°C of global warming above pre-industrial levels, with a likely range of 0.8°C to 1.2°C. Global warming is likely to reach 1.5°C between 2030 and 2052 if it continues to increase at the current rate. (high confidence)....Warming from anthropogenic emissions from the pre-industrial period to the present will persist for centuries to millennia and will continue to cause further long-term changes in the climate system, such as sea level rise, with associated impacts (high confidence), but these emissions alone are unlikely to cause global warming of 1.5°C (medium confidence)* (IPCC 2018 https://www.ipcc.ch/sr15/chapter/spm/).

The major role of IPCC is to find the fact from trustworthy sources, document them with appropriate references and prepare reports in a way that concerned stakeholders understand the factual truth of climate change and use produced knowledge in their policy directives. Another role is creating and combining scattered knowledge base in a single source, so concern

[2]IPCC, (2018), Summary for Policymakers. In: Global Warming of 1.5°C. An IPCC Special Report on the impacts of global warming of 1.5°C above pre-industrial levels and related global greenhouse gas emission pathways, in the context of strengthening the global response to the threat of climate change, sustainable development, and efforts to eradicate poverty [Masson-Delmotte, V., P. Zhai, H.-O. Pörtner, D. Roberts, J. Skea, P.R. Shukla, A. Pirani, W. Moufouma-Okia, C. Péan, R. Pidcock, S. Connors, J.B.R. Matthews, Y. Chen, X. Zhou, M.I. Gomis, E. Lonnoy, T. Maycock, M. Tignor, and T. Waterfield (eds.)]. World Meteorological Organization, Geneva, Switzerland, 32 pp. https://www.ipcc.ch/sr15/chapter/spm/

stakeholders can see the climate change scenarios within IPCC portal (in its series of assessment reports). The reliability of IPCC depends on how knowledge source is chosen, how validity of chosen resources is tested, what is the procedure of validation and reliability of used such resources. It is hard to point answers of these, normally raised general questions in IPCC documentation.

The IPCC reports are the product of combined efforts of many people from the member countries. However, the IPCC itself is a political organization, because its governing authority (the chair, deputy chair etc.) is chosen or nominated by the government with their own interests. The IPCC chair and his/her deputy have facilitating role in production of assessment reports; they are the main responsible for the quality production. However, there is no checking and rechecking mechanisms of their roles. The people who are invited/nominated to involve in the assessment preparation process also have no any obligations and there is no a political evaluation system in the IPCC assessment report production procedures. Because of the lack of reliability and validity test mechanism, IPCC assessment reports get the valid questionable critics. Errors in IPCC 2007 report is an example of validation problems. Cause of these errors are the organizational problems. "climate change as an organizational problem. It is an organizational problem in the sense that acknowledging and addressing the problem are the result of organizing practices – practices that coordinate and control the activity of a wide array of individual and collective actors. And organizing practices that seek to coordinate and control others' activity revolve around authority (Porter *et al* 2018:876).

9.4 The IPCC is a Popular Organization in Climate Change Domain

There are hundreds of climate change knowledge producing organizations in terms of its size, coverage and ground reach. However, still it is most favored organization by the climate change domain. Table below shows the occurrences in search

The table shows the occurrences in thousands in google search combinedly- universities and climate change has the highest number, followed by the governments; World Bank; Governments; Scientists and United Nations respectively. It is obvious because of the coverage and presence. The occurrence number of IPCC is low, however, the percentage of occurrence on climate change is 73.13% which shows that, it is kind of synonyms of climate change. Second popular stakeholder in climate change is university and third is IUCN. This simple search indicates that, IPCC is not a sole organization

Table 9.1 The most popular organizations on climate change

Concern Stakeholders	General (in thousands)	With Climate Change (in thousands)	% of climate change occurrences
IPCC	22,700	16,600	73.13
World Bank	4,310,000	317,000	7.35
IUCN	21,200	11,700	55.19
WWF	65,400	12,500	19.11
United Nations	861,000	346,000	40.19
Universities	10,430,000	654,000	62.70
Government	7,920,000	775,000	9.79
Scientist	1,010,000	175,000	17.33

Source: Google search on (2/13/20- 4.30AM). Occurrences in thousands[3]

in climate change knowledge production, and it needs strong collaboration with other organizations who are also producing trustworthy knowledge on climate science.

9.5 Future of Organization-Sociological Perspective

The future of any organization depends, how organization strictly follows the rules and regulation stated in its constitution; does organizational product lies within the vision, mission and objectives of the organization, does the leadership has controlling mechanism to its product?

Organizational theory can be described as the study of formal groups organized to achieve or attain specific goals in efficient manner. Organizations have been leading the sociopolitical scenarios of the world for at least the last three centuries. In other words, organizations have been changing the world's socioeconomic landscape (Perrow 1991). As Perrow notes "organizations are the key to society because large organizations have absorbed society. They have vacuumed up a good part of what we have always thought of as society and made organizations once a part of society into a surrogate of society (1991: 726, as cited by Scott and Davis 2007: 340). This influence, which has increased gradually over time, mostly in the developed regions of the world during the twentieth century, can be found in our everyday lives. Now, we are in organizational firms from birth to joining the workforce, as well as in our prayers, and we even die in organizations; along the way, we derive our identities from our associations with them. Organizations are related to every aspect of our daily life. Organizations are not only the building blocks of our societies, and a basic vehicle for collective action, but they are also our life form. They bring into being the social structure of our societies and basically form our futures. Organizations are a fundamental part of contemporary societies; we enthusiastically turn to them or create them when a need or crisis exceeds our own personal abilities or resources. Moreover, organizations are at the core spirit of every society and nation. Through organizations we make differences in our society and achieve collective goals. Theoretically, organizations are dedicated to extending the idea of creating a new base to create new opportunities. These broader contexts of organizations theories apply not only to the domestic organizational environment; but also, equally apply in the case of international organizations like IPCC.

Organizations are formed by the contexts or environments in which they are established. Modern organizations replicate the impact of their historical origins in societies characterized by growing privileged circumstances and conflicts over the control and distribution of products and services. Organizations come in many puzzling forms because they have been clearly designed to deal with a wide range of social, cultural, economic and political problems. Because they have emerged under widely varying environmental conditions, they have to deal with complexity within and emerged externalities. The nature of IPCC holds different levels of complexity because of its stand as intergovernmental organization.

'First organizations are complex because of complex adaptive systems, differentiations in agents, variations on decision making and problem-solving techniques and networks, information technology and algorithmic complexity, second, organization hold complex adaptive systems, loose coupling and models, edge of chaos, simple rules and complex behavior, emergence and recombination and evolution and thirdly, organizational interdependence, cellular automata, micro-behavior and macro-structure complex inter-organizational dynamics, sensitivity to initial conditions and path dependence (Baum and Rowley, 2005). Adaptation of Climate Science – in multisectoral scenario has to pass/experience through complex adaptive systems, because each country has difference social, cultural, geographical and political limitation or boundaries. The complexity also exists in decision making system because each countries definition of public participation may differ due to their own governance system (Bhandari 2019:107).

"The progress of humanity has been predominantly due to the effectiveness of organizations to achieve human pursuits. Organizations have always made use of knowledge and technology to survive. With the rise of large corporations in the early twentieth century came a strong interest in research in management and organizational theory. The awareness of the importance of information and knowledge, followed by a constant search for ways to create, store, integrate, tailor, share, and make available the right knowledge

to the right people at the right time, led to the birth of knowledge management in the 1990s. Knowledge organizations, currently in their embryonic form, focus on networking and knowledge creation, sharing, and application. The ultimate challenge of the future is to liberate and amplify the knowledge and creativity of all organizational members, which will enable the rise of the knowledge organization" (Bennet and Bennet 2004:1). Bennet and Bennet (2004) argument captures the direction of the knowledge production organizations like IPCC; whereas IPCC relies on secondary information, own creativity could be new role to be added in future.

Similarly; Nadler, and Tushman, (1999) notes that, "There are elements of today's organization design that are timeless, but there are also new strategic imperatives that flow from the reshaped environment that will raise design issues for the organization of the future. .... environment will continue to drive the strategic architecture or an enterprise and the variety of ways in which the enterprise manages the work of its people in pursuit of strategic objectives. At the same time, the changing environment will create 6 strategic imperatives for future organizations: focus portfolios, with various business models; abbreviate strategic life cycles; create "go-to-market" flexibility; enhance competitive innovation; and manage intra-enterprise cannibalism. Eight core competencies in which organizations will have to become proficient in light of these challenges are suggested: increased organization clock speed, design structural divergence, organizational modularity, hybrid distribution channels, asymmetrical research and development, conflict management processes, organizational coherence, and team management" (Nadler, and Tushman, 1999:45).

In the IPCC case, roles are clear, it bonded within the organizational rules, norms, procedures; however, because of its nature the controlling and monitoring mechanism has been questionable. The IPCC is brand of climate change knowledge production; the maintenance- here maintaining the objective of producing science of climate change fact, it needs a nonpolitical evaluation body or mechanism.

9.6 The Future of the Scientific Organization Depends on How Organizational Mechanism Values the Knowledge of Scientists/Practitioners

In the case of IPCC, this part is largely questionable. Many scholars particularly from the developing world, still think that, the IPCC is western world organization which rarely try to involve scholars from developing world (base on field interviews- India, Nepal, Bangladesh, Pakistan, Thailand, Kenya.....). Even women from the developing world have very nominal representation in the IPCC process (IPCC 2019).

9.7 What is the Meaning of Scientific Outcome?

The social science scholars value the knowledge validation process of IPCC; however, they fear about the policy relevancy. The prominent environmental sociologist, Prof. Riley Dunlap states:

> "*The IPCC has been and will continue to be the essential source for knowledge about climate change, with its reports offering excellent summaries and syntheses of the latest scientific understanding as it is mandated to provide Unfortunately, I fear that its policy relevance will continue to decline, as major nations such as the USA, Brazil and Australia increasingly ignore its recommendations and set poor examples for other nations Efforts to agree on international accords promoting carbon reduction measures is harder than ever in an era of nationalism, with populist leaders promoting national self-interest and attacking the very idea of international cooperation I am not optimistic* (Riley Dunlap 2020[4])."

Similar observation is outlined by Urmi A Goswami (2018[5]): "The United States, Russia, Saudi Arabia, and Kuwait succeeded in preventing the inclusion of a key and the most recent scientific assessment of the possible impact of global warming in the final agreement at the UN climate change talks in Poland. Despite recognizing that the October assessment by UN climate science panel Intergovernmental Panel on Climate Change (IPCC) encapsulates the "best available science" in the final decision, countries opted to focus on the IPCC's ability to be punctual while putting together a work program to

---

[4]Prof. Riley Dunlap, Regents Professor of Sociology, and Dresser Professor Emeritus, Department of Sociology, Oklahoma State University, Stillwater, OK 74078, USA (in personal email conversation- Monday, February 10, 2020)

[5]By Urmi A Goswami, ET BureaulLast Updated: Dec 18, 2018, 11.38 AM IST US, Russia succeed in keeping IPCC report out of climate deal IPCC report calls for rapid reduction of greenhouse gases to limit adverse impacts of global warming. https://economictimes.indiatimes.com/news/politics-and-nation/us-russia-succeed-in-keeping-ipcc-report-out-of-climate-deal/articleshow/67140360.cms?utm_source=contentofinterest&utm_medium=text&utm_campaign=cppst

make the 2015 Paris climate pact operational. US, Russia succeed in keeping IPCC report out of climate deal-IPCC report calls for rapid reduction of greenhouse gases to limit adverse impacts of global warming". What to say? What is the meaning of scientific outcome if the power game continues as noted above??

9.8 There's Never Been a Coherent Response

Similarly, another worldly known environmental sociologist Prof. Timmons Roberts shares his experince: "*The IPCC's mega assessment reports always have been the go-to source for those seeking to characterize the consensus on key climate science issues. I believe it will continue to be, and the AR6 is well underway and will be impressive. The downside is the lack of any analysis of power, and a downplaying and very selective use of political economy and sociology. This leaves a huge blind spot in the reports, in which they meticulously describe our problem but not the actors blocking the very solutions they advance. The executive summaries have to be approved by every government on earth, so that's not entirely surprising. But an equal or far greater influence on the creation and persistence of this blind spot is the ignorance and discomfort of physical scientists and even some social scientists with the bare knuckles' manipulation of the whole process We know that fossil fuel companies actively infiltrated and steered the IPCC processes in the past. There's never been a coherent response, and the world continues to stumble along without a clear understanding of what's in the way of dealing with climate change (I was a contributing author on the SR1.5 at the late stages of addressing comments, and sought to add more on this topic, including the work of Riley Dunlap, Aaron McCright, and Robert Brulle. What I wrote was pruned, garbled, and soft-pedaled in confusing language in the final draft*)". (Timmons Roberts[6] 2020).

To be a real knowledge producing organization the IPCC, needs to incorporate the points raised above. What to add here- where IPCC is going and who is guiding this international intergovernmental panel? The future of the IPCC depends on how it addresses these true problematic issues.

[6]Ittleson Professor of Environmental Studies and Sociology, Department of Sociology, Brown University, Providence, RI 02912, USA (in personal email conversation- Monday, February 10, 2020 7:57 AM)

9.9 Social Scientists Say is Often Ignored

Social thinker Professor Robert Brulle have similar response as above: "*Your research sounds extremely interesting. I've actually published quite a bit about the shape of climate change research and its neglect of sociological perspectives, and the ideological character of this research as exemplified in the IPCC I guess my comment on the IPCC is that its failure to robustly include social science perspectives and the domination of the post-political ideology in its reports makes the IPCC reports increasingly marginal and less relevant over time. The core scientific questions have been answered, and now it is really just a process of further refining the scientific questions But left unanswered are the core political and cultural battles over how to address climate change Here sociology has made a number of contributions But these have been pretty much ignored or buried deep in the report For example, probably the largest obstacle to climate action comes from organized opposition to climate change Yet, you would be hard pressed to learn about this from reading the IPCC reports. With little new to offer in the natural sciences, I see the IPCC reports becoming less and less relevant to public policy development until it engages with the structural and organizational barriers to climate action*" (Robert Brulle 2020[7]). "Probably the largest obstacle to climate action comes from organized opposition to climate change" who are they- no need to expose, so, whose weakness is this? Absolutely, it is the IPCC and its management system. They should be able to maintain their stand to validate themselves.

9.10 Accepting Weak Points – Address them with the Acknowledgement – Repair from the Route: Could Pave New Directions to the Knowledge Base Organizations: The IPCC

What are the week points of IPCC?

1. Errors on assessments reports- Errors in assessments reports and its impact on the reputation of IPCC is discussed on the chapter IPCC and deniers. Particularly, based on the IPCC 2007 assessment report error. United Nations intervened to resolve the problem and Enteracademy

[7]Robert Brulle, Professor Emeritus, Department of Sociology, Drexel University Philadelphia, USA (in personal email conversation- Monday, February 10, 2020)

Institute was appointed to evaluate the IPCC's errors. Academy suggested few points and give some direction to overcome such problems. However, IPCC did not/have not make clear that, such errors will not occur again. IPCC presented 2014 assessment report and in 2018 a special report on the impacts of global warming of 1.5°C. Why general public should trust IPCC, how one knows that there are no any errors in the reports? What is the evaluation mechanism of valid information? IPCC is silent. This silence is boasting the deniers. There are thousands of scholars who are with IPCC, accept that errors can happen and some time it is normal too, because of too many people involved, and eye slip in common, because we trust each other. As we know, IPCC does not do own research, IPCC documents have no space to acknowledge to those scholars, who are actually the core knowledge producers. Adding few pages to the core scholars of both physical sciences and social sciences should not harm to IPCC.

There are hundreds of international reginal and national organizations, individuals and groups who have been advocating on climate change impact, warning governments, policy makers to utilize the IPCC recommendations. These advocating organizations, individuals, groups are using valid case study embedded with the ground facts. However, the IPCC which is the most important governments body on this domain have not incorporate the people's say.

The media houses, newspapers, social medias are one of the major stakeholders to bring the right facts to the public domain. The climate change knowledge flow through media is more powerful and important for the general public. In fact, they are the knowledge drivers advocates, activists, and societal change maker; however, IPCC has even not valued and gave the space for those knowledge flow even in annex or in acknowledgements. The role of media is explained in the chapter related Deniers.

2. Representation of scientists- composition and hegemonic western boys club model: Details of scientist's representation is presented in a chapter "Developing World's scientists' representation". The representation of scientists is still mainly dominated by male developed countries scientists. It seems that, the representation of developing countries scientists seems in growth if we see from first assessment report to fifth assessment reports. However, if we one evaluates who the one from developing countries – are they really represent the developing countries? Not really, their nationalities might be from developing countries but?? ….. It is

important to note that the scientists who are listed as the representatives of developing countries actually have worked in their respected countries and aware of the ground truth of the reality.

3. IPCC information is too much secondary- As noted earlier the composition of IPCC should be proportionally represented with the all members organization as much as possible. As seen in the previous chapters, the developing world representation was very minimal in the first assessment report, slightly increased 2nd, 3rd, 4th and 5th. However, it is still insignificant in terms of population, country's geographic location, economic and social structure. The participants are not equally distributed. The domination is always from western developed countries. Several COP conferences recommended for the proportional distribution/involvement for the IPCC assessment process. The existing embodied hegemonic relationship among the continuously participating contributors have no intention of encouraging the new participants, who are mostly from developing countries. The IPCC stands as knowledge producing, knowledge validating, and organizing for policy recommendations. The IPCC does policy recommendations based on produced knowledge through the assessment reports or through the special reports. There is no question about the validity of produced knowledge; however, who has the firsthand information about the devastating impact of climate change? For example, the pain of floods in Bangladesh- who knows better than the Bengali scholar who has faced the Ganges and Brahmaputra rivers flood in his or her house from childhood or the person who studies floods in Bangladesh in academic papers? In relation to cyclone impact, or impact of sea level rise, who knows better – the scientists of small island countries or scientists of the mainland (USA, Western Europe etc.)? Only the wounded one knows where the pain is and what makes him or her feel less pain. In another example, who knows the intensity of labor pain- the woman who is delivering a baby or the nurses who are helping her to deliver? Does this scenario not provide the room to state that the policy recommendations of IPCC are too generic?
The climate change induced problems are not equal- some areas of planet are suffering more than other areas. Therefore, IPCC which is only fully the most trusted knowledge producing organization, still needs to seek the floating knowledge- which holds the reality of the victims of climate change. IPCC should include more social scientists from the climate vulnerable countries, who can bring the factual truth

not only base on the quantitative analysis but by analyzing through qualitative reasoning. Nursing or proving medication is wonderful approach to sicken world due to climate induced problems, however, it is essential to know the cause of the problem through ground truth. Ground truth can be acquired only by involving scientists who hold firsthand information and experiences.

4. No proper room for non-western languages- In the IPCC citation list, one can rarely find any literature or knowledge listed or acknowledged other than western world scientists speak or write (please see IPCC reports citation list to verify). In fact, the large part of knowledge is ignored in the IPCC process, which allows to critical thinkers that, IPCC assessment reports, are not complete and insofar, not able to capture the real essence of this climate change challenges. IPCC is still within the larger box of western, male dominated circumstances. This western hegemonic model has a such a shield system which does not allow to consider the value of traditional, indigenous, as well the climate science or research written other than (English or whatever is their languages are) they speak. The women particularly from the developing world or non-English speaking countries are not being able to even to suggest their knowledge within in the group meetings of the IPCC, because, the circumstances itself created by within western mode, and there is no way to penetrate that even being within the discussion forum (base on discussion with participant).
5. Ignoring to ignorance- the deniers-skeptics and politics- as such by principle IPCC can be considered as a political organization founded by politicians by the government organizations- United Nations agencies- UNEP and WMO. The administrative body is appointed by governments through voting. The process involves power exercise. So, the composition of IPCC- including selection or nomination of experts is part of political strategy, which is governed by unseen western power of knowledge (noted earlier as well as previous chapters). There is no question of contribution of the experts; they all deserve full appreciation. However, whoever are involved do not represent member state equally. In another words, insofar, from first assessment report to fifth assessment report, IPCC has not focus on proportional representation of scientists- from developed and developing world. There are some improvements- however not significant. Why? Is it the political strategy? Or, are there no skilled, knowledgeable people in the developing world- (it is not true), so why, need to explore from insiders (IPCC).

> *"The climate change denial campaign has been successful. Organized and well-funded denial efforts have convinced many policy-members and citizens that the scientific evidence for human caused atmospheric warming remains so uncertain that regulating carbon emissions is not urgent. As a result, the United States remains a laggard, and often an obstacle, in international efforts to ameliorate this grave threat to the planet and the wellbeing of many Americans"* (Dunlap and McCright 2013).

Ignoring ignorance- or deniers- without any doubt most of the stakeholders of climate science accept the truth that the major causes of climate change is due to anthropogenic disturbances on nature or humans ambition to win the nature, which occurred or occurring due to disturbances on earth ecosystems or over exploitation of natural resources. Surprisingly, we have deniers who reject the science proven fact. No problem- they have freedom of expression. However, is there any limitation of freedom of expression? Absolutely yes. Otherwise, it can create social disorder, and no one has right to distract social harmony to fulfil the personal greed, ego, anger and self-satisfaction. The IPCC and the all concern stakeholders need to be aware with those who may able to inject deviant behavior in the society (there are already few such cases- see following conversations).

So, the IPCC and all concern stakeholders need to make sure, the epidemics of personal greed, ego, anger and self-satisfaction should not penetrate to our education, economic, political, social system (cultural, traditional and religious) as well as harmonious relationship with nature.

6. **Woman representation- Minimal representation in the IPCC- even who have contributed have no good experiences due to the domination of male counterparts**.

   "Gender inequalities intersect with climate risks and vulnerabilities. Women's historic disadvantages — their limited access to resources, restricted rights, and a muted voice in shaping decisions — make them highly vulnerable to climate change." (Human Development Report 2007/08. Fighting climate change: Human solidarity in a divided world)

   ... "in essence, a term used to emphasize that sex inequality is not caused by the anatomic and physiological differences that characterize men and women, but rather by the unequal and inequitable treatment socially accorded to them. In this sense, gender alludes to the cultural, social, economic and political conditions that are the basis of certain standards, values and behavioral patterns related to genders and their relationship" (Riquer 1993, as cited in Aguilar, Granat, and Owren 2015).

"Differences in vulnerability and exposure arise from non-climatic factors and from multidimensional inequalities often produced by uneven development processes. These differences shape differential risks from climate change... People who are socially, economically, culturally, politically, institutionally, or otherwise marginalized are especially vulnerable to climate change and also to some adaptation and mitigation responses... This heightened vulnerability is rarely due to a single cause. Rather, it is the product of intersecting social processes that result in inequalities in socio-economic status and income, as well as in exposure. Such social processes include, for example, discrimination on the basis of gender, class, ethnicity, age, and (dis)ability" (IPCC 2014 Summary for Policymakers, as cited in Aguilar, Granat, and Owren 2015).

As noted in the above quotes (also noted below), in principle, the IPCC accepts the notion of discrimination to *gender, class, ethnicity, age, and (dis)ability*"; in the same time, it has not been able to avoid discriminative structure within its system. In the IPCC structure, selection of participants is guided by the western hegemonic (Gramsci 1920; Robinson 1972; Bates 1975; Hoare 1977; Salamini 1984; Adamson 1985; Scott 1997; Kendie 2006; Riley 2011) male centric system "a chain of western mostly white color male club" knowingly or unknowingly, where, overall woman participation is still nominal and woman participation from developing world is almost none (see table titled- Participant in the IPCC process by gender and developing countries).

"One mechanism through which the universal norms of world hegemony are expressed is the international organization. Indeed, international organization functions as the process through which the institutions of hegemony and its ideology are developed. Among the features of international organization which express its hegemonic role are the following: (1) the institutions embody the rules which facilitate the expansion of hegemonic world orders; (2) they are themselves the product of the hegemonic world order; (3) they ideologically legitimate the norms of the world order; (4) they co-opt the

elites from peripheral countries; (5) they absorb counterhegemonic ideas" (Cox 1983:172 as cited in Peng 2018:57). To some extent as an international governmental organization, the IPCC is still dominated by the western supremacy and not yet able to penetrate or break the chain of hegemonic order; in which woman and man of the south are still considered incompetent in knowledge production.

And whoever women got a chance of participation, insofar, have not raise any voice except very few (few examples are listed below). The example of labor pain applies here too. The major victims of climate change are the women and children of developing world (Women's Environmental Network 2010). "Since Kyoto, the idea of climate change as not simply an environmental issue, but one of social justice, has been given increasing recognition. However, the way in which climate change affects groups that face discrimination and under-representation, such as women, ethnic minorities and indigenous people, has not received sufficient attention" (Women's Environmental Network 2010:5).

The following finding of Women's Environmental Network 2010, shows a true outline of women suffering:

This report by Women's Environmental Network examines the distinct impacts of climate change on women in both developed and developing countries, women's contribution to climate change, and their involvement in decision making about tackling climate change.
It finds that, because of their increased likelihood of living in poverty, and their gendered social roles, women are more likely than men to:

- die in climate change-related disasters, and suffer from increased workload, loss of income,
- health problems, and violence and harassment in the aftermath of such events;
- be displaced, or encounter problems when other (usually male) family members migrate for economic reasons;
- experience increased burden of water and fuel collection, and resulting health problems,
- due to increased incidence of drought or other changes in climate; feel the effects of rising food prices most acutely, and be the first to suffer during food,
- shortages; suffer exacerbated health inequalities;

- suffer from violence, including sexual violence, in resource conflicts;
- be expected to, and need to, adapt to the effects of climate change, increasing their workload;
- suffer as a result of intended solutions to the problem of climate change, such as forestry projects and biofuel production.

OPPOSITE TO THIS: *women always care environment one way or another.* on average, women tend to contribute less to climate change. This is because of their poverty - lower consumption roughly equates with lower greenhouse gas emissions - and their social roles - for example, women are less likely to fly for business reasons, and less likely to hold a driving license or own a vehicle. It is also observed that women are more inclined towards pro-environmental behavior, such as recycling and energy efficiency, taking part in citizen actions, and are more likely to favor policies that reduce greenhouse gas emissions, such as limiting airport expansion, or taxing activities with a large climate change impact.

Source: Women's Environmental Network (2010:5).

As illustrated in the Women's Environmental Network (2010) reports as well as many academic papers (Bian, Leslie, & Cimpian, 2017; Raymond, 2013; Shen, 2013b, Handley, Brown, Moss-Racusin, & Smith, 2015, Blickenstaff, 2005; Budig & England, 2001; Clancy, Nelson, Rutherford, & Hinde, 2014; Johnson, Widneall, & Benya, 2018; Nelson, Rutherford, Hinde, & Clancy, 2017; Shen, 2013a, Sardelis, Oester, & Liboiron, 2017; Settles, Cortina, Stewart, & Malley, 2007 etc.) as well the major UN Development agencies (UNDP, UNEP, UNESCO, UNFCCC, WMO, WHO, UNHCR, IPCC etc.) as well as international development agencies (World Bank and Regional Development Banks etc.), international nongovernmental organizations (IUCN, WWF, WRI, ICIMOD) and other INGOs and NGOs have hundreds of publications, which talk about the necessity of equal participation male and female. However, practically efforts to strengthening the capacity of women are remain questionable.

An example from ECOSOC – "The process of assessing the implications for women and men of any planned action, including legislation, policies or programs, in all areas and at all levels. It is a strategy for making women's as well as men's concerns and experiences an integral dimension of the design, implementation, monitoring and evaluation of policies and programs in all political, economic and societal spheres so that women and men benefit equally, and inequality is not perpetrated. The ultimate goal is to achieve gender equality" (The UN's Economic and Social Council -ECOSOC- 1997; Women 2000; as cited in Aguilar, Granat, and Owren 2015:27).

> *"No consideration of indigenous knowledge who have been adjusting with nature without hampering; Ignoring the contrasting views of proponent and prominent scholars of who analyze climate change with social sciences perspectives (particularly scholars of social sciences- sociology, anthropology, culture, philosophy, media studies, development economic etc.)"* (UNEP 2014).

In the policy documents of most of the agencies listed above (including in the abbreviation) have produced many directives to increase women in the mainstream development arena, however, even women lead organizations have not incorporated own directives. The simple reason is the organizational bureaucratic frame holds the traditional male centric mechanism, where women leadership begin to enjoy, so empowering another women concept become invisible. Even, it has been heard and noticed (during the data collection phase), that, suppression to the subordinate take place, so empowering women became out of the box agenda (base on discussion with the research participant of INGO- name of organization could not be exposed due to confidentiality).

It is noteworthy to illustrate that; **the IPCC seems serious to increase women participants in its assessment process;** however, it is not being able to do so.

For example, "Enhancing Gender Balance in the Intergovernmental Panel on Climate Change Event organized by the Government of Canada on 5 September 2017 7:00 PM – 9:00 PM at the International Civil Aviation Organization (ICAO) in Montreal, Canada. Information note for delegates of IPCC-46- https://www.ipcc.ch/site/assets/uploads/2018/08/Enhancing-Gender-Balance-in-the-IPCC-Montreal.pdf

According to the press release "The event focused on best practices, achievements and recent work on the inclusion of gender balance principles under the United Nations structure to explore ways in which to strengthen the contribution of women's involvement in climate change science and in the IPCC. An invitation to participate in this discussion was extended to all registered delegates of IPCC" (IPCC 2017:46).

According to Elena Manaenkova: "Women further hold key knowledge in natural resources management and are key actors in climate adaptation and mitigation. Women are a minority in science, technology, engineering and mathematics (STEM). They constitute only 30% of the world's researchers in these fields. Women are also under-represented in climate negotiations. At COP21 women accounted for 32% of party delegations. There is a lack of mentoring and support networks. There is a lack of outreach at schools and universities and not many policies that addresses work-life balance" (IPCC 2017:2).

Research suggests that women in science continue to face discrimination, lower pay, fewer citations, and less access to funding (Ceci & Williams, 2011; Shen, 2013a). The barriers to women include the graduate-level environment; the maternal wall/glass ceiling; performance evaluation criteria that do not account for maternal leave; less access to research funding, inadequate recognition; inadequate support for leadership bids; and unconscious gender bias (UNESCO, 2015). The scientific journal Nature notes that only 22% of their reviewers and 20% of their first author invited commenters were women (2015-2016) ("Slow progress," 2017). Women are also underrepresented across fields as authors of single-authored papers or as first author (West, Jacquet, King, Correll, & Bergstrom, 2013). IPCC-XLIX/Doc. 10, p.5 Another large body of work explores the factors that contribute to the imbalance of women in research. Both men and women are documented to have gender biases against women in the sciences, which begin at a very early age (Bian, Leslie, & Cimpian, 2017; Raymond, 2013; Shen, 2013b), and men within STEM have been demonstrated to be more sceptical of evidence of gender bias in science (Handley, Brown, Moss-Racusin, & Smith, 2015). Other hypothesized factors include the lack of female role models at higher career levels, childbearing and rearing responsibilities falling more heavily on women, wage discrepancies, and a "chilly climate" for women in male dominated fields that can include sexual harassment (Blickenstaff, 2005; Budig & England, 2001; Clancy, Nelson, Rutherford, & Hinde, 2014; Johnson, Widneall, & Benya, 2018; Nelson, Rutherford, Hinde, & Clancy, 2017; Shen, 2013a). Even when gender balance improves, with more women represented in science and leadership, women may not have equal opportunity to speak or influence decision making (Sardelis, Oester, & Liboiron, 2017; Settles, Cortina, Stewart, & Malley, 2007). (IPCC 2019:5).

Source: IPCC 2019:5

Similar event was also organized in 2019, Forty-Ninth Session of the IPCC, Kyoto, Japan, 8 - 12 May 2019, IPCC-XLIX/Doc. 10 16.IV.2019),

Table 9.2 Authors- distribution by Gender Working Group- one to six and Women %

Report No		WG1	WG2	WG3	Total	Women %
1	Male	34	21	21	76	
	Female	0	2	6	8	8
	Unknown	0	9	9	18	18
2	Male	70	202	51	323	
	Female	3	28	7	38	10
	Unknown	3	12	3	18	5
3	Male	112	164	119	395	
	Female	14	28	14	56	12
	Unknown	4	9	3	16	3
4	Male	137	180	163	480	
	Female	28	38	26	92	16
	Unknown	0	0	0	0	
5	Male	208	213	220	641	
	Female	47	79	52	178	22
	Unknown	0	0	0	0	
6	Male	169	156	157	482	
	Female	63	106	71	240	33
	Unknown	0	0	0	0	

Source: IPCC 2019, IPCC-XLIX/Doc. 10:13

Which was completely focused on how woman participants can be increased[8].

However, throughout in the IPCC history, and till yet, there have not been proportional distributions in IPCC bureau. For example, - during first assessment report, there were no women representation in all working groups (WG), in second -WG1 has 0%, WG2 10%, WG3 25% respectively. Similarly, in third assessment –has no women representation; in forth, there was WG1 29%, WG2 13%, WG3 0% representation of women. In fifth assessment- WG1 13%, WG2 13%, WG3 22% and in sixth assessment – women representation in bureau was relatively low, whereas, WG1 33%, WG2 20%, WG3 11% respectively (IPCC 2019, IPCC-XLIX/Doc. 10:13[9]).

Similar, can be found on the Authors- distribution by Gender Working Group (see table below)

As illustrated in the table, there is no women authors in Working Group (WG) one in first assessment, likewise in WG2, about 10%, WG3 28% women participants, total about 8% only. In second assessment total women participation was only 10%; in third 12%, forth 16%, fifth 22% and in

[8]https://www.ipcc.ch/site/assets/uploads/2019/01/170420190604-Doc.10Gender.pdf

[9]ibid

the sixth assessment report woman participation was 33% respectively. The scenario shows some increment of women participation; however, developing country's women participation is still insignificant (IPCC 2019:14).

The women who were the participants on the IPCC processes have not a good experience working with the male colleagues (text box below is taken directly from the IPCC 2019, IPCC-XLIX/Doc. 10:13)

"As a CLA and late-career scientist (and white, male), I am in a privileged position so tend to be respected and listened to etc. by default."
"I am probably something like a "dinosaur" in IPCC terms. I notice that it is easy for me personally to say what I want to say, and I do find that I am treated with respect. If I was younger and there for the first time, then things might be radically different."
"I was the only woman in my chapter team. I felt I was invisible and not listened to. Classic situation of me saying something and then being ignored, followed by a man saying the same thing and being listened to. Moreover, my experience of past IPCC involvement was ignored."
"My chapter Chair was particularly aggressive with me, contradicting most of what I said, to the point that I just ended up saying less in the hope that at least what I thought was important would not be cut out. I never really had the feeling that I was knowledgeable during IPCC meetings. Rather I had the feeling I was rather unknowledgeable. The IPCC very much fed my impostor syndrome. My feeling is that climate science is very macho in general, and the IPCC is a regrouping of climate scientists."
"The most active CLA of my chapter... was a woman, and deliberately and effectively inclusive. It made a huge difference."
"The sensitivity of the research teams is outstanding. I feel the IPCC is a best example of considering diversity in participant's experiences."
IPCC-XLIX/Doc. 10, p.19

These texts in the box are self-explanatory, but very powerful- *"I was the only woman in my chapter team. I felt I was invisible and not listened to" "I am probably something like a "dinosaur" in IPCC terms. I notice that it is easy for me personally to say what I want to say, and I do find that I am treated with respect" "My chapter Chair was particularly aggressive with me, contradicting most of what I said, to the point that I just ended up saying less in the hope that at least what I thought was important would not be cut out. I never really had the feeling that I was knowledgeable during IPCC meetings"*. What does these women participants view? Why such a bios treatment? In these two major events focus to increase women participants somehow give a clear picture that, IPCC is not yet being able to make equal treatment women and man, or man are dominating women counterparts.

And here once again, I repeat it has not been able to avoid discriminative structure within its system. In the IPCC structure, selection of participants is guided by the western hegemonic male centric system "a chain of western mostly white color male club" knowingly or unknowingly, where, overall woman participation is still nominal and woman participation from developing world is almost none (see table above titled- Participant in the IPCC process by gender and developing countries). To be of trustworthy knowledge producing international governmental organization the IPCC needs to evaluate it male centric approach and include more women participants particularly from developing and developed world.

References

Adamson, Walter. (1985), Hegemony and Revolution: A Study of Antonio Gramsci's Political and Cultural Theory. Berkley, CA: University of California Press.

Aguilar, L., Granat, M., and Owren, C. (2015), Roots for the future: The landscape and way forward on gender and climate change. Washington, DC: IUCN & GGCA.

Bates, Thomas (1975), Gramsci and the Theory of Hegemony, Journal of the History of Ideas, 36 (2): 351-366

Baum J A and Rowley T J. (2005), Companion to Organizations: An Introduction. In: Baum J A. Editor. The Blackwell Companion to Organizations. Wiley-Blackwell, UK.

Bennet D., and Bennet A. (2004), The Rise of the Knowledge Organization. In: Holsapple C.W. (eds) Handbook on Knowledge Management 1. International Handbooks on Information Systems, vol 1. Springer, Berlin, Heidelberg

Bhandari, Medani P. (2020), The Phobia Corona (COVID 19) - What Can We Do, Scientific Journal of Bielsko-Biala School of Finance and Law, ASEJ 2020, 24 (1): 1-3, GICID: 01.3001.0014.0769, https://asej.eu/resources/html/article/details?id=202946

Bhandari, Medani P. (2019), Sustainable Development: Is This Paradigm the Remedy of All Challenges? Does Its Goals Capture the Essence of Real Development and Sustainability? With Reference to Discourses,

Creativeness, Boundaries and Institutional Architecture. Socio Economic Challenges, 3(4), 97-128. https://doi.org/10.21272/sec.3(4).97-128.2019. https://essuir.sumdu.edu.ua/bitstream/123456789/76484/1/Bhandari_Sustainable_Development.pdf

Bian, L., S.J. Leslie, and A. Cimpian (2017), 'Gender Stereotypes about Intellectual Ability Emerge Early and Influence Children's Interests', Science 355(6323): 389–91.

Brooks, N. and W.N. Adger, (2005), Assessing and enhancing adaptive capacity, Adaptation Policy Frameworks for Climate Change: Developing Strategies, Policies and Measures, B. Lim and E. Spanger-Siegfried, Eds., Cambridge University Press, Cambridge, 165-182

Budig, M.J. and P. England (2001), 'The Wage Penalty for Motherhood', American Sociological Review 66: 204–25.

Button, S. B. (2001), Organizational efforts to affirm sexual diversity: a cross-level examination. The Journal of Applied Psychology, 86(1), 17–28.

Campbell, L.G., S. Mehtani, M.E. Dozier, and J. Rinehart (2013), 'Gender-Heterogeneous Working Groups Produce Higher Quality Science', PLoS ONE 8(10): 1–6.

Ceci, S. J., and Williams, W. M. (2011), 'Understanding current causes of women's underrepresentation in science', Proceedings of the National Academy of Sciences, 108(8), 3157–3162. https://doi.org/10.1073/pnas.1014871108- https://www.pnas.org/content/108/8/3157

Clark, M.J., (2002), Dealing with uncertainty: adaptive approaches to sustainable river management. Aquat. Conserv., 12, 347-363

Clancy, K.B.H., K.M.N. Lee, E.M. Rodgers, and C. Richey (2017), 'Double Jeopardy in Astronomy and Planetary Science: Women of Color Face Greater Risks of Gendered and Racial Harassment', Journal of Geophysical Research: Planets 122(7): 1610–23.

Clancy, K.B.H., R.G. Nelson, J.N. Rutherford, and K. Hinde (2014), 'Survey of Academic Field Experiences (SAFE): Trainees Report Harassment and Assault', PLoS ONE 9(7): 1–9.

Cox, Robert W. (1981), "Social Forces, States and World Orders: Beyond International Relations Theory." Millennium: Journal of International Studies. 10 (2): 126–55.

Cox, Robert W. (1983), "Gramsci, Hegemony and International Relations: An Essay in Method." Millennium: Journal of International Studies. 12 (2): 162–75.

Dunlap R, and Brulle R (eds) (2015), Climate change and society: sociological perspectives. Oxford University Press, New York 460

Dunlap, Riley E. and Aaron M. McCright, (2011), "Organized Climate Change Denial," in The Oxford Handbook of Climate Change and Society, edited by John S. Dryzek, Richard B. Norgaard, and David Schlosberg, Oxford University Press, 144-160,

Dunlap, Riley E. and Aaron M. McCright, (2010), "Climate Change Denial: Sources, Actors and Strategies," in Routledge Handbook of Climate Change and Society, edited by Constance Lever-Tracy, Routledge, 240-259.

Dunlap, Riley E. and Aaron M. McCright (2013), The Climate Change Denial Campaign, Scholars Network, https://scholars.org/sites/scholars/files/ssn_key_findings_dunlap_and_mccright_on_climate_change_denial.pdf

Gay-Antakia, Miriam and Diana Livermana (2018), Climate for women in climate science: Women scientists and the Intergovernmental Panel on Climate Change, PNAS, vol 115, no. 9: 2060–2065

Gramsci, Antonio. (1971), Selections from the Prison Notebooks. Translated and edited by Quintin Hoare and Geoffrey Nowell Smith. New York: International Publishers.

Handley, I.M., E.R. Brown, C.A. Moss-Racusin, and J.L. Smith (2015), 'Quality of Evidence Revealing Subtle Gender Biases in Science Is in the Eye of the Beholder', Proceedings of the National Academy of Sciences 112(43): 13201–6.

Ide, T. J. Schilling, J.S.A. Link, J. Scheffran, G. Ngaruiya and T. Weinzierl. (2014), On exposure, vulnerability and violence: Spatial distribution of risk factors for climate change and violent conflict across Kenya and Uganda. Political Geography 43: 68-81.

IPCC (2019), Forty-Ninth Session of The IPCC, Kyoto, Japan, 8 - 12 May 2019, IPCC-XLIX/Doc. 10 16.IV.2019), Agenda Item: 5, WMO 7bis, Avenue de la Paix C.P. 2300 1211 Geneva 2 Switzerland, https://www.ipcc.ch/site/assets/uploads/2019/01/170420190604-Doc.10Gender.pdf

IPCC, (2018), Summary for Policymakers. In: Global Warming of 1.5°C. An IPCC Special Report on the impacts of global warming of 1.5°C above pre-industrial levels and related global greenhouse gas emission pathways, in the context of strengthening the global response to the threat of climate change, sustainable development, and efforts to eradicate poverty [Masson-Delmotte, V., P. Zhai, H.-O. Pörtner, D. Roberts, J. Skea, P.R. Shukla, A. Pirani, W. Moufouma-Okia, C. Péan, R. Pidcock, S. Connors, J.B.R. Matthews, Y. Chen, X. Zhou,

M.I. Gomis, E. Lonnoy, T. Maycock, M. Tignor, and T. Waterfield (eds.)]. World Meteorological Organization, Geneva, Switzerland, 32 pp. https://www.ipcc.ch/sr15/chapter/spm/

IPCC. (2014), Climate change 2014: Impacts, adaptation and vulnerability. Summary for policymakers. Retrieved: https://ipcc-wg2.gov/AR5/images /uploads/WG2AR5_SPM_FINAL.pdf 44.

Janus, Steffen Soulejman. (2016), Becoming a Knowledge-Sharing Organization: A Handbook for Scaling Up Solutions through Knowledge Capturing and Sharing. Washington, DC: World Bank. doi:10.1596/978-1-4648-0943-9.

Johnson, P., Widneall, S. E., and Benya, F. (2018), Sexual Harassment of Women: Climate, Culture, and Consequences in Academic Sciences, Engineering, and Medicine. Washington, D.C. https://doi.org/10.4135/9781452218595.n210

Karlsson, S., T. Srebotnjak, and P. Gonzales (2007), 'Understanding the North-South Knowledge Divide and Its Implications for Policy: A Quantitative Analysis of the Generation of Scientific Knowledge in the Environmental Sciences', Environmental Science and Policy 10(7–8): 668–84.

Keels, E. (2017), Oil Wealth, Post-conflict Elections, and Postwar Peace Failure. Journal of Conflict Resolution 61 (5): 1021-1045.

Kendie, Daniel (2006), "How Useful is Gramsci's Theory of Hegemony and Domination to the Study of African States?" African Social Science Review: Vol. 3: Iss. 3, Article 5. https://digitalcommons.kennesaw.edu/assr/vol3/iss3/5

Kniveton, D., K. Schmidt-Verkerk, C. Smith and R. Black. (2008), Climate Change and Migration: Improving Methodologies to Estimate Flows. Migration Research Series No. 33. Geneva: International Organization for Migration.

Loder, N. (1999), 'Gender Discrimination "Undermines Science"', Nature 402(6760): 337.

Macgregor, S. (2010), A stranger silence still: the need for feminist social research on climate change. Sociological Review, 57(2), 124–140.

MacIntyre, A. (1985), After Virtue, 2nd edn., London, UK: Duckworth

MacPhee, D. and S.S. Canetto (2015), 'Women in Academic Atmospheric Sciences', Bulletin of the American Meteorological Society 96(1): 59–67.

Maddrell, A., Thomas, N. J., and Wyse, S. (2019), 'Glass ceilings and stone floors: an intersectional approach to challenges UK geographers face across the career lifecycle'. Geografiska Annaler, Series B: Human Geography, 101(1), 7–20. https://doi.org/10.1080/04353684.2018.1555670

Miriam Gay-Antakia and Diana Livermana (2018), Climate for women in climate science: Women scientists and the Intergovernmental Panel on Climate Change, PNAS, 2060–2065 I PNAS I February 27, 2018 I vol. 115 I

Moss-Racusin, C. A., Van Der Toorn, J., Dovidio, J. F., Brescoll, V. L., Graham, M. J., and Handelsman, J. (2014), Scientific Diversity Interventions. Science, 343. https://doi.org/10.1126/science.1245936

Nadler, D. A., and Tushman, M. L. (1999), The organization of the future: Strategic imperatives and core competencies for the 21st century. Organizational Dynamics, 28(1), 45–60. https://doi.org/10.1016/S0090-2616(00)80006-6

Nelson, R.G., J.N. Rutherford, K. Hinde, and K.B.H. Clancy (2017), 'Signaling Safety: Characterizing Fieldwork Experiences and Their Implications for Career Trajectories', American Anthropologist 119(4): 710–22.

Pearse, R. (2017), 'Gender and climate change'. Nature Climate Change, 8(April), 1–16. https://doi.org/10.1002/wcc.451

Peng, Bo (2018), China, Global Governance, and Hegemony: Neo-Gramscian Perspective in the World Order JCIR: VOL. 6, No. 1

Raymond, J. (2013), 'Most of Us Are Biased', Nature 595: 33–4.

Riley, Dylan J., (2011), Hegemony, Democracy, and Passive Revolution in Gramsci's Prison Notebooks, California Italian Studies, 2(2) https://sociology.berkeley.edu/sites/default/files/faculty/Riley/hegemonydemocracy.pdf

Riquer, F. (1993), Población y género. México: Consejo Nacional de Población (CONAPO).

Robinson, Roland. (1972), Non-European Foundations of European Imperialism: Sketch for a Theory of Collaboration:' In Roger Owen and Bob Sutcliffe, (eds.), Studies in the Theory of Imperialism (140–141). London, England: Longman.

Salamini, Leonardo, (1984), Gramsci and the Marxist Sociology of Knowledge, Sociological Quarterly, 1984,15(3): 259–380.

Sardelis, S., S. Oester, and M. Liboiron (2017), 'Ten Strategies to Reduce Gender Inequality at Scientific Conferences', Frontiers in Marine Science 4(July): 1–6.

Scott, James (1997). Hegemony and the Peasantry, Politics and Society, 7(3): 267–296.

Settles, I.H., L.M. Cortina, A.J. Stewart, and J. Malley (2007), 'Voice Matters: Buffering the Impact of a Negative Climate for Women in Science', Psychology of Women Quarterly 31(3): 270–81.

Shen, H. (2013), 'Mind the Gender Gap', Nature 495: 22–4.

Shen, H. (2013), Women's Work. Nature, 495, 21.

Smooth, W. G. (2016), Intersectionality and women's advancement in the discipline and across the academy. Politics, Groups, and Identities, 4(3), 513–528. https://doi.org/10.1080/21565503.2016.1170706

Smooth, W. G. (2016), Intersectionality and women's advancement in the discipline and across the academy. Politics, Groups, and Identities, 4(3), 513–528. https://doi.org/10.1080/21565503.2016.1170706

Solomon, Susan and Martin Manning (2008), The IPCC Must Maintain Its Rigor, Science, Vol. 319, Issue 5869, pp. 1457, DOI: 10.1126/science.1155724 https://science.sciencemag.org/content/319/5869/1457.full

Terry, G. (2009), 'No Climate Justice without Gender Justice: An Overview of the Issues', Gender & Development 17(1): 5–18.

The World Bank Group. (2011), Gender and climate change: Three things you should know. Washington, DC: The World Bank Group.

The World Bank. (2011), World development report, Gender equality and development. Washington, DC: The World Bank.

Tsoukas, H., and Shepherd, J. (2004). Introduction: Organizations and the Future, From Forecasting to Foresight, Wiley-Blackwell

UNEP. (2014), UNEP Gender equality Policy and Strategy, UNEP, Nairobi.

UNESCO. (2015), UNESCO Science Report Towards 2030. Paris.

UNESCO. (2018), Women in Science. Paris.

United Nations Development Program (2007), Human Development Report 2007: Climate Change and Human Development–Rising to the Challenge 5th Edition, Palgrave Macmillan http://hdr.undp.org/sites/default/files/hdr_20072008_summary_english.pdf

UN-Women, and Mary Robinson. (2013), The Full View: Advancing the goal of gender balance in multilateral and intergovernmental processes,Climate JusticeFoundation

West, J.D., J. Jacquet, M.M. King, S.J. Correll, and C.T. Bergstrom (2013), 'The Role of Gender in Scholarly Authorship', PLoS ONE 8(7).

Williams, Mariama (2015), Gender and Climate Change Financing: Coming out of the margin, Routledge IAFFE Advances in Feminist Economics, Series, Routledge, London

Women 2000. (1997), Gender mainstreaming. Retrieved from http://www.un.org/womenwatch/daw/ csw/GMS.PDF

Yamineva, Y. (2017), 'Lessons from the Intergovernmental Panel on Climate Change on Inclusiveness across Geographies and Stakeholders', Environmental Science and Policy 77: 244–51.

Yohe, G.W., R.D. Lasco, Q.K. Ahmad, N.W. Arnell, S.J. Cohen, C. Hope, A.C. Janetos and R.T. Perez, (2007), Perspectives on climate change and sustainability. Climate Change 2007: Impacts, Adaptation and Vulnerability. Contribution of Working Group II to the Fourth

Assessment Report of the Intergovernmental Panel on Climate Change, M.L. Parry, O.F. Canziani, J.P. Palutikof, P.J. van der Linden and C.E. Hanson, Eds., Cambridge University Press, Cambridge, UK, 811-841. https://www.ipcc.ch/site/assets/uploads/2018/02/ar4-wg2-chapter20-1.pdf

10

Case Study: Bashudaiva Kutumbakka – The Entire World is Our Home and all Living Beings are Our Relatives; Why We Need to Worry About Climate Change? With Reference to Pollution Problems in the Major Cities of India, Nepal, Bangladesh and Pakistan*

In this chapter, I have directly or indirectly revealed the interconnected impact of geographical, and socio-cultural environment on personal motivation building. As such I have shared the story of why and how I became interested in the conservation of nature and natural resources, what was the problems and how I overcome and continuously working on the same track with same focus in his entire life; however, it might be the story of each environmentalist who have tried to continue environment conservation action

*Part of this chapter was first published as an interview at the Akamai University, website; and Advance Journal of Agriculture and Environ Science. (2019);2(1): 8–35. DOI: 10.30881/aaeoa.00019- http://ologyjournals.com/aaeoa/aaeoa_00019.pdf and Pellam Journal Of Science-Institute for Positive Global Solutions – Bibliotheque: World Wide Society,MMXIX, NO. 2, ISSN#: 1544-5399. http://bwwsociety.org/journal/science/archive/2019/climate-change-interview.htm Author has modified to fit with this chapter. Author would also like to declare that the part of this chapter has been appeared more than fifteen different international journals, News Paper (The Asia Environmental Daily, https://asiaenvdaily.com/index.php/editorials/71-editorials/26837-bashudaivakutumbakkam-the-entire-world-is-our-home-and-all-living-beings-are-our-relatives-why-we-need-to-worry-about-climate-change-with-reference-to-pollution-problems-in-the-major-cities-of-india-nepal-bangladesh-and-pakistan), Social Medias like LinkedIn, Facebook, Tweeter etc.. As well as in the University Websites of Akamai University and Sumy State University. The texts of this chapter are also illustrated in opinion papers of several scholars as an example of cultures contributions on environment leadership building.

and activism and academic scholarships together. I have spent my entire life as a conservationist, evaluator of climate change impact, social empowerment and educationalist, and I have devoted entire life for the conservation of nature and social services. This true story tells how personal background makes people's perceptions on nature and society and what role a spiritual / tradition, Indigenous knowledge can motivate himself or herself to devote on conservation of nature and social empowerment and make aware of impact of climate change, how organizations like IPCC can bring experts in the mainstream climate science. I have shared why I think **"Vashudhiva Kuttumbakam"** the entire planet is our home and all living beings are our relatives. I have shared the essence of education; I use the term EDUCARE-education for life. I have tried to explain the severity of current environmental impact on human and other living beings and explicate how and why we need to worry. I have drawn evidences of negative impact of climate change on cities and how dangerous is the pollution condition in major cities of India, Pakistan, Bangladesh and Nepal. *"My intention, of life is to pay back; give or contribute to the society in fullest whatever I have, earned, or experienced."* Hopefully readers will enjoy reading the entire book- "Getting the Facts Right: The IPCC and the Role of Science in Managing Climate Change- With reference of personal reflection on pollution in major cities of Asia". I am hopeful that readers (all concerned stakeholders who care the health of planet) will be benefited from this true an intrinsic motivational story with the evidences of scientifically grounded facts.

Q: Speak about your career in environmental protection and climate change. These are obviously serious and pressing concerns for people throughout Asia, as well as around the world.

A: Acknowledgement ["It is my pleasure to share knowledge and expertise on the issues of environmental protection and climate change. My family, all relatives, my network friends and colleagues, communities, and various societies (wherever I have been), including the nature and culture, traditions combinedly nurtured, taught me, without any expectations. *My intention, of life is to pay back; give or contribute to the society in fullest whatever I have, earned, or experienced. And I have tried to answer, why I am motivated to devote myself within the domain of environmental protection and climate change.* I would be more than happy, if readers find this information useful. I am open to engage in any kind of collaborative research, or any other tasks which can contribute to overcome or minimize the devastating impact of climate change. I would like to clearly state that, most of the information,

I have stated in this interview/discussion are based on web-search, secondary sources, as well as based on my published papers or books etc. I declare that, *most of the pictures, graphs, tables are taken from the websites. I have tried to provide proper sources, citations, and links of the original sources. However, if I missed to note any source, I request to forgive me and apologize in advance to the all concern authors, journalists, government agencies and any other stakeholders whom I have cited in this note"*[1].

Thank you very much for encouraging me to discuss and share my experience in this very critical issue. Environment protection and climate change is a global concern and global challenge. I think the COVID 19 kind of epidemic has connection with the changing global environment (still we need extensive research to validate this proposition). It is established notion that major environment concern/problems are Ozone layer depletion; Global warming; and Loss of biodiversity. As such each of these problems has negative impact on life on earth. *"IPCC-Fifth Assessment Report provides a comprehensive assessment of sea level rise, and its causes, over the past few decades. It also estimates cumulative CO_2 emissions since pre-industrial times and provides a CO_2 budget for future emissions to limit warming to less than 2°C. About half of this maximum amount was already emitted by 2011....There is alarming evidence that important tipping points, leading to irreversible changes in major ecosystems and the planetary climate system, may already have been reached or passed. Ecosystems as diverse as the Amazon rainforest and the Arctic tundra, may be approaching thresholds of dramatic change through warming and drying. Mountain glaciers are in alarming retreat and the downstream effects of reduced water supply in the driest months will have repercussions that transcend generations"* (United Nations 2020[2]). For example, Ozone layer depletion has Effects on Human and Animal Health (i.e. eye diseases, skin cancer, infectious diseases); Effects on Human and Animal Health (increases radiation could change species composition, or change in plant forms); Effects on Aquatic Ecosystems (affect the distribution of phytoplankton's, reproduction system alteration); Effects on Bio-geo-chemical Cycles (affect terrestrial and aquatic bio-geo-chemical cycles); Effects on Air Quality (can increase both production and destruction of ozone and related oxidants such as hydrogen peroxide-which

---

[1]Bhandari Medani P. (2019). "Bashudaiva Kutumbakkam"- Adv Agr Environ Sci. (2019);2(1): 8–35. DOI: 10.30881/aaeoa.00019 (second part) http://ologyjournals.com/aaeoa/aaeoa_00019.pdf

[2]United Nations (2020)- https://www.un.org/en/sections/issues-depth/climate-change/ (retrieved 3/20/20)

can have direct adverse effects). Similarly, another problem is global warming which more severe. Evidence shows that Global temperature increase 0.3–6 within last 100 years, and major contributor for this change are we the human. Globe is warming, and main cause is greenhouse gases (GHG) causing global warming is carbon dioxide. The evidence of global warming is Rise in global temperature and Rise in sea level which ultimately affecting ecosystems of the planet. *Documented increases in global air and sea temperatures over the last century have demonstrated unequivocally that our planet is warming. Most climatologists agree that the warming trend will continue, and at an accelerating pace unless the causes of global warming are addressed immediately. This reality, and the urgent need for action, is finally being recognized by society and governments around the world (Kleinschmit 2009:1).* There is inevitable relationship between agriculture and environment [the surroundings or conditions in which a person, animal, or plant lives or operates (web[3]), it also covers, climatic condition, its pattern and change). Human existent and civilization began with the development of agriculture which fully relies on climatic variations of specific location of Earth Surface. All of us, have been witnessing the variation and change on climate one way or another. With no doubt, We, the humans are directly or indirectly responsible and have been contributing for this climate change, wherever we live or do; however, the degree of contribution and impact may vary. As a matter of fact, many initiatives and steps have been taken at the international to individual levels [various international treaties, policies, actions, etc.). However, still there are unmeasured and unexplored issues on "how climate change has already impacted and will affect the lives support ecosystems". On the other hand, there are also people, who actually do not support the statement that, *"Human influence on the climate system is clear, and recent anthropogenic emissions of greenhouse gases are the highest in history. Recent climate changes have had widespread impacts on human and natural systems. Warming of the climate system is unequivocal, and since the 1950s, many of the observed changes are unprecedented over decades to millennia. The atmosphere and ocean have warmed, the amounts of snow and ice have diminished, and sea level has risen" (IPCC 2014:2). "Anthropogenic greenhouse gas emissions have increased since the pre-industrial era, driven largely by economic and population growth, and are now higher than ever. This has led to atmospheric concentrations of carbon dioxide, methane and nitrous oxide that are unprecedented at least the last 800,000 years. Their effects, together*

[3]http://www.businessdictionary.com/definition/environment.html (access on 2/3/2018)

with those of other anthropogenic drivers, have been detected throughout the climate system and are extremely likely to have been the dominant cause of the observed warming since the mid-20th century" (IPCC 2014:4[4]).

In my opinion, the people or group, only externally, oppose the statement "the climate change is occurring due to anthropogenic disturbance in natural environment" in the inner heart they know, the fact that, globe is warming and we human are responsible. I think, every one of us have witnessed or experienced the change in global environment or heard about the recent past and current situation of our environment.

Q: *We would like to start off with asking you about how you became interested in protecting the environment in the first place? What led you to making this the focus of your academic studies and your eventual long-term career?*

A: How, I became, interested protection of environment?? This is very important question. To respond this, I need to go back to my childhood.

I born in the most beautiful place of the Planet, at the bank of the **'Hyatung Jharana'** waterfall *(One of the highest waters falls in South Asia with 365-meter height situated in the Eastern, Nepal. It is located at the confluence of VDC Ishibu and Samdu, Terathum District*), a very remote village. I grew up listening of the music of water falls and rivers. At the age of one and half, I was taken to my Grandparent (mother side) to Hwaku, where I had direct view of Tamor River, one of the tributaries of the Saptakoshi River. My Grandparent's house is located at the top of the Hill, from where one can see the several tops of Mountains, two Evergreen Rivers Phoguwa khola and **Iwa khola and Tamor River where** Phoguwa and **Iwa merge**.

I raised in the spiritual environment. My Grandparents taught me to consider all living creatures (plant and animal) and non-living objects (like stone, soil, air, and cloud) as deities, creation of God. I learn to appreciate the nature, before I learn to speak and walk. I grew in the primitive kind of environment if we compare the current developed world context. I think base of my orientation towards conservation of nature and nature, is the culture, which taught me to appreciate nature. According to Eastern (particularly Hindu) mythology, each living and non-living structure are interconnected

---

[4]IPCC (2014) Climate Change 2014: Synthesis Report. Contribution of Working Groups I, II and III to the Fifth Assessment Report of the Intergovernmental Panel on Climate Change [Core Writing Team, R.K. Pachauri and L.A. Meyer (eds.)]. IPCC, Geneva, Switzerland, 151 pp. https://www.ipcc.ch/pdf/assessment-report/ar5/syr/SYR_AR5_FINAL_full_w cover.pdf

and supportive to each other. For example, there are millions of gods and goddesses and each of them have association with animal and plant.

Name of God/Goddess	Symbol	Vehicle/plant associated
Goddess Bagabati	goddess of power	Lion/tiger Red flowering tree
Lord Shiva	god of law/Protection	Bull/Poisonous plant
Vishnu	god of Creation	Stroke (Garuda)/*Ficus religiosa*
Laxmi	Goddess of wealth	pair of Elephant/Lotus
Saraswati	Wisdom and knowledge	Swan and peacock/Lotus
Ganesh	the combination of power Wealth and wisdom	Rat/*Sandilon dectilon*
Yama	God of death	Wild water buffalo/Bamboo

Similar principle may apply with other religious mythologies (Baha'i, Buddhism, Christianity, Daoism, Hinduism, Jainism, Judaism, Islam, Shinto, Sikhism and Zoroastrianism etc.) in different form.

The first thing I learn from the childhood was the appreciation to "whatever we see, and feel, including, air, the surrounding, trees, birds, animal and structure of rock, terrain, land escape", because they were all included in the daily morning prayers. Some of them include:

Om Bhadram Karnebhih Shrinuyama Deva Bhadram Pashyemakshabhiryajatrah?

Sthirairangaistushtuva Sastanubhirvyashemahi Deva-Hitam Yadayuh?

Meaning: Gods, may we with our ears listen to what is good, and with our eyes see what is good, ye Holy Ones. With firm limbs and bodies, may we extolling you attain the term of life appointed by the Gods.

Om Aa No Bhadrah Kratavo Yantu Vishvatoadabdhaso Aparitasa Udbhidah?

Deva No Yatha Sadamidvridhe Asannaprayuvo Rakshitaro Dive-Dive?

Meaning: May powers auspicious come to us from every side, never deceived, unhindered, and victorious. That the Gods ever may be with us for our gain, our guardians' day by day unceasing in their care.

[Om Dyauh Shantirantariksha Shantih Prithivi Shantirapah Shantiroshadhayah Shantih?

Vanaspatayah Shantirvishvedevah Shantirbrahma Shantih Sarva Shantih Shantireva Shantih Sa Ma Shantiredhi?]

Meaning: "May peace radiate there in the whole sky as well as in the vast ethereal space everywhere. May peace reign all over this earth, in water and in all herbs, trees and creepers. May peace flow over the whole universe. May peace be in the supreme being Brahman. And may there always exist in all peace and peace alone." https://www.drikpanchang.com/puja-vidhi/shanti-patha/shanti-patha.html

Most probably, repetedly reciting such prayers, might have given some inner intrinsic motivation to me to devote for the conservation of nature and natural resources.

I, think, in the scholarly term, the home, tradition, culture and belief help to build my mindset to appreciate, protect nature and also motivate others to do the same. While, I was still in forth grade my uncle (maternal), Ram Chandra Gautam, who was a school teacher, begin to teach me about the world culture and religions and how they all have motive of public welbeing. The first profounding concept I got enternalize was "*Vasudhaiva Kutumbakam" (the entire world or earth is your familty).*

I think the lession and essence of "Vasudhaiva Kutumbakam" begin to build within my inner self, on which I strongly follow and believe. This philosophy is written in Hitopadesha, 1.3.71: *'ayam nijah paroveti ganana laghuchetasam; udaracharitanam tu vasudhaiva kutumbhakam'* [**Meaning:** The distinction "This person is mine, and this one is not" is made only by the narrow-minded (i.e. the ignorant who are in duality). For those of noble conduct (i.e. who know the Supreme Truth) the entire world is one family (one Unit).]. Be detached, be magnanimous, lift up your mind, enjoy the fruit of .... Freedom (Maha Upanishad 6.71–75).

When I was at high school, I studied world religions and culture. I was so amazed when, I found that, the purpose of all traditions (including indigenous), cultures, religions were the same "this earth belongs to all of us and human task is to protect the mother earth and fellow living and non-living beings, who help before our birth to death and even after death".

Here I would like to list some of important quotations from the great epics (all are cited from the Faith in conservation : new approaches to religions and the environment by Martin Palmer with Victoria Finlay published by the International Bank for Reconstruction and Development/The World Bank, http://documents.worldbank.org/curated/en/570441468763468377/pdf/269751Faith0in0Conservation010paper.pdf) on environment conservation issues:

Christianity: "*The Bible: No one lighting a lamp puts it under a basket, but on a lampstand and it gives light to all in the house. In the same way, let your light shine before others, so that they may see your good works and give glory to your Father in heaven*". -The Bible (Matthew 5: 15–16) (page 37). "*Nature and Nature's laws lay hid in night: God said, "Let Newton be!" and all was light*". (Alexander Pope around 1718) (page 18).

Islam: "*O children of Adam!. .. eat and drink: but waste not by excess for Allah loveth not the wasters.* -Holy Qur'an, Surah 7:31" (page 3).

Jain: "*Ahimsa-this is a fundamental vow and runs through the Jain tradition like a golden thread. It involves avoidance of violence in any form, through word or deed, not only to human beings but to all nature. It means reverence for life in every form including plants and animals. Jains practice the principle of compassion for all living beings at every step in daily life. Jains are vegetarians*". -The Jain Statement on Ecology (page 15).

http://documents.worldbank.org/curated/en/570441468763468377/pdf/269751Faith0in0Conservation010paper.pdf

Daoism: "*Daoism has a unique sense of value in that it judges affluence by the number of different species. If all things in the universe grow well, then a society is a community of affluence. If not, this kingdom is on the decline" -The Daoist statement on ecology (chapter 10) (page 23).*

Chinese medicine has traditional root and philosophy: "Its worldview is based on belief in the Dao-the nature of the universe-which is best described in a famous series of verses in the Dao De Jing, written in the fourth century B.C.:

The Dao gives birth to the One: The One gives birth to the Two: The Two gives birth to the Three: The Three gives birth to every living thing. All things are held in yin and carry yang:

And they are held together in the qi of teeming energy.

The One is the universe, which gives birth to the two primal forces of yin and yang, which are the natural forces of opposites. Yin, for example, is cold, wet, winter, female, and earth, while yang contrasts to this by being hot, dry, summer, male, and heaven". (page 26).

Baha'i: "*Baha'i scriptures teach that, as trustees of the* ***planet's vast resources and biological diversity, humanity must seek to protect the "heritage [of] future generations***"; *see in nature a reflection of the divine;*

approach the earth, the source of material bounties, with humility; temper its actions with moderation; and be guided by the fundamental spiritual truth of our age, the oneness of humanity. The speed and facility with which we establish a sustainable pattern of life will depend, in the final analysis, on the extent to which we are willing to be transformed, through the love of God and obedience to His Laws, into constructive forces in the process of creating an ever-advancing civilization" (page 76).

Buddhism: Nature as a way of life: "*The Buddha commended frugality as a virtue in its own right. Skillful living avoids waste and we should try to recycle as much as we can. Buddhism advocates a simple, gentle, nonaggressive attitude toward nature-reverence for all forms of nature must be cultivated" (page 81).*

"We need to live as the Buddha taught us to live, in peace and harmony with nature, but this must start with ourselves. If we are going to save this planet, we need to seek a new ecological order, to look at the life we lead and then work together for the benefit of all; unless we work together no solution can be found. By moving away from self-centeredness, sharing wealth more, being more responsible for ourselves, and agreeing to live more simply, we can help decrease much of the suffering in the world" (page 82).

Judaism: *"Love your neighbor as yourself" (Lev. 19:18) (page 113).*

"When God created Adam, he showed him all the trees of the Garden of Eden and said to him: "See my works, how lovely they are, how fine they are. All I have created, I created for you. Take care not to corrupt and destroy my universe, for if you destroy it, no one will come after you to put it right." (Ecclesiastes Rabbah 7) (page 111). In Jewish sources, the rationale for humanity's obligation to protect nature may be found in the biblical expression, "For the earth is Mine" (Lev. 25:23). The Bible informs us that the Earth is not subject to man's absolute ownership but is rather given to us "to use and protect" (Gen. 2:15)" (page 111).

Shintoism: "*Shinto regards that the land, its nature, and all creatures including humans are children of Kami. Accordingly, all things existing on this earth have the possibility of becoming Kami. . . . The ancient Japanese considered that all things of this world have their own spirituality, as they were born from the divine couple. Therefore, the relationship between the natural environment of this world and people is that of blood kin, like the bond between brother and sister*" (page 127).

Sikhism: "*The Sikh scripture, Guru Granth Sahib, declares that the purpose of human beings is to achieve a blissful state and to be in harmony with the earth and all creation" (page 131).*

Men, trees, pilgrimage places, banks of sacred streams, clouds, fields.

Islands, spheres, universes, continents, solar systems. The sources of creation, egg-born, womb-born, earth-born, sweat-born, oceans, mountains and sentient beings. He, the Lord, knows their condition, 0 Nanak. Nanak, having created beings, the lord takes care of them all. The Creator who created the world, He takes thought of it as well". (466) (page 134).

Zoroastrianism: "*Whoever teaches care for all these seven creations, does well and pleases the Bounteous Immortals; then his soul will never arrive at kinship with the Hostile Spirit.*

When he has cared for the creations, the care of these Bounteous Immortals is for him, and he must teach this to all mankind in the material world. –Shayasht ne Shayast (15:6)1

These actions, according to Zoroastrianism, will lead toward "making the world wonderful," when the world will be restored to a perfect state. In this state the material world will never grow old, never die, never decay, will be ever living and ever increasing and master of its wish. The dead will rise, life and immortality will come, and the world will be restored to a perfect state in accordance with the Will of Ahura Mazda" (World Bank 2010[5]). These above listed illustrations clearly indicate that, all world religions appreciate nature and encourage and try to empower human to nourish or contribute to protect the mother earth. The world religions, cultures, traditions and my own surroundings were major motivators for me to devote myself for conservation of nature and natural resources formally or informally. I began to think and act on conservation of nature practically like planting trees, campaigning against hunting birds from the school time.

This intrinsic motivation became more visible and action oriented from 1985. During 1984-1985, I began research on indigenous knowledge and religious believe on plant conservation, particularly the Banyan (*Ficus benghalensis*) and Peepal (*Ficus religiosa L)* trees as a student of sociology & anthropology. Hindu people plant these trees on important moments in

[5]World Bank (2010) (Lord of Wisdom) (page 145) [Once again, above text are copied from the book, "Faith in conservation: new approaches to religions and the environment by Martin Palmer with Victoria Finlay published by the International Bank for Reconstruction and Development/The World Bank, available at: http://documents.worldbank.org/curated/en/570441468763468377/pdf/269751Faith0in0Conservation010paper.pdf]

their lives, like the birth of a child or the death of a relative. I learned a lot about the importance of trees and how local people are adjusting with nature. In the meantime, I was also fortunate to attend lecture of Indian Environmentalist Sunderlal Bahuguna [is a noted Garhwali environmentalist, Chipko movement leader and a follower of Mahatma Gandhi's philosophy of Non-violence and Satyagraha]. His topic of lecture was on how we are related to nature and why we have to protect forest. His lecture took me back to my childhood, on the essence of *Bsudaiva Kutumbakam* (listed above). My father (late) Loknath Bhandari, always tried to boast my instructive motive. My father was great admirer of Sunderlal Bahuguna, who also called as forester, BAN RAJA (Lion of Forest- who try to protect forest from tree loggers) (voluntarily).

In the deep, inner mind, I thought, should do whatever I can. From the childhood, I started to motivate the local people to plant Banyan and Peepal trees as well as other religiously important plant species on their important event of lives. It was not very difficult task to do so, because, they already knew the importance of these trees and broadly forest itself. These species are supposed holly trees in Hindu Mythology, so normally people do not cut these plant species. However, this believe is not very powerful as it was before, but still elder people feel sin if they cut these species.

	Tulasi is cultivated for religious and supposed traditional medicine purposes, and for its essential oil. It is widely used as a herbal tea, commonly used in Ayurveda, and has a place within the Vaishnava tradition of Hinduism, in which devotees perform worship involving holy basil plants or leaves. Ocimum tenuiflorum (synonym Ocimum sanctum), commonly known as holy basil, tulasi (sometimes spelled thulasi) or tulsi, is an aromatic perennial plant in the family Lamiaceae. It is native to the Indian subcontinent and widespread as a cultivated plant throughout the Southeast Asian tropics. Source: https://en.wikiped ia.org/wiki/Ocimum_tenuiflorum
Another plant species considered as secret, comonly seems planted in most of the houses of Hindus	

	Kush (Dharbham) – Eragrostis Cynosuroides It is a tradition of keeping Kush home from centuries. It is scientifically proved that this grass absorbs radiations and bad rays. Even the x-ray can be absorbed by this grass, imagine what else it can do. Source: http://www.weallnepali.com/nepali-festivals/babu-ko-mukh-herne-din Desmostachya bipinnata, commonly known in English by the names Halfa grass, big cordgrass, and salt reed-grass, [5] is an Old World perennial grass, long known and used in human history. In India it is known by many names, including: daabh, darbha, kusha, etc.[6] source: https://en.wikipedia.org/wiki/Desmostachya_bipinnata
	Pterocarpus santalinus (Red Sandal): *Pterocarpus santalinus* is used in traditional herbal medicine as an antipyretic, anti-inflammatory, anthelmintic, tonic, hemorrhage, dysentery, aphrodisiac, anti-hyperglycaemic and diaphoretic. Source: https://en.wikipedia.org/wiki/Pterocarpus_santalinus *Photos by Hari Bhandari (Rangeli -3, Morang, Nepal)*
	Rudraksha is the seed of a fruit from a tree which belongs to the Elaeocarpus family. https://www.gemsratna.com/benefit-of-rudraksha **Rudraksha**, Sanskrit: *rudrâkşa* (Sanskrit: रूद्राक्ष) ("Rudra's [Shiva's] teardrops"), is a seed traditionally used as prayer beads in Hinduism. The seed is produced by several species of large evergreen broad-leaved tree in the genus Elaeocarpus, with Elaeocarpus ganitrus being the principal species. Source: https://en.wikipedia.org/wiki/Rudraksha 

<table>
<tr>
<td></td>
<td>Buddha chitta plant. Ziziphus budhensis, the scientific name for buddha chitta plant is a species of plant in the Rhamnaceae family endemic to the Timal region of Kavreplanchok in Central Nepal.[1] 'Buddha chitta' or 'ZiziphusBudhensis' is clearly different from the species of 'Ziziphus' already known to grow in Nepal, and it did not match with the Ziziphus species reported to be found in India, Pakistan, Bhutan, Bangladesh, and China. Source: https: //en.wikipedia.org/wiki/Buddha_chitta_mala
</td>
</tr>
<tr>
<td>
Photo: Laxmi Gurung, Pokhara, Nepal</td>
<td>
Source: http://hindi.webdunia.com/astrology-articles/importance-of-five-113122600061_1.html (panch pallav-five secret plants (Peepal, Gular (*Ficus racemosa*), *Ashoka, Mango, Banyan)*</td>
</tr>
</table>

Saraca asoca (the Ashoka tree; lit., "sorrow-less") is a plant belonging to the Caesalpinioideae subfamily of the legume family. It is an important tree in the cultural traditions of the Indian subcontinent and adjacent areas. Source: https://en.wikipedia.org/wiki/Saraca_asoca *Source are listed in the picture itself – Here I would like to declare that these pictures are used only to show the condition of pollution – they are not official and no relation with the political activism.*

Banyan and Peepal species are also very good food and habitat source of many bird species. Therefore, these species have a significant importance from the bio-diversity conservational point of view. Banyan and Peepal plant species need feverish temperature for germination, therefore, to establish a nursery is expensive. However, these plants are common and could be found in the old house wall, roof as well as other places where normally birds came and stay or drop their fecal. From the bird's fecal seed spray in different and difficult place, where the seedlings appear. I thought, to establish a nursery by collecting seedlings.

I along with Miss Prajita started to collect those plant species by riding wall or wherever we saw the seedlings. Later on, Mr. Pradeep Banskota also joined this work and we manage to establish a nursery in my house as well as in a small land provided by the Biratnagar Municipality. This plant collection created awareness in the Biratnagar City in Nepal and people started to give new name to them like MAD Conservationist. This aspiration of madness for conservation of nature and natural resources is still my identity (please see Than Medani got mad- by Govinda Luiten- ehimalayantimes- http://ehimalayatimes.com/bichar-sahitya/40627-2020).

Firstly, from January 1, 1985, we started survey the marginal land and started plantation as well with available saplings of any plant species. In 1986, we stared awareness campaign on tree plantation and environment conservation. We also started the survey and monitor the forest degradation as well as recording the wild animal, birds and plant species in Jhapa, Udayapur, Dhankuta and Morang districts through interviews with local residence. We also continued to visit same place again and again to monitor the forestry encroachment by the locals illegally and ligulae. In 1987, 10 people from different village development council also joined our group. In this year, I got married to Prajita. People were increasing in the group, so Prajita Bhandari proposed to establish an official forum to work regularly and systematic way. All group members supported his new idea, said that something should be done with it. Together with my wife Prajita, we publicly announced and the decision to devote our lives for conservation of nature and natural resources. In 1988 we cofounded the APEC group (Association for the Protection of Environment & Culture-http://www.geocities.ws/ngo_apec/apec2.html), in my leadership. Since then, I have been working in this field together with my wife Prajita. Her father late Dwarika Nath Devkota and mother Durga Devi Devkota, were very supportive financial and socially to us. They were very spiritual personalities. They used to always remind us the meaning of "*Basudaiva Kutumbakkam*". When, we finished all my property in conservation work, she gave me a piece of land, with no rights to sell. They wanted us to survive with any severe situation. When people named us as MAD Conservationist, actually, they were only happy person, "they always told us, do not worry, it is good symbol to reach to the destiny. I would like to acknowledge that, in the conservation mission entire family, were with us and are with us even now. In those time many people stand with us like Narayan Paudel, Bhaktalal Upadhyaya, Krishna Busan Bal, Govinda Luitel, Gyanedra Bahdur Thapa, Hari Bhandari, Narayan Poudel-Department of National Park, Biratnagar Municipality, Koshi Tappu Wildlife Reserve, Govinda Pandeya, Hari Gyawali, Gularia Municipality, about 125 Schools of Eastern Nepal, and many more. In reality, without their supports, we would not be able to continue our conservation journey. Wherever, we go, whatever, we do, our major motive of conservation has not changed and will not change.

The environment conservation movement, including many action projects were running with to some extend smoothly under my leadership (1985-2002). However, there was a turning point during 2000-2002. In Nepal, there were unseen problems for social activists, environmentalists, etc. (I prefer not to disclose). I felt like I needed to change my direction from core conservationist to academia. There are many people who helped me to move forward. I feel blessed from all individuals with I come cross during my academic journey. I always feel fresh, exited, thrilled; and feel that, a new journey is just began and going on. The destination is still bit far, however, I feel I am still learning about the world. I want to contribute for the societal harmony, peace and freedom. I see several severe problems i.e. the deteriorating peace in the world due conflicts and violence; environmental degradation- climate change-loss of biodiversity and unsustainable development. Additionally, I

(Medani and Prajita Bhandari, in green dresses, with school's team during conservation campaign)

see the poverty, and the lives in Slums in the developing world. My whole, objective of the life is not to have any prestigious or easy life but devote myself in complete to empower people who could contribute to protect the environment and contribute for the social wellbeing of South and North. I have acquired knowledge, expertise and experience, which I want to share with the young and old, or anyone who thrive for knowledge and want to use knowledge to build new world of social and environmental harmony.

Moreover, the mode if problems in the developing world and developed world seems different; however, they are omnipresent. To resolve them, both north and south needs to join hands.

I would like to assure, that these are the basis, which led me to making this the focus of my academic studies and in my long-term career. Basically, the as I said, I was, and still I am, aware of the importance of environment for human existence. I am convinced that, if I say, but not act, it does not make any sense. As a matter of fact, my motto is to continue on this work until my last breath. My request to readers, is to learn to love themselves first and think, how, we are loved by nature. If, we feel we are loved by nature (if not loved and nurtured- we were not in current situation), we have to love and contribute whatever we can, not for nature shake but shake of ourselves.

As a scholar of social sciences, and an advocate for social justice, I continuously search for the links between my scholarship and the local, national, and global realities we face and with which we live. I try to bring those connections, those links between theory and real-world events with each and every day.

I go so far as to label this approach "EDUCARE"—a way to nurture constructive dialogue that cultivates the creation and recognition of a collective knowledge base even among people who may not always see eye to eye. To my mind, this is an absolutely essential aspect of living in a rapidly populating world, fraught with the dangers of viewing people who think differently as competitors or, worse, enemies. I see this approach arising naturally out of my academic interests and professional experience, which center on the intersection of identity with local, national, and global social, cultural, political, economic, and environmental change.

I strive to achieve these aims first by shaping my curriculum to enable scholarships to relate theoretical currents to their lives, for example, by having them integrate eco-friendly practices into their daily routines, such as recycling and using low-carbon transportation options. The foundation of my work is the disproportionate impact that global environmental changes are having on communities and nations that are the most vulnerable on the planet—vulnerability that has been decades, if not centuries, in the making. This vulnerability is owed to the marginalization of, if not institutionalized of, discrimination against the resident populations.

At personal level, I am trying to devote myself to find the way to resolve these problems individually and as a group. Every problem has a solution. However, the problems of the contemporary world have very long roots; history and connections. We need to resolve these problems as they have emerged. In reality, through the research, conferences, experiences and the ground truth, we know some cause of the problems. However, more rigorous research is needed to find the root of anthropogenic hampers in nature. It is also necessary to resolve the humanitarian crisis, and the way to minimize the gaps between north and south. These long-rooted problems cannot be resolved overnights. We need to develop a global strategy, which can be implemented within given time. There have been many international commitments to resolve the global crisis on environment, health and food crisis. However, the implementation part depends on the leadership of individual countries. The individual countries are not being able to overcome from the neighborhood blame culture *(the problem created by south, they should solve, or by north they should pay for that etc.).* In addition to that, at the global level there is no specific program for the citizen empowerment. Until or unless, citizens do not have any stake in resolving the crisis, we cannot imagine implementing any program at the ground level, where actual problem is situated. There is a strong need of long-term commitments, devotions, and actual work at the ground level. The fact is "world always dominated by the noble and kind-hearted people". I am optimistic, problems are everywhere, individual to community, nations and at the globe; however, they can be solved, with collective efforts.

Q: We know that South Asia has been an area of special expertise in your studies. In your opinion, what are the most serious environmental issues facing India, Pakistan, Nepal, and Bangladesh today?

A: Yes, originally, my area of practical work and research was focused Asia, particularly in Nepal, where I born, extended to India where I got my intermediates and bachelor's degrees. And, I was attracted and motivated explore more about Bangladesh and Pakistan who belong to similar social and environmental problems. However, I am equally concerned and working to other geographical locations like east Africa, Australia, Japan, middle east, Europe, Eurasia, post-Soviet countries and north America etc. As I noted above, one way or other, the entire world has similar problems, in terms of environmental issues, therefore, in my research, always I see the global scenario and narrow down to specific geo-locations.

Primarily, environmental issues emerged due to anthropogenic disturbances on nature and natural resources or we can state that it is human impact on the living environment (ecosystems). Worldwide, there are many environmental problems, few of them can be listed as follows:

Table 10.1 The major environment Problems in South Asia

Contamination of Drinking Water	Water Pollution	Air Pollution
Wildlife Conservation and Species Extinction	Loss of Tropical Rainforests	**Climate Change/Global Warming**
Biological pollutants	**Carbon footprint**	**Consumerism**
Dams and their impact on the environment	**Ecosystem destruction**	**Energy conservation**
Fishing and its effect on marine ecosystems	**Food safety**	**Genetic engineering**
Intensive farming	**Land degradation**	**Land use**
Deforestation	**Mining**	**Nanotechnology and future effects of Nano pollution/nanotoxicology**
Natural disaster	**Nuclear issues**	**Other pollution issues**
Overpopulation	**Resource depletion**	**Soil contamination**
Sustainable communities	**Toxins**	**Waste**

Source: By Vijayalaxmi Kinhal https://greenliving.lovetoknow.com/Top_30_Environmental_Concerns

Basically, these all and many more are the serious environmental issues in India, Pakistan, Nepal, and Bangladesh as well as to South Asia. However, **Pollution**; Climate Change; Global Warming; Deforestation and Overpopulation impacts are commonly known environmental issues in the region.

The impacts of these issues impact depend on the geo-locations. For example, degree of pollution can be reduced in short period of time if emission reduction policy implemented and used the modern tools to reduce pollutants but global warming –which is causing sea level rise, habitat change, disasters etc. are very severe also difficult issues to solve. Global warming and climate change are interconnected issues and there is no quick fix and short-term solutions. The impacts of such issues emerged due to long anthropogenic interventions in the natural environment, and as they are cumulative impacts, to overcome from such issues also need collective and continuous efforts from local to international levels.

The impacts depend on their geographical locations, size, economic status, and coping mechanism to deal with the crisis. For example, Bangladesh and Nepal belongs same region and have almost same geographical size, however, the degree and the modality of problems are different, but interconnected. For example, one of the impact of global warming "have long lasting effects which can result in melting of glaciers, climate change, droughts, diseases and increase in hurricanes frequency" (https://www.conserve-energy-future.com/current-environmental-issues.php). Here, if we examine the scenario, Nepal's glaciers melting helping sea level rise and Bangladesh, which is belong to low land has direct effects of sea level rise. As such this interconnectedness applies to entire planet. This also brings us to the point of interdependency of chain of ecosystems. As social scientist I also see interconnections of social system and our behavior with our environment as well as to the society. As known notion, industrialization process occurred during 1760-1840 in western industrialized world and continues to date and will continue to future (modality might be different, but objectives remains the same (efficiency and production). We need to think who are the victims? We know, what is the answer; however, it is too late to blame to the past or even to present. It time to consider the fact, the entire planet is victim and only collectively we can minimize the severity of the environmental problems.

The Impact of Pollution on Planetary Health: Emergence of an Underappreciated Risk Factor

Pollution is a massive, overlooked cause of disease, death and environmental degradation. ...... Pollution was responsible in 2015

Table 10.2 A brief comparative account of four countries (territory, demographics, economics, and health profiles)

Category	Indicators	IND	PAK	BGD	NPL
Population	Territory (land surface in 000 Sq. Km.)	3287	796	144	147
	Population density (per Sq.km.) 2008	361	210	1120	195
	2016	445	251	1252	202
	Increased in 8 years	84	41	132	7
	Total (million) 2016	1,324.2	193.2	163	29
	Rural (% of total)	71	64	73	83
	Over 65 years (% of total)	5	4	4	4
	Young	50	63	50	63
	Old	8	7	6	7
Economy	GNI/capita (US$)	1,220	1,000	580	440
	PPP GNI/capita (US$)	3,280	2,680	1,550	1,180
	Annual growth GDP (%)	7.3	2	6.2	5.3
	% male 15 years and older	81	85	84	76
	% female 15 years and older	33	21	58	63
	Extreme Poverty (% <US$1.25 PPP)	41.6	22.6	49.6	55.1
Health Indicators	Mortality rate, infant (per 1,000 live births)	52	73	43	41
	Maternal mortality ratio (per 100,000 live births)	450	320	570	830
	Crude death rate (per 1,000 population)	7	7	7	6
	Life expectancy (years)	64	67	66	67
Health Services[a]	Hospital beds (per 10,000 populations)	7	10	3	2
	Physicians (Density per 10000 population)	6	8	3	2
Health Financing	Total expenditure on health (% of GDP)	4	2.9	3.5	4.9
	General government expenditure on health (% of total)	28	29.7	35.7	39
	Per capita total expenditure on health (US$)	43	24	17	20

[BGD= Bangladesh, IND=India, NPL=Nepal and PAK=Pakistan; the population data source

Sources: World Bank, World Development Indicators 2010. Data for health services are from World Health Statistics 2009. Original data sources include ILO, WHO, UNICEF, UNFPA, and World Bank, Maternal Mortality in 2005 (maternal mortality ratio); and WHO National Health Accounts (health financing data). Note: Data are for 2008 except for extreme poverty (2002–2005), maternal mortality ratio (modeled estimates, 2005), hospital beds (2000–2008), and physicians (2000–2007). a. per 10,000 populations as in World Bank 2011:21; 2016) Bhandari 2012]

for 9 million premature deaths – three times as many deaths as caused by AIDS, tuberculosis and malaria combined. 92% of PRD occurs in low and middle-income countries (LMICs), and in the hardest hit countries, PRD is responsible for more than 1 death in 4. Household air and water pollution, the traditional forms of pollution, are decreasing, and deaths from pneumonia and diarrhea are down. But ambient air, chemical and soil pollution are all on the rise, and non-communicable diseases (NCD) caused by these forms of pollution are increasing. Pollution and climate change are closely linked; both arise from the same sources, and both can be controlled by similar solutions. PRD causes great economic losses. These include productivity losses that reduce gross domestic product in LMICs by up to 2% per year as well as health care costs that account for 1.7% of health care spending in high-income countries and up to 7% in LMICs. Welfare losses due to pollution are estimated to amount to $4.6 trillion per year, 6.2% of global economic output.
Source: The impact of pollution on planetary health: emergence of an underappreciated risk factor: By Philip J. Landrigan, and Richard Fuller, ISSUE NO. 29: United Nations Environment Program (2018-01-08); https://wedocs.unep.org/bitstream/handle/20.500.11822/22416/Perspective_No_29_web.pdf?sequence=1&isAllowed=y

Going back to four Asian countries mostly, I consider ten environmental issues of Environmental Performance Measurement Index developed by Yale University (Yale Center for Environmental Law and Policy) and Columbia University (Center for International Earth Science Information Network) in collaboration with the World Economic Forum and the Joint Research Centre of the European Commission (https://en.wikipedia.org/wiki/Environmental_Performance_Index). In the index they use ten categories (1) environmental burden of disease; (2) water resources for human health; (3) air quality for human health; (4) air quality for ecosystems; (5) water resources for ecosystems; (6) biodiversity and habitat; (7) forestry; (8) fisheries; (9) agriculture; and (10) climate change (Bhandari 2012); which I consider major environmental issues of global importance. However, I would like to restate, that Pollution; Climate Change; Global Warming; Deforestation and Overpopulation are the severe environmental issues of South Asia. The degree of problems are different in each countries. For example, Nepal's major issues are, Deforestation (need to fulfill the demand growing population); Soil erosion (due to elevation, landslides are common); Pollution (crowding cities- air, water, soil pollutions are common); Climate change impact on glaciers and melt. Bangladesh- except glacier all applies; Pakistan: all applies and for India all above applies in full phase. However, degrees can be different.

Sewage and waste management problem (http://www.yourarticlelibrary.com/environment/10-major-environmental-challenges-faced-by-india/9862).*Source are listed in the picture itself – Here I would like to declare that these pictures are used only to show the condition of pollution – they are not official and no relation with the political activism.*

Q: *We would like to briefly discuss Bangladesh for a moment. All countries in Asia and throughout the world are facing the effects of climate change and will be affected in various ways. However, Bangladesh, with approximately 46% of its population living in areas a mere 10 meters above sea level, would seem to be one of the most vulnerable in Asia. Can you briefly discuss how climate change will affect the lives of these 75 million people, what steps the govt. is taking to mitigate these effects, and whether you think Bangladesh is doing enough? Can the government do enough?*

A: Yes, you are absolutely right. We have been witnessing the effects of climate change, even in our day to day lives. As EDF notes "A warming Earth disturbs weather, people, animals and much more". There are ample of evidences that *More heat alters ice, weather and oceans* (The cryosphere – the frozen water on Earth – is melting; Weather is getting more extreme; The oceans are getting hotter, expanding and becoming more acidic); *Human life and prosperity suffering* [(Climate change is a major threat to agriculture: Farms are more likely to face attacks from weeds, diseases and pests, which reduce yield; Warmer, polluted air affects our health; Infrastructure and transportation are at risk]; *natural habitats become hostile-(*The ice Arctic animals need is vanishing; Coral and shellfish are suffering; Forests are more prone to deadly infestations) (Environmental Defender Fund – 2018;

https://www.edf.org/climate/how-climate-change-plunders-planet). I am sure, we all have seen many environmental changes one way or another. We have been witnessing very frequent extreme disastrous even within decades. Such events are more severe in the climate prone country like Bangladesh.

The climatic hazards have always leaded the human casualties, poverty and extreme economic damage in Bangladesh. In 1970 about 500,000 people were killed due to cyclone alone in Bangladesh. For example, from 1971 to 2001, about 505,378 people were killed 147, 5994 were insured and 3625, 1500 people became homeless. During these 30 years about 33 million people of Bangladesh were affected directly from the climatic hazards, which damaged more than US$10 billion worth of property (Government of Bangladesh 2001:113-117).

I think, most worrisome problem of Bangladesh is impact of sea level rise (which is already visible), over population (most densely populated country), and people dependency on agriculture (which is also very fragile). The following brief notes from Rezaul Karim provides a nice snap of the major problems.

Climate Change & its Impacts on Bangladesh Bangladesh is one of the largest deltas in the world which is highly vulnerable to Natural Disasters because of its Geographical location, Flat and low-lying landscape, Population density, Poverty, Illiteracy, Lack of Institutional setup etc. In other words, the Physical, Social as well as Economic conditions of Bangladesh are very typical to any of the most vulnerable countries to Natural Disasters in the world. The total land area is 147,570 sq. km. consists mostly of Floodplains (almost 80%) leaving major part of the country (with the exception of the north-western highlands) prone to flooding during the rainy season. Moreover, the adverse effects of Climate Change – especially High Temperature, Sea-level Rise, Cyclones and Storm Surges, Salinity Intrusion, Heavy Monsoon Downpours etc. has aggravated the overall Economic Development scenario of the country to a great extent.	Bangladesh experiences different types of Natural Disasters almost every year because of the Global Warming as well as Climate Change impacts, these are: **Floods/Flash Floods** (Almost 80% of the total area of the country is prone to flooding). **Cyclones and Storm Surges** (South and South-eastern Parts of the country were hit by Tropical Cyclones during the last few years). **Salinity Intrusion** (Almost the whole Coastal Belt along the Bay of Bengal is experiencing Salinity problem). **Extreme Temperature and Drought** (North and North-western regions of the country are suffering because of the Extreme Temperature problem). Notes from A.K.M. Rezaul Karim, Source: https://www.ncdo.nl/artikel/climate-change-its-impacts-bangladesh

	Source: **Landscape full of trash in Bangladesh** https://www.boredpanda.com/environmental-pollution-overdevelopment-overpopulation-overshoot-global-population-speakout/
	Source: Dhaka Tribune- WB: **Bangladesh losing 1% of its GDP every year due to air pollution**: Md Amjad HossainPublished at 10:37 PM December 10, 2017Last updated at 11:33 PM December 10, 2017-http://www.dhakatribune.com/bangladesh/dhaka/2017/12/10/wb-bangladesh-losing-1-gdp-every-year-due-air-pollution/

Source are listed in the picture itself – Here I would like to declare that these pictures are used only to show the condition of pollution – they are not official and no relation with the political activism.

Similarly, the IPCC in its first assessment report estimated that a 1-meter rise in sea level could inundate 17 percent of Bangladesh ..... and that this could decrease the agricultural productivity of many delta countries that can least afford losses (IPCC 1990: 6.2). The IPCC further noted, in the case of tropical Asia that the projected climate changes ... include strengthening of monsoon circulation, increases in surface temperature, and increases in the magnitude and frequency of extreme rainfall events. ... These changes could result in major impacts on the region's ecosystems and biodiversity; hydrology and water resources; agriculture, forestry, and fisheries; mountains and coastal lands; and human settlements and human health (IPCC 1998: 385)[6]. Similarly, in the additional remarks IPCC noted that the individual countries, regions, resources, sectors, and systems will be affected by climate change not in isolation but in interaction with one another (IPCC 1998: 403). Likewise, in the third assessment report (TAR) it states that the deltas will be exposed to potential inundation both due to climate change and to human-induced stresses (IPCC 2001a: 343-380; 533-590; 843-876)[7].

[6]IPCC (1998) The Regional Impacts of Climate Change. An Assessment of Vulnerability, Cambridge University Press, NY

[7]IPPC (2001) Climate Change 2001, the Scientific Basis, Cambridge University Press, NY
IPPC (2001a) Climate Change 2001, Impacts, Adaptation and Vulnerability and Mitigation, Cambridge University Press

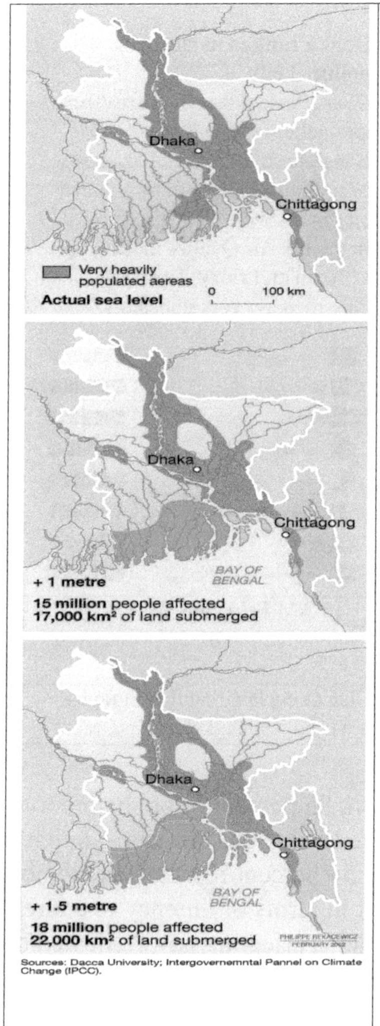

Source: Impact of sea level rise in Bangladesh Impact of sea level rise in Bangladesh. three maps in a time relapse resulting in 18 million people affected, 22,000 km2 of land submerged by flooding. For any form of publication, please include the link to this page: www.grida.no/resources/5648
Year: 2009
From collection: Vital Water Graphics 2
Cartographer: Philippe Rekacewicz, February 2006

"Bangladesh's exposure to the growing hazard of sea-level rise in the 21st century needs to be seen in the perspective of its exposure to current environmental hazards and its growing development needs. If sea-level is currently rising at 1.3 mm/year, that is by only 13 mm (= 0.5 inch) in 10 years. Even if the rate is 3 mm/year, that is by only 30 mm (=1.2 inches) in 10 years. But Bangladesh's population of 150 million is currently growing at ca 2 million a year: i.e., it could grow by 20 million in the next 10 years. That will generate much greater pressure on the country's land and water resources and its economy than will a slowly-rising sea-level. The country's agricultural land is already fully developed; in fact, considerable areas of valuable farmland are being lost to expansion of settlements and infrastructure each year (Brammer, 2010)" [(Hugh Brammer (2014) Bangladesh's dynamic coastal regions and sea-level rise, Author links open overlay panel Climate Risk Management Volume 1, 2014, Pages 51-62. Text taken from page 61)]

Source are listed in the map itself – Here I would like to declare that these maps are used only to show the general from of country – they are not official and no relation with the political boundaries of the countries- these are used just as rough sketch.

In sum-up, Bangladesh is having serious environmental problems, both natural and anthropogenic. Bangladeshi people are aware of overall climate change scenarios and severity of its impact in their day-to-day lives. For example, according to World Bank public survey report of 2009, in reposes

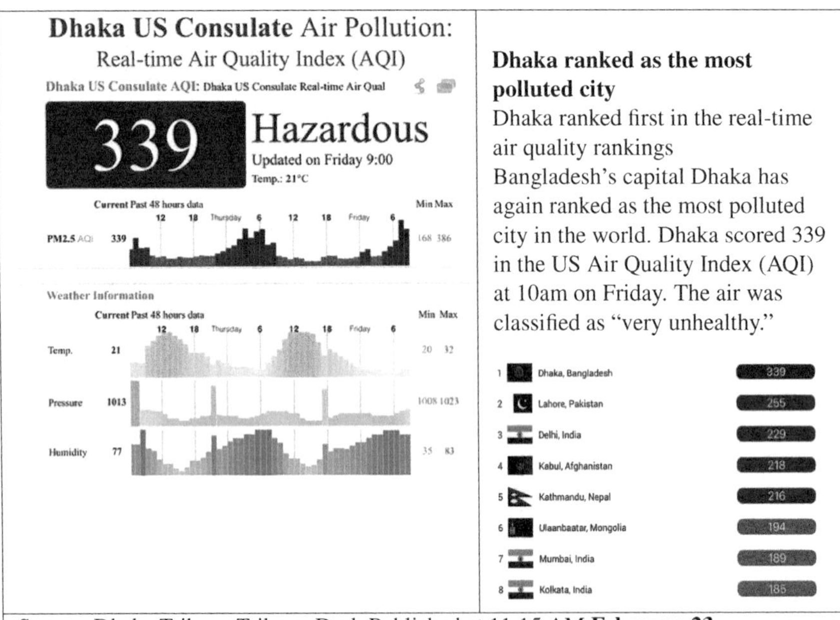

Dhaka ranked as the most polluted city

Dhaka ranked first in the real-time air quality rankings

Bangladesh's capital Dhaka has again ranked as the most polluted city in the world. Dhaka scored 339 in the US Air Quality Index (AQI) at 10am on Friday. The air was classified as "very unhealthy."

Source: Dhaka Tribune,Tribune Desk Published at 11:15 AM **February 23, 2018**-http://www.dhakatribune.com/bangladesh/environment/2018/02/23/poor-air-quality-strangles-life-dhaka-residents/

Source are listed in the graphs itself – Here I would like to declare that these graphs are used only to show the general scenarios - these are used just as sketch.

on seriousness of climate change as a problem, 85 percent of public reported as very serious problem and 14 percent reported as serious problem; similarly in response to the climate change as priority 54 percent reported as topmost priority and 34 percent reported and priority; in terms of urgency to address 67 percent reported that climate change impact is affecting now, followed by 26 percent reported that it will affect within 10 years (The World Bank 2009: xiii)[8].

The government of Bangladesh is also very serious in addressing the climate change induced problems. For example, the Government of Bangladesh

[8]The World Bank (2009) The World Bank World Development Report 2010, "Public Attitudes toward Climate Change: Findings from A multi-Country Poll"; (15 countries released 12-3-09), The World Bank, Washington, DC. (The poll was carried out by WorldPublicOpinion.org , a project managed by the Program on International Policy Attitudes (PIPA) at the University of Maryland) http://siteresources.worldbank.org/INTWDR2010/Resources/Background-report.pdf http://www.worldpublicopinion.org/pipa/articles/btenvironmentra/649.php (accessed on 07/18/2010)

report to its citizens and the global community disseminated on September 2007, states that the rapid global warming has caused fundamental changes to our (Bangladesh) climate. ***No country and people know this better than Bangladesh,*** where millions of people are already suffering. Sudden, severe and catastrophic floods have intensified and taking place more frequently owing to increased rainfall in the monsoon......Bangladesh is recognized worldwide as one of the country's most vulnerable to the impacts of global warming and climate change.

What government is doing:

The phrase **"No country and people know this better than Bangladesh"** used by the government actually spell out the truth which they have been experiencing the risk, insecurity, conflicts and the severity of climate change impact, for which they say the ***innocent victims*** of global warming.

The major step began from the approval of the UNFCCC on June 09, 1992 which was ratified on April15, 1994. To fulfill the commitments to the UNFCC and other international obligations the government of Bangladesh also prepared several legal and policy instruments.

The government of Bangladesh has taken major step to cope with the national environmental crisis and shown its strong commitment by preparing and submitting the National Adaptation Plan of Action (NAPA) (2005[9]) plan to UNFCC in 2005. The NEPA is the policy guidance which outlines the major program and plan in addressing the climate change issues in Bangladesh. This action plan was prepared as a response to the decision of the Seventh Session of the Conference of the Parties (COP7) of the United Nations Framework Convention on Climate Change (UNFCCC) (NEPA 2005: xvi). With the consideration of adverse effects of climate change including variability and extreme events based on existing coping mechanisms and practices, the NEPA (2005) recommended adaptation strategies.

However, the environmental risk factors are not only associated with the climate change but also associated with the deep-rooted poverty, population growth, insecurity and also the instable government and its ineffectual

[9]Government of Bangladesh (2005) National Adaptation Program of Action (NAPA) Final Report November 2005, Ministry of Environment and Forest Government of the People's Republic of Bangladesh http://www.moef.gov.bd/bangladesh%20napa%20unfccc%20version.pdf (accessed on 07/29/2010)

administrative system. So far, there are no any visible achievements, however, generally, the Government seems convinced to materialize the aim "to eradicate poverty and achieve economic and social well-being for its entire people, through a pro-poor, climate resilient and low-carbon development by adaptation to climate change, mitigation, technology transfer and adequate and timely flow of funds for investment, within a framework of food, energy, water and livelihoods security (GOB 2008:2)".

Bangladesh is actively participating and with the international Treaty Obligations related to environment conservation including climate change and its initiation for the forests and environment management since 1972 clearly indicates that the country's seriousness to address the environment issue internationally and domestically. In the Bangladesh environmental conservation history of global relations another significant milestone step was signing of the Convention on Biological Diversity, at the Earth Summit in 1992, and preparation of policy guidelines to manage the biodiversity in the country. In the Bangladesh environmental conservation history, the international organizations have been playing the major role those includes the UN agencies, multilateral donor agencies, development banks, embassies of developed world, conservation international organizations and the NGOs and general concern citizens, in the specific areas of their expertise.

On the basis of available literature, my interaction with locals, NGOs, INGOs and government official, I conclude that, most of the concerned stakeholders (general public, NGOs communities, government et.) are aware of the adverse impact of environmental change due to global warming and climate change. Particularly, the residents of coastal area are also worried, who witnessing increasing hazardous weather pattern. I find them, brave, courageous and optimistic. They have developed own way of survival. The international agencies, NGOs, national and local governments are also working hard to save this vulnerable areas' lives, through early warning systems (cell phones, radio, TV, information stations), and by supplying the essential tools and equipment. However, in comparison to the severity and intensity problems, the available coping mechanism are too minimal. Bangladesh needs global help and resources even to minimal minimization of this problem.

Q: *Many of our readers focus on the larger, more industrialized, wealthier, and probably more polluted countries: China, Vietnam, Malaysia, Indonesia, Pakistan, and India. Nepal is somewhat forgotten. Can you tell us what major environmental issues are facing Nepal?*

A: I do not think, Nepal is less focused or forgotten country in terms of climate change. Even in terms of pollution, Nepal's capital Kathmandu is considered among the list of most polluted cities in the world.

Nepal's Kathmandu ranks 5th in Pollution Index 2017
KATHMANDU: Nepal's capital city Kathmandu has ranked 5th in Pollution Index 2017 mid-year as published by the Numbeo.com recently. .... Kathmandu slumps two spots down to 5th with 96.57 pollution index. Numbeo said it included relevant data from World Health Organization and other institutions for the ranking. The cities were listed on the basis of air pollution and then the water pollution/accessibility followed by other pollution types.
According to the Department of Environment of Nepal, the particulate matter (PM 2.5) of Ratnapark is 107 μg/m^2 *marking Kathmandu as one of the unhealthy cities to live in.*
PM 2.5 indicates the matter present in the air that are 2.5 microns or below. These particles include dust, coal, particles exited from power plants and home heating, car exhaust and pollen plants among others. Kathmandu's downfall was heralded due to the snail-paced road expansion projects in the Kathmandu Valley and delay in the underground installation of Melamchi Drinking Water pipes in the city. The government's failure to replace the old and outdated vehicles plying on the roads of the city have also added air pollution in Kathmandu. Source: The Himalayan Times (Published: July 05, 2017 10:56 pm On: Nepal) https://thehimalayantimes.com/nepal/nepals-kathmandu-ranks-5th-in-pollution-index-2017/

Similarly, in recent years, dust pollution has been another big threaten and subject of major concern for the public health.

Above images and media named the synonyms DUSTMANDU or MASKMANDU for Kathmandu actually depicts the reality of pollution in Kathmandu. One aspect, I would like to restate that, people in South Asia are very active and concerned on the issues of public concern, whether, it is the case of Dhaka, Delhi, Karachi, or Kathmandu.

Going back to the question on major environmental problems in Nepal, they are similar to Bangladesh (except direct impact of sea level rise and coastal related problems). However, Nepal's geographical situation is different than any country in the world, due to its altitudinal variations, which can be divided as a) Low land 60 to 900 meters from sea level; b) Midland 900 to 2500 meter from the sea level; c) Highland 2500 to 2750 meter from the sea level and d) Trans-Himalayan 2750 to 8848 meter from the sea level.

The extreme altitudinal gradient of Nepal results the occurrence of 10 bio-climatic zones from tropical to naval within a horizontal span of less than 180 km virtually making Nepal a treasure house of biological and cultural diversity. A total of 118 ecosystems, 75 vegetation and 35 forest

Kathmandu Valley: Dust bowl of pollution, by Surendra Bahadur Sijapati
Pollution is more vicious than modern-day terrorism. Leave it unchecked and it will strangulate every living thing on this planet. So all who love or live in the Kathmandu Valley must work hand in hand to deal with and eradicate pollution................Between the words, however, it literally hid the day to day sufferings of the inhabitants of Kathmandu Valley blended in blinding and suffocating dust, excruciating noise and nasty smell. The so-called capital city of Nepal and a national melting pot is rapidly turning into a "dust bowl" of pollution, unhindered.
It is easy to tag Kathmandu as **"Dustmandu" or "Maskmandu"** which merely illustrates the tip of the iceberg. But underneath, a huge white core is hidden and we must responsively act before it is too late. Just imagine, one little spark of a pandemic can cause a huge loss. Source: *The Himalayan Times,* Published: February 17, 2017 2:51 am On: Opinion https://thehimalayantimes.com/opinion/kathmandu-valley-dust-bowl-pollution/

World Environment Day: Campaigners lay dead demanding right to clean air, Youths initiate **#Maskmandu campaign,** seek to declare Kathmandu Valley 'Mask-Must Zone'
- Anuj Kumar Adhikari, Kathmandu

Source: The Kathmandu Post 'Capital,' Anuj Kumar Adhikari, Published: 05-06-2016 17:21, http://kathmandupost.ekantipur.com/news/2016-06-05/world-environment-day-campaigners-symbolically-lay-dead-demanding-clean-air.html

Source are listed in the picture itself – Here I would like to declare that these pictures are used only to show the condition of pollution – they are not official and no relation with the political activism.

types have been identified. Nepal comprises seven ecological zones which occur in the following order from south to north: (a) Terai, (b) Siwalik zone, (c) Mahabharat Lekh, (d) Midlands, (e) Himalaya, (f) Inner Himalaya, and (g) Tibetan marginal mountains. The diversity, threats, strategies and action plans of Nepal have been discussed under six broad categories *viz.* forests, rangelands, protected areas, agro-ecosystems, wetlands, and mountain ecosystem (Lillesø; Shrestha; Dhakal; Nayaju and Shrestha 2005; Dahal 2004; Bhattarai 2005). The impact of changing environment varies; we still do not know the degree of impact and severity different ecosystems. In general, the identified *(those are similar to Bangladesh, India, Pakistan)* major environmental problems of Nepal can be summarized as: (1) Degradation of air quality, (2) Degradation of drinking water, (3) Degradation of natural resources, (4) Lack of solid waste management, (5) Degradation of surface water quality, (6) Diminishing of water resources, (7) Release of toxic pollutants, (8) Loss of biodiversity, (9) Impacts of climate change, and (10) Improper land use [(Environment Statistics Nepal (2015) http://cbs.gov.np/image/data/2016/Compendium%20of%20Environment%20Statistics%20Nepal%202015.pdf]

Q: *Climate change tends to hit low-lying areas, as well as naturally more arid areas, more severely. Is climate change a threat to Nepal, and if so, how?*

A: The impact of climate change has been noticed globally, however, it is more visible in the climate sensitive areas such as in Himalayas or the lowland and coastal islands including countries like Nepal (high Himalaya), Bangladesh other countries located in climate prone areas. In this context Nepal belongs to the vulnerable to climate change due to its extreme elevation variations.

Its elevation increases from south to north and is accompanied by decreasing temperatures. It is the home of the 8 of the10 highest mountain peaks of the world, including Mount Everest (at 8848 m), with the lowest area of about 60 (Kechanakalan, Jhapa) meter from sea level. Nepal contains a climatic variation of tropical to arctic within the distance of about 200 kilometers south to north containing lower land Terai plain, Siwalik Hills, Middle Mountains, and High Mountains (OECD 2003[10]; Karki 2007[11]; Lamichhane and Awasthi 2009).

Several international institutions, bilateral and multilateral agencies and independent authors including the government of Nepal have highlighted the impact of the climate change in Nepal; however, still it has limited information regarding the impacts of climate change in economic growth, development, resource conservation and basic livelihood (UNEP

[10]OECD (2003) Development and Climate Change in Nepal: Focus on Water Resources and Hydropower, Environment Directorate, Working Party on Global and Structural Policies, Working Party on Development Co-operation and Environment, Organization for Economic Co-operation and Development, Publications Service, OECD, Paris, France (by Agrawala, Shardul; Raksakulthai, Vivian; Aalst, Maarten van; Larsen, Peter; Smith; Joel and Reynolds, John 2003) http://nepaldisaster.org/download/Development.pdf (accessed on 10/31/2010)

[11]Karki, M.B. (2007) Nepal's Experience in Climate Change Issues, Fourteenth Asia Pacific Seminar on Climate Change, Sydney, Australia. Available at: www.apnet.org/docs/14th_seminar/karki.pdf. (accessed on 10/31/2010) http://www.adb.org/Documents/Books/emerging-issues-challenge/

2001; OECD 2003; Karki 2007; ADB and ICIMOD 2006; IPCC 2007a, b; Lamichhane and Awasthi 2009; USAID 2009; GON 2010[12]). For example, USAID in its Country Assistance Strategy Nepal for 2009-2013, highlights that Nepal is extremely vulnerable to natural disasters due to its elevation. The strategy states that Nepal has an extraordinarily high vulnerability to natural disasters – ***including major floods, landslides, drought and earthquakes – due to its geographic location, low levels of development, minimal infrastructure and institutional capacity, and dependence on rain-fed agriculture***. Severe rural poverty causes populations to inhabit marginal lands in areas at high risk for natural disasters, which is then aggravated through unsustainable practices (deforestation, over-farming or over-grazing) ... There are several critical gaps and impediments to addressing Nepal's vulnerability to disasters. Many of the most vulnerable populations are also the most physically remote, impoverished, and the least-educated (USAID 2009:18[13]).

"Nepal's diverse geography makes it vulnerable to various climatic impacts, including ***extreme temperatures, erratic rainfall, drought, floods, melting snow and glacier retreat****. The mountain areas are also vulnerable to glacial lake outburst floods (GLOFs) from melting glaciers, a risk" [Source: Nepal-How the people of Nepal live with climate change and what communication can do (BBCMEDIA http://dataportal.bbcmediaaction.org/site/assets/uploads/2016/07/Nepal-Report.pdf].*

Nepalis feel the impact of changes in climate now

Nepalis believe that temperatures have risen, rainfall has become less predictable and floods and droughts have increased during the last 10 years. People also feel the environment has changed and nearly nine in ten say that insects and pests have increased.

They feel the impact of these changes in climate now – over half say that agricultural production is decreasing and the majority think that these changes are having an impact on their health. The majority of people (66%) feel very worried about the impact these changes will have on their lives in the future – more than in any of the other six

[12]Government of Nepal – GoN (2010) The Future of Nepal's Forests Outlook for 2020 (Asia Forestry Outlook Study 2020: Country Report Nepal), Submitted By: Ministry of Forests and Soil Conservation (MOFSC), Singh Durbar, Kathmandu, Submitted to: Food and Agriculture Organization of the United Nations Regional Office for Asia and the Pacific, Bangkok, Thailand http://www.forestrynepal.org/images/publications/Forestry%20Outlook%20study%20Country%20Paper%20Nepal%202008.pdf (accessed on 10/19/2010)

[13]USAID (2009) Country Assistance Strategy Nepal (2009–2013) U.S. Agency for International Development, U.S. Mission to Nepal, Kathmandu, Nepal http://pdf.usaid.gov/pdf_docs/PDACN451.pdf (accessed on 10/31/2010)

Climate Asia countries. This worry comes both from their personal experiences and from media coverage.
People are taking action and responding to these changes – one in five have made changes to their livelihoods, including changing job, supplementing their income or migrating. People are also making smaller changes; for instance farmers are rotating crops and growing different crops. However, people want to take more action. In fact, among all the Climate Asia countries, Nepalis are the most willing to make more changes to adapt. The majority strongly want to make changes to improve agricultural production, prepare more for floods and droughts and cope better with water shortages. Most people are struggling to take as much action as they would like for a number of reasons. They feel they need more money, government backing and information on how to respond. Others, such as some housewives from the Eastern and Central Terai, feel isolated within their communities and don't think their actions would make a difference. [Source: Nepal- How the people of Nepal live with climate change and what communication can do (BBCMEDIA http://dataportal.bbcmediaaction.org/site /assets/uploads/2016/07/Nepal-Report.pdf].

The above conclusion of BBC survey actually reveals the major threats of climate change in Nepal, which are already visible. Most of the concerned stakeholders are aware of climatic threats and worried.

The Government of Nepal has been actively involved in mitigating the impact of climate change and trying to draw international attention and help. Nepal signed the most of international treaties, agreements and active participants, supporter of major climate change deals.

Nepal still lacks the proper instrumental arrangements in addressing the changing scenario of climate regime. However, there is new hope because, now, there is new elected body from local, federal and national level. The current government has full majorities in both houses including federal levels. Hopefully, stable government will be able to fulfill the people aspiration in addressing and minimize the impact of climate change.

Q: *In South Asia, Pakistan and India obviously grab the most attention. I would like to start out discussing Pakistan. Most readers know that Pakistan faces severe energy and water shortages, which are a major concern for the Abbasi govt. Can you briefly discuss these issues for our readers and what steps PM Abbasi is taking to address these issues?*

Aside from energy and water, what, in your opinion, are the most pressing environment/climate change issues facing Pakistan?

A: The Islamic Republic of Pakistan shares its boundary with India to the east, Iran and Afghanistan to the west, and China to the north. The country lies between 24°N to 37°N Latitude and 61°E to 75.5°E Longitude. Pakistan has a land area of 880,000 square kilometer including Azad Jammu & Kashmir (AJK) and the Northern Areas (without AJK 796,095 square kilometers) that, along with the coastline along with the Arabian Sea about 1,046 kilometers long, and 22,820 square kilometers of territorial waters and an Exclusive Economic Zone covering about 196,000 square kilometers in the Arabian Sea (borders with Afghanistan 2,430 kilometers; China 523 kilometers; India 2,912 kilometers; and Iran 909 kilometers). Pakistan's dominant geomorphic features include the Indus River and its drainage basin. From the mouths of Indus near the Tropic of Cancer, Pakistan extends about 1,700 km to the river's sources in the Himalayan, Hindu Kush, and Karakorum mountains, where several peaks exceed 8,000 meters in height (ADB 2008)[14].

Pakistan has varieties climatic variation due to its unique location. It constitutes a broad latitudinal spread, and immense altitudinal range, and number of the world's broad ecological regions, as defined by various classification systems. It contains areas that fall under three of the world's eight biogeography "realms" (Indo-Malayan, Pale arctic, and Afro-tropical); four of the world's ten "biomes" (desert, temperate grassland, tropical seasonal forest, and mountain); and three of the world's four "domains" (polar or mountain, humid-temperate, and dry). The great variety of landscapes, including rangeland, forest, wetland, and other wildlife habitats has generated a rich diversity of life forms. However, among the south Asian countries Pakistan holds the least varieties of biodiversity (ADB 2008:13-14; UNEP 1995[15]; Government of Pakistan 2000:8[16]; ADB 2008:14; Government of Pakistan 2009).

As it has verities of climatic zones, Pakistan faces severe energy and water shortages as well as various climatic hazards throughout the history (Iqbal 2009)[17]. It is highly vulnerable to disasters caused by Climate Change, and especially prone to floods and droughts. Sandstorms, dust storms,

---

[14]The Asian Development Bank-ADB (2008) Islamic Republic Of Pakistan: Country Environment Analysis, ADB, Manila. http://www.adb.org/Documents/Assessments/Country-Environmental/PAK/Country-Environment-Analysis.pdf (accessed on 05/10/2010)

[15]United Nations Environment Program-UNEP (1995) Global Biodiversity Assessment, Cambridge: Cambridge University Press

[16]Government of Pakistan (2000) (with the collaboration of IUCN and WWF) Biodiversity Action Plan for Pakistan, A framework For Conserving Our Natural Wealth, Government of Pakistan, World Wide Fund for Nature, Pakistan and International Union for Conservation of Nature and Natural Resources, Pakistan. http://www.iucn.pk/publications/Biodiversity%20Action%20Plan.pdf

[17]Iqbal, M. Mohsin (2009) Vulnerability of Pakistan to Climate Change Hazards, Global Change Impact Studies Centre (GCISC) Islamabad, National Disaster Awareness Day -2009" presentation, at Convention Centre, Islamabad (on October 08, 2009) http://ndma.gov.pk/Do

Table 10.3 Climatic Hazards in Pakistan

Natural Hazards	Human Induced Hazards
Earthquakes	Transport accidents
Floods	Oil spills
Tsunami	Urban fires
Avalanches	Civil conflicts
Landslides	International displacements
Cyclones/Storms	Radiological (CNR)
Glacial Lake Outburst Floods (GLOF)	Accidents
Droughts	
River erosion	
Pest attacks	

Source: IUCN and Government of Pakistan (2009:4)

micro-cloudbursts, cyclones and tsunamis are additional threats (IUCN and Government of Pakistan 2009:4)[18]. The major climatic hazards in Pakistan can be listed as.

Pakistan has been facing severe problem due to the climatic extreme events such as Earthquakes, Floods, Landslides, Cyclones/Storms, Droughts, River erosion, Glacial Lake Outburst Floods (GLOF) etc. and experiencing human causality, displacement and loss of property. For example, in July and August 2010 due to extreme flood more than 1600 people were killed and 14 million became homeless and 170 million people were affected, destroyed 1,226,678 houses (National Disaster Management Authority 2010) and damaged about a trillion-dollar worth of property and infrastructures of the country (United Nations Office for the Coordination of Humanitarian Affairs-OCHA 2010). This flood has been considered one of the worst natural disasters of Pakistani history which affected to entire nation. As published in various news sources: The United Nations has rated the floods in Pakistan as the greatest humanitarian crisis in recent history with more people affected than the South-East Asian tsunami and the recent earthquakes in Kashmir and Haiti combined. Maurizio Giuliano, a spokesman for the UN Office for the Coordination of Humanitarian Affairs (OCHA) said: This disaster is worse than the tsunami, the 2005 Pakistan earthquake and the

cuments/DMEC%2009/Climate%20Change%20Hazards.ppt.ndma.gov.pk/Documents/.../Climate%20Change%20Hazards.ppt (accessed on 08/20/2010)

[18]IUCN and Government of Pakistan (2009) Climate Change Disaster Management in Pakistan, IUCN Pakistan Country Office, Karachi 75530, Pakistan and National Disaster Management Authority (NDMA), Government of Pakistan, Prime Ministers Secretariat, Islamabad, Pakistan http://cmsdata.iucn.org/downloads/pk_cc_dm_vul.pdf (accessed on 08/20/2010)

Haiti earthquake (Tweedie 2010[19]; BBC 2010[20]; OCHA 2010). According to Pakistan's National Disaster Management Authority (NDMA)[21], the affected area covers 132,421 km, including 1.4 million acres of cropped land (The Current Affairs 2010)[22].

Source: Duniyanews.tv-A National Communications Services Company; Last Updated On 18 October,2017 11:35 am http://dunyanews.tv/en/Pakistan/410284-

Source are listed in the picture itself – Here I would like to declare that these pictures are used only to show the condition of pollution – they are not official and no relation with the political activism.

As other Asian countries, Pakistan is facing similar environmental problems such as deforestation, air pollution, water pollution, noise pollution, climate change, pesticide misuse, soil erosion, natural disasters and desertification. However, noticeable problems are waste management and water crisis. The few pictures (from various news sources) show the reality of the fact.

---

[19]Tweedie, Neil (2010) Pakistan floods: disaster is the worst in the UN's history, The Teligraph UK, (Neil Tweedie in Charsadda-Published: 6:07PM BST 09 Aug 2010) http://www.telegraph.co.uk/news/worldnews/asia/pakistan/7935485/Pakistan-floods-disaster-is-the-worst-in-the-UNs-history.html (accessed on 08/21/2010)

[20]BBC (2010) Pakistan flood aid not getting through – UN http://www.bbc.co.uk/news/world-south-asia-10997669 (accessed on 08/20/2010) BBC (2010) BBC News South Asia August 17, 2010 last updated at 11:56 ET http://www.bbc.co.uk/news/world-south-asia-10997669?print=true (accessed on 08/20/2010)

[21]National Disaster Management Authority (2010) News Update of current flooding, National Disaster Management Authority (NDMA), Government of Pakistan

[22]The Current Affairs (2010) Pakistan Floods 2010, Floods in Pakistan, Pakistan Floods Donating and Organizational Relief Campaign Information, The Current Affairs, Pakistan TheCurrentAffairs.com (By admin at August17, 2010, 12:22 pm) http://thecurrentaffairs.com/pakistan-floods-2010-floods-in-pakistan-pakistan-floods-donating-and-organizational-relief-campaign-information.html (accessed on 08/21/2010)

In pictures: Karachi's garbage problem

Source: Duniyanews.tv-A National Communications Services Company; Last Updated On 18 October,2017 11:35 am http://dunyanews.tv/en/Pakistan/410284-

Source are listed in the picture itself – Here I would like to declare that these pictures are used only to show the condition of pollution – they are not official and no relation with the political activism.

People & Society **Keeping trash alive**

A garbage-strewn canal in Board Bazaar, Peshawar | Shahbaz Butt, White Star
Source: Amna Chaudhry, Herald, Updated Sep 04, 2017 09:00pm https://herald.dawn.com/news/1153824/keeping-trash-alive

Source are listed in the picture itself – Here I would like to declare that these pictures are used only to show the condition of pollution – they are not official and no relation with the political activism.

Pakistan's Water Crisis: Why a National Water Policy is Needed

(Recently, the Pakistan Council of Research in Water Resources (PCRWR) delivered a grave warning: if the government does not take action, the country will run out of water by 2025. Severe water scarcity is already having a negative impact on the country's public health and the economy. Over 80 percent of water supplied is considered unsafe, and water scarcity and water-borne diseases are resulting in a loss of up to 1.44 percent of GDP.)

Women gather water at a central drinking pump. Severe water scarcity in Pakistan is already having a negative impact on the country's public health and the economy, with over 80 percent of water supplied in the country is considered unsafe. Source: Photo/Hisaar Foundation, November 1, 2017, By Brayshna Kundi https://asiafoundation.org/2017/11/01/pakistans-water-crisis-national-water-policy-needed/

Source are listed in the picture itself – Here I would like to declare that these pictures are used only to show the condition of pollution – they are not official and no relation with the political activism.

Arsenic in drinking water threatens up to 60 million in Pakistan

Villagers collect water from a broken water pipeline in the outskirts of Islamabad. B. K. BANGASH/ASSOCIATED PRESS

Source: By Giorgia Guglielmi Aug. 23, 2017, 4:15 PM: Science: 2018 American Association for the Advancement of Science. All rights Reserved. AAAS is a partner of HINARI, AGORA, OARE, CHORUS, CLOCKSS, CrossRef and COUNTER. http://www.sciencemag.org/news/2017/08/arsenic-drinking-water-threatens-60-million-pakistan

Source are listed in the picture itself – Here I would like to declare that these pictures are used only to show the condition of pollution – they are not official and no relation with the political activism.

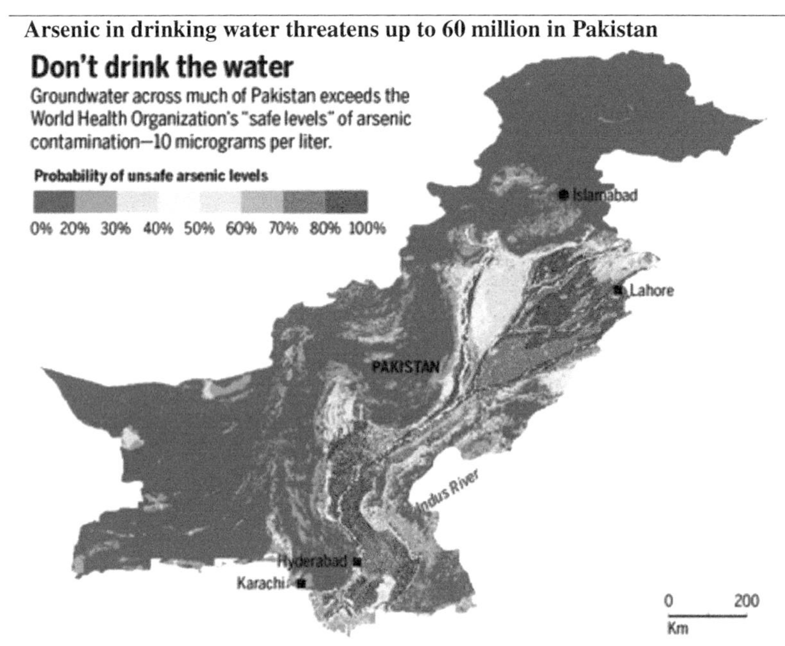

Source: By Giorgia Guglielmi Aug. 23, 2017, 4:15 PM: Science: 2018 American Association for the Advancement of Science. All rights Reserved. AAAS is a partner of HINARI, AGORA, OARE, CHORUS, CLOCKSS, CrossRef and COUNTER. http://www.sciencemag.org/news/2017/08/arsenic-drinking-water-threatens-60-million-pakistan Posted in: Asia/PacificClimateEarthHealth- doi:10.1126/science.aap7590- Declaration: Source are listed in the map itself – Here I would like to declare that these maps are used only to show the general from of country – they are not official and no relation with the political boundaries of the countries- these are used just as rough sketch.

The above pictures nicely depict the severity of the climate change in Pakistan. The government, the general public, the international organizations (all concern stakeholders) are aware of the problems and also working to minimize the devastating impacts. The following will show what people are doing to overcome with this devastating problem through plantations.

One Billion Trees Planted in Pakistan's NW Province

ISLAMABAD —

Pakistan's northwestern province, Khyber Pakhtunkhaw (KPK), has planted an unprecedented 1 billion trees in just more than two years and surpassed an international commitment of restoring 350,000 hectares of forests and degraded land.

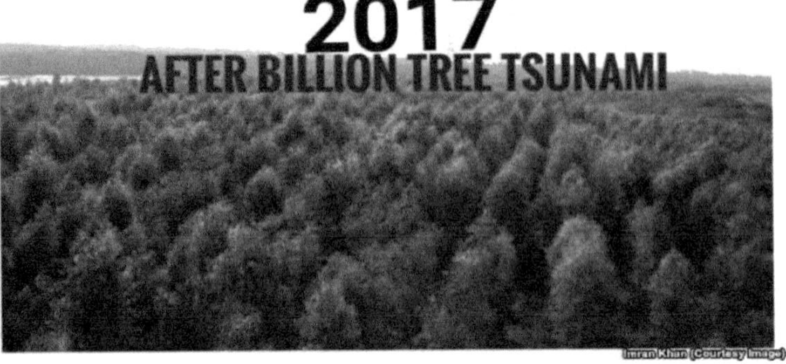

Source: August 13, 2017 5:10 AM; By Ayaz Gul- https://www.voanews.com/a/one-billion-trees-planted-in-pakistan-nw-province/3983609.html

Source are listed in the picture itself – Here I would like to declare that these pictures are used only to show the condition of pollution – they are not official and no relation with the political activism.

"**Share** this article to let the world know about this amazing environment initiative by Pakistan" Source: http://pakiholic.com/billion-tree-tsunami-huge-contribution-pakistan-stop-climate-change/.

Billion Tree Tsunami a Huge Contribution of Pakistan to Stop Climate Change: The initiative was launched in 2013 and since then more than 75% of the 1 billion trees have been planted in the province.

We often hear about the global climate change and its negative impacts on our planet. New initiatives are being taken all over the world in an effort to reverse this climate change. One such initiative is taken by the Khyber Pakhtunkhwa province of Pakistan. This mega initiative is given the name of Billion Tree Tsunami. According to the details of this initiative, the government of Khyber Pakhtunkhwa province in Pakistan is planting 1 billion trees. The best thing about this project is that more than 75% trees have already been planted and are growing rapidly. Source: http://pakiholic.com/billion-tree-tsunami-huge-contribution-pakistan-stop-climate-change/

Source are listed in the picture itself – Here I would like to declare that these pictures are used only to show the condition of pollution – they are not official and no relation with the political activism.

"**Share** this article to let the world know about this amazing environment initiative by Pakistan" Source: http://pakiholic.com/billion-tree-tsunami-huge-contribution-pakistan-stop-climate-change/.

Source are listed in the picture itself – Here I would like to declare that these pictures are used only to show the condition of pollution – they are not official and no relation with the political activism.

These illustrated initiatives are only few examples. There are such programs in each province and districts.

Government of Pakistan has been taking climate change issue seriously. In its report to the UNFCC in 2001 and taskforce report of 2010, it repeats that, because climate change is posing a direct threat to its water security, food security and energy security (Government of Pakistan 2001; 2010:1)[23]. There are several national level policy commitments in addressing the degrading

[23]Planning Commission, Government of Pakistan (2010) Final Report of the Task Force on Climate Change, Planning Commission, Government of Pakistan, Islamabad http://115.186.133.3/pcportal/usefull%20links/Taskforces/TFCC%20Exec%20Summary%20and%20%20Recomm.pdf (accessed on 06/03/2010) Ministry of Environment, Local Government and Rural Development (2001) National Report of Pakistan on the Implementation of United Nations Convention to Combat Desertification (UNCCD) Government of Pakistan, Islamabad http://unfccc.int/resource/docs/natc/paknc1.pdf (accessed on 06/03/2010)

environmental situation of the country. Pakistan has been a party to various Environmental Conventions and Protocols. To evaluate the climate change issue regularly the Government of Pakistan launched a comprehensive program called " SMART" (self-Monitoring and Reporting Tool) with the help of Sustainable Development Policy Institute (SDPI), which was founded by the government to address the issues of sustainable development, environment and climate change, which helps to monitor release of effluents and emissions from the Industries initially with 50 industrial Units later expending to 200 and 400 Industrial Units all across the country under phase-I program (SDPI 2006; Khan 2009)[24]. There have been several projects running in Pakistan to address the climate change issue in the country with the financial and technical support of various international organizations i.e. the World Bank, Asian Development Bank, GTZ, DFID, UN agencies, the IUCN, the WWF etc. (Khan 2009; World Bank 2006; SDPI 2006)[25]. In addition to Government's plan, policies and projects, the government established institutions, agencies as well as international organizations also have been working to address the climate change issues in Pakistan.

The scenario clearly indicates that the government of Pakistan is serious in addressing the degrading environmental situation and impact of climate change, through its national and international commitments. The following section elaborates on Pakistan's involvement on international treaties, convention and conferences including climate change. However, the global context, Pakistan has signed only about 16 percent of bounding treaties related to environment and climate change. In addition to showing its commitments in the international and national policy instruments, the government of Pakistan has been maintaining it memberships with the numerous of International Organizations.

Q: *We can't discuss South Asia without talking about India. It's the largest, wealthiest, most industrialized nation in South Asia. Yet, it does have severe*

---

[24]SDPI (2006) Sustainable Industrial Development/National Environmental Quality Standards, Sustainable Development Policy Institute, Islamabad http://www.sdpi.org/research_Programme/environment/sustainable_industrial_development.htm#6 (accessed on 06/03/2010) Khan, Javed Ali (2009) Multilateral Environmental Agreements (MEA's) Ministry of Environment, Government of Pakistan, Islamabad. http://www.lead.org.pk/c11-1nts/faculty%20Presentation/Jawed%20Ali%20Khan.ppt. (accessed on 06/03/2010) Jawed Ali Khan is the Director of Pakistan Environmental Protection Council (PEPC).

[25]The World Bank (2006) Pakistan Strategic Country Environmental Assessment South Asia Environment and Social Development Unit, Document of the World Bank, World Bank Office Islamabad and Washingrton, DC. http://www.environment.gov.pk/NEW-PDF/PK-SCE-FText-Oct-2006%20.pdf

problems. Smog blankets most cities, sporadic electricity, troubles with its rail network and mass transit, the list is almost endless. In your opinion, what are the most serious issues facing India today, and what steps is the Modi government taking to address these issues?

A: As I noted, in the first section of this interview, India holds second largest population in the world and seventh largest in terms of geographical territory. Yes, India has ample of opportunities and threats. I think, the geographical wealth and abundance and its large population is also opportunity, strength and key of innovation, for industrialization and growth. It is established notion that, there is negative side effects of over population and industrialization. The pollution, waste, increase of environmental health risk, transit and traffic issues can be count few examples of such side effects. However, it has long history to tackle issues, which still remain strongly influential in Indian rich culture and traditions.

"India has a civilizational legacy which treats Nature as a source of nurture and not as a dark force to be conquered and harnessed to human Endeavour. There is a high value placed in our culture to the concept of living in harmony with Nature, recognizing the delicate threads of common destiny that hold our universe together. The time has come for us to draw deep from this tradition and launch India and its billion people on a path of ecologically sustainable development" (Dr. Manamohan Singh. Prime Minister of India, June 30, 2009)[26].

> *"India is the cradle of human race, the birthplace of human speech, the mother of history, the grandmother of legend, and the great grandmother of tradition. Our most valuable and most astrictive materials in the history of man are treasured up in India only!"* Mark Twain (1835-1910).

Environmental protection was cast as a citizen's duty and as a part of the people's spiritual lives, which can be found in all major religious epic, grew in India. Examples of the spiritual importance of environmental protection can be drawn from the Hindu, Islamic and Buddhist theologies. The five principles of Buddhism are as follows: abstinence from killing all forms of life; abstinence from sexual misconduct; abstinence

[26]http://www.pmindia.nic.in/lspeech.asp?id=690 Downloaded on January 3, 2020

from lies and deceit; abstinence from theft and abstinence from intoxicants. The first principle clearly points to the importance of the conservation of all living beings. Similarly, in Islam *"the conservation of the environment is based on the principle that all the individual components of the environment were created by God, and that all living things were created with different functions, functions carefully measured and meticulously balanced by the Almighty Creator. Although the various components of the natural environment serve humanity as one of their functions, this does not imply that human use is the sole reason for their creation. . . .* (Engel and Engel 1990; as cited by Deen 2002:2)[27]. In the major Hindu epics, such as the Vedas, Puranas, Upanishads, and others, there are detailed descriptions of the trees, plants and wildlife and their importance to the people. "*The Rig Veda highlighted the potentialities of nature in controlling the climate, increasing fertility and improvement of human life emphasizing on intimate kinship with nature. Atharva Veda considered trees as abode of various gods and goddesses. Yajur Veda Emphasized that the relationship with nature and the animals should not be that of dominion and subjugation but of mutual respect and kindness. Many animals and plants were associated with Gods and Goddesses so that they were preserved for the future generations. As they were associated with supernatural powers, no one dared to misuse the resources and therefore there was a check on the excess utilization of resources*" (Wilson 1977; Bryant2001; Lal 2005 as cited by Budholai 2009:1)[28].

India holds the most influential role in South Asia: it has the oldest and most extensive history of civilization, colonization, and struggle for independence, as well as an extended democratic system and the second largest population on the globe. Pakistan and Bangladesh were part of India until they became independent; therefore, these two countries have been following a similar tradition of conservation bureaucracy as India. India's influence in modern Nepal is even stronger than in Pakistan and Bangladesh because India and Nepal share an open border and citizens of both countries do not require work permits. Similarly, Nepal and India have the largest Hindu population

[27]Engel J. R. and Engel J. G. (1990) Ethics of Environment and Development, London: Bellhaven Press Deen, Mawil Y. Izzi (Samarrai) (2000) Islamic Environmental Ethics, Law, and Society, King Abdul Aziz University, Jeddah http://www.mbcru.com/Texas%20Tech%20Mypage/Conservation%20Biology/Assignment%202/IzziDeenIslamicEcol.pdf (accessed on 04/20/2010).

[28]Budholai, Bharat (2009) Hidayatullah National Law University, Raipur (C.G): Environment Protection Laws in the British Era: http://www.legalserviceindia.com/articles/brenv.htm Legal Service India.com Bryant, Edwin (2001) The Quest for the Origins of Vedic Culture: The Indo-Aryan Migration Debate, Oxford: Oxford University Press, ISBN 0195137779 Lal, B.B. (2005) The Homeland of the Aryans; Evidence of Rigvedic Flora and Fauna & Archaeology, New Delhi, Aryan Books International. Wilson, H. H. (1977) Rig-Veda-Sanhitā: A Collection of Ancient Hindu Hymns. 6 vols. (London, 1850-88); repring: Cosmo Publications (1977)

and therefore have strong cultural ties. Because of these facts, to some extent an understanding of India in general reflects an understanding of the other three nations.

As India has all kind of ecosystems and diversities, it has also verities of environmental problems, the following table, (from Mahesh Chandra, 2015) nicely summarizes.

Some of the major environmental concerns confronting India include:

- Air pollution from industrial effluents and vehicle emissions;
- Energy-related environmental problems such as, chemical & oil pollution and Greenhouse Gas (GHG) emissions (Greenstone and Hanna, 2014);
- Water pollution from raw sewage, the lack of adequate sanitation, and non-portable water throughout the country;
- Municipal solid waste management (MSWM) remains a challenge for India due to the rising population and the resultant infrastructural needs (Dube, Nandan, and Dua, 2014);
- Over-population and its strain on natural resources; and
- Agricultural factors such as, runoff of agricultural pesticides, overgrazing, short cultivation cycles, slash and burn practices, destructive logging practices, and deforestation of timber reserves for fuel, all contribute conjointly to the decimation of the subcontinent's environmental system.

Source: (Chandra, Mahesh (2015) "Environmental Concerns in India: Problems and Solutions," Journal of International Business and Law: Vol. 15: Iss. 1, Article 1. Available at: http://scholarlycommons.law.hofstra.edu/jibl/vol15/iss1/1)

There are area specific problems in terms of pollution, waste, sewages, over population, poverty. The following few pictures (published in various newspaper illustrate few examples).

The key environmental challenges facing India today include (i) air pollution; (ii) poor management of waste; (iii) growing water scarcity; (iv) declining levels of groundwater; (v) water pollution; (vi) forest preservation and quality; (vii) loss of biodiversity; (viii) land and soil degradation; and (ix) increasing frequency of natural disasters, including droughts and floods. Coupled with the demands of India's increasing population, these challenges place mounting pressure on India's environmental resources. Growth of India's economy will place further pressures on India's natural resource base, and reinforce the need to sustainably exploit and manage these resources. This is particularly important because India's poor suffer most from declining natural resource productivity. Source: ADB- https://www.adb.org/sites/default/files/linked-documents/cps-ind-2013-2017-ena.pdf

As there are countless many problems, however, as its culture and tradition indicate that, there are numerous efforts are also taking place to overcome

How Delhi became the most polluted city on Earth

Breathing in the Indian capital this month was like smoking 50 cigarettes a day.

On November 8, pollution surged so high that some monitoring stations reported an Air Quality Index of 999, way above the upper limit of the worst category, Hazardous. (An extra-sensitive air quality instrument at the US embassy got a reading of 1,010, as you can see in the chart below.)

When Delhi became the most polluted city on Earth

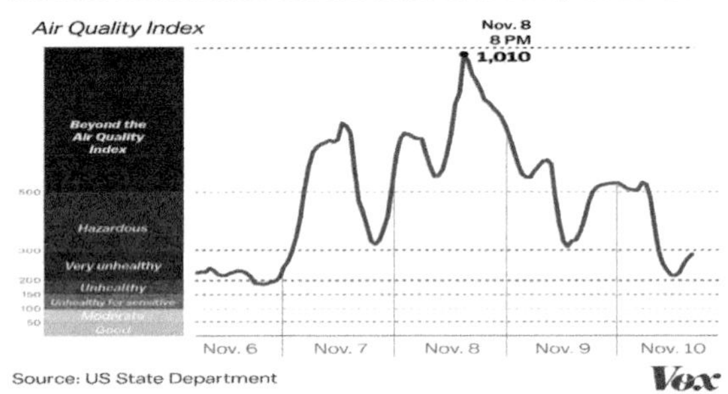

Source: VOX: By Umair Irfan Updated Nov 25, 2017, 4:32pm EST https://www.vox.com/energy-and-environment/2017/11/22/16666808/india-air-pollution-new-delhi

Source are listed in the graph itself – Here I would like to declare that these graphs are used only to show the condition of pollution – they are not official and not accurate.

75% of India's air pollution-related deaths are rural, study finds

(CNN)Rural Indians, who make up about two-thirds of the country of 1.3 billion people, are disproportionately at risk of breathing polluted air, according to new research. India's air pollution has been making headlines for years, with attention focused on Delhi, the capital, once named the most polluted city in the world by the World Health Organization.

Source: CNN: By Huizhong Wu, CNN: Updated 3:32 AM ET, Mon January 15, 2018 https://www.cnn.com/2018/01/15/health/india-air-pollution-study-intl/index.html

India's air pollution crisis risks becoming humanitarian catastrophe

Source: PHOTO: Vehicles drive through smog in New Delhi. (Reuters: Saumya Khandelwal)

ABC News-Correspondents Report By South Asia correspondent Siobhan Heanue Posted 23 Feb 2018, 5:52pm http://www.abc.net.au/news/2018-02-24/air-pollution-in-india-risks-becoming-a-humanitarian-catastrophe/9469922

Source are listed in the picture itself – Here I would like to declare that these pictures are used only to show the condition of pollution – they are not official and no relation with the political activism.

from such problems. For example, India is signatory of most of the environment related treaties and conventions. India's Constitution, Article 51(c), sets the directive principle requiring the State to foster respect for International law and treaty obligations (Divan, 2002)[29]. India first accepted the impact of climate change after the 1947 publication of Sisir Kumar Mitra's book titled The Upper Atmosphere, published by Calcutta, Royal Asiatic Society of Bengal. *"He considered, for the first time, the atmospheric environment as a whole-neutral and ionized - its thermal structure and distribution of constituents, its motions, the interaction of the solar radiation and the particle streams from the sun with these constituents. He also considered the atmosphere from the surface to the fringe of the upper atmosphere*" (Indian National Science Academy, 2001:144[30]).

As guided and accepted by the constitution of India, there are many initiates, government of India and even general public have been working to minimize the impact of environmental change, throughout the history, with some disputes on natural resource management.

> "*Nature-based conflicts have increased in frequency and intensity in India. They revolve around competing claims over forests, land, water and fisheries, and have generated a new movement struggling for the rights of victims of ecological degradation. The environmental movement has added a new dimension to Indian democracy and civil society. It also poses an ideological challenge to the dominant notions of the meaning, content and patterns of development*" (Gadgil and Guha1994:100)[31].

However, together with government, general public are looking options to overcome with the increasing problems due to environment degradation.

Government has Taken Series of Steps to Address Pollution-Related Issues: Environment Minister
Pollution is a matter of concern in cities and towns and is caused due to introduction of contaminants into the environment viz. air, water and soil that may cause

[29]Divan, Shyam (2002) International Environmental Law (On line publication on July, 12th, 2002) http://www.nls.ac.in/CEERA/ceerafeb04/html/documents/internationalenvtlawshyamdiwan.htm#N_1_ (accessed on 05/01/2010).

[30]Indian National Science Academy (2001) Pursuit and Promotion of Science – The Indian Experience, Indian National Science Academy, New Delhi. http://www.insaindia.org/INSA-book.pdf (accessed on 05/02/2010)

[31]Gadgil, Madhav and Guha, Ramchandra (1994) Ecological conflicts and the environmental movement in India, Development and Change Vol. 25: 101-136.

adverse change in ambient conditions. The Government has taken a series of steps to address issues related to water pollution, air & vehicular pollution, industrial pollution, improper waste disposal etc. in cities, towns and metropolises.
The major steps being taken by the Government to control pollution inter alia include the following:-
(i) Notification of National Ambient Air Quality Standards;
(ii) Formulation of environmental regulations/statutes;
(iii) Setting up of monitoring network for assessment of ambient air quality;
(iv) Introduction of cleaner/alternate fuels like gaseous fuel (CNG, LPG etc.), ethanol blend etc.;
(v) Promotion of cleaner production processes.
(vi) Launching of National Air Quality index by the Prime Minister in April 2015;
(vii) Implementation of Bharat Stage IV (BS-IV) norms in 63 selected cities and universalization of BS-IV by 2017;
.........(xix) Implementation of National River Conservation Plan for abatement of pollution in identified stretches of various rivers and undertaking conservation activities...
Source; Business Standard- Delhi Last Updated at July 20, 2016 00:21 IST http://www.business-standard.com/article/government-press-release/government-has-taken-series-of-steps-to-address-pollution-related-issues-116071901137_1.html

The government has taken steps to mainstream both climate change mitigation and adaptation within its developmental framework through policies and action plans. India's Eleventh Five Year Plan (2007–2012) recognized the urgent need to balance the growth–environment tradeoff, given the dangers posed by India's large population, economic growth, and ever-increasing demands on natural resources including water. India has made admirable efforts on its climate change program through the National Action Plan on Climate Change (NAPCC). In response to the NAPCC, all Indian states have been directed to prepare a state climate change action plan (SCAP), detailing sector-specific plans to adapt to and mitigate climate change. A few states have already initiated studies to prepare their state's climate action plans and aim to identify the most vulnerable sectors and regions to projected climate change and to develop adaptation projects. These action plans also aim to assess GHG emissions and identify potential mitigation programs and projects.
India launched its NAPCC in June 2008, and has established eight national missions to address various aspects of climate change mitigation and adaptation: (i) solar energy, (ii) enhanced energy efficiency, (iii) sustainable habitats, (iv) water, (v) sustaining the Himalayan ecosystem, (vi) national mission for a green India, (vii) national mission for sustainable agriculture, and (viii) strategic knowledge for climate change. ADB has been at the forefront of supporting some of these missions and mainstreaming climate change adaptation and mitigation concerns into its operations, in line with India's Twelfth Five Year Plan. The latter emphasizes mainstreaming of climate change interventions across sectors and supports the development of SCAPs that may be dovetailed into the NAPCC by developing specific action programs for sector operations to facilitate mitigation and adaptation action against the challenge of climate change.
Source: ADB- https://www.adb.org/sites/default/files/linked-documents/cps-ind-2013-2017-ena.pdf

The listed texts and figures give some general outlines of what government is doing to address the severity of the environment induced problems in India. There may be less practical work, than, whatever is advertised in the government publications or in the newspapers to support or oppose the government; however, it clearly indicates that, the government and other concerned stakeholders are aware of the severity of the problems and trying their best to overcome, / minimize the problems.

India plants 66 million trees in 12 hours as part of record-breaking environmental campaign
More than 1.5 million volunteers were involved in the huge operation

"Volunteers in India planted more than 66 million trees in just 12 hours in a record-breaking environmental drive. About 1.5 million people were involved in the huge plantation campaign, in which saplings were placed along the Narmada river in the state of Madhya Pradesh throughout Sunday. India committed under the Paris Agreement to increasing its forests by five million hectares before 2030 to combat climate change".
Source: (Chris Baynes Monday 3 July 2017 09:45 BST) http://www.independent.co.uk/news/world/asia/india-plant-66-million-trees-12-hours-environment-campaign-madhya-pradesh-global-warming-climate-a7820416.html

Source are listed in the picture itself – Here I would like to declare that these pictures are used only to show the condition of pollution – they are not official and no relation with the political activism.

Theoretically, and in practice to some extent, India is committed to environment conservation, which includes addressing climate change issues, and has been working to develop collaborative plans and projects with the Multilateral Donor Agencies, Development Agencies, Civil Societies and the private sector. India has also shown its firm commitment both domestically and internationally by signing and ratifying the major conventions and

6 Important Eco-Friendly Steps Taken by Narendra Modi Government

(1) Swachh Bharat Abhiyan: The Swachh Bharat Abhiyan is India's biggest cleanliness drive ever. The campaign covers as many as 4041 towns and aims at cleaning streets, roads, and infrastructure.

(2) Clean Ganga Mission: Modi's Clean Ganga Plan involves five ministries working in close co-operation to see the dream project through.

(3) National Air Quality Index (NAQI): The NAQI will simplify air quality rendition and will help raise awareness about alarming levels of air quality across the country.

(Image Credits: blogs.wsj.com)

(4) Toilets Before Temples: Modi government is working tirelessly to ensure that affordable sanitation reaches the people who need it. In addition to this, Modi government has also focused on the spread of e-toilets in rural as well as urban India.

(Image Credits: abc.net.au)

(5) Mount Everest Ascent: Although the Mount Everest cleaning drive is essentially an Indian Army undertaking, authorities have made it clear that they draw inspiration from Narendra Modi's Swacch Bharat Abhiyan. A team of Indian Army climbers have set off on a mission to bring back at least 4000 kg of non-biodegradable waste from the world's highest peak. This includes waste materials left behind by climbers over decades.

(Image Credits: scoopwhoop.com)

(6) Water Conservation: In a bid to raise awareness about water conservation, the Narendra Modi government directed the states of India to ensure that 50% of the work taken up by MNREGA, should be for the improvement of water conservation.

(Image Credits: timesofindia.com)

Source: Skymet Weather Services Pvt. Ltd. 2018-9 April 2015 03:09 PM https://www.skymetweather.com/content/weather-news-and-analysis/6-important-eco-friendly-steps-taken-by-narendra-modi-government /(Main Image Credits: narendramodiblog.wordpress.com) (Featured Image Credits: indiatimes.com) all texts and pictures are taken from Skymet Weather Services Pvt. Ltd. 2018, https://www.skymetweather.com/content/weather-news-and-analysis/6-important-eco-friendly-steps-taken-by-narendra-modi-government/ on 3/18/2018

Source are listed in the picture itself – Here I would like to declare that these pictures are used only to show the condition of pollution – they are not official and no relation with the political activism.

treaties related to the environment and climate change, most of which were from the first United Nations Conference on Environment in Stockholm in 1972 (Kishwan, Panday, Goyal and Gupta, 2007)[32]. Since the Conference, India has prepared and implemented approximately 40 Acts and Policies. Consequently, India has also established close working networks with the International Network Organization, which includes: The United Nations, IUCN, WWF, Winrock International and climate change[33]. In addition, India is enforcing the establishment of institutional capacity to incorporate the new policies and acts as needed to address the seriousness of the global environment crisis, which is activating the role of local agencies with the recognition of their contributions.

Q: *All governments face difficult policy choices, having to balance environmental/health issues versus national development. This is not unique to Asia. But these policy choices are probably felt more acutely in Asia as the "west" has already achieved a certain level of national development that most nations in Asia are still trying to achieve. How do you feel about the policy choices being made in Nepal, Bangladesh, Pakistan, and India? Are these the optimal choices?* Are they the only choices? Or do you think that the governments of South Asia can do better?

A: As I noted in the earlier sections; the major environmental problems in the region include, land degradation & desertification loss of biodiversity; fresh water depletion & degradation; solid waste management; degradation of air quality; environmental health issues; degradation and depletion of coastal and marine resources; and natural disasters and their consequences (SACEP 2010). Specifically, they can be summarized as:

Most of the issues and causes noted in the table are related to the depletion of forest canopy. The major underlying cause is deforestation, which is associated with the public dependency on natural resources for the livelihood. There are also increasing trend of environmental hazards related human health. The environmental health problems can either be due to the lack of access to essential environmental resources (clean air, water, shelter adequate food etc.,) and due to unhealthy and unsafe work environments. Health issues in the form of premature death, chronic bronchitis and other

[32]Kishwan, Jagadish; Panday, Devendra; Goyal, AK and Gupta AK (2007) India's Forests, Government of India, Ministry of Environment and forest, New Delhi.

[33]India's major donors and collaborative partners in addressing the Environment and Climate change issues (we resources found at) http://www.karmayog.com/lists/foreigndonors.htm (accessed on 05/02/2010).

Table 10.4 Key environmental issues and causes

Common problems Climate change and associated Natural disasters; Increasing population pressures; Pollution due to land based activities; Intensive agriculture development; Coral mining and Increased pressure from tourism		
Country Specific Problems		
Country	Key Issues	Key Causes
Bangladesh	Marginalized populations forced to live on and cultivate flood-prone land; loss of biodiversity; limited access to potable water; water-borne diseases prevalent; water pollution, especially of fishing areas; arsenic pollution of drinking water; urban air pollution; soil degradation; deforestation; severe overpopulation: natural disasters (especially floods and cyclones which kill thousands of people and causes heavy economic losses every year); food security risks; industrial pollution; import of hazardous waste.	High population density and urban primacy; reliance on private transport; urbanization and deficits in urban infrastructure (including one of the world's 30 largest cities – Dhaka); increases in unmanaged marine-based tourism; green revolution/agrochemicals and run-off; high demand for bio-fuels; lack of controls on industrial effluent; over exploitation and/or pollution of groundwater.
India	Deforestation; soil erosion; overgrazing; desertification; loss of biodiversity; air pollution; water pollution; huge population base and large growth rate is overstraining natural resources; natural disasters such as floods, cyclones and landslides are common; high death rates and ailments associated with indoor air pollution.	High rates of urbanization and deficits in urban infrastructure (including in four of world's 30 largest cities); reliance on private transport; industrial effluents and vehicle emissions; increases in unmanaged marine- based tourism; green revolution/ agrochemicals and run-off; reliance on biofuels.
Nepal	Deforestation; soil erosion and degradation; loss of biodiversity; water pollution; natural disasters such as floods and landslides in rural areas; food security risks	High rates of urbanization; reliance on private transport; increased demands for timber; increased population density and cultivation of marginal lands.
Pakistan	Water pollution; seasonal limitations on the availability of natural freshwater resources; majority of the population lacks access to potable water; deforestation; soil erosion; coastal habitat loss and degradation of marine environment; desertification; loss of biodiversity: natural disasters, mainly due to floods.	High rates of urbanization and deficits in urban infrastructure; industrial wastes; population increases in coastal areas and rise in tourism; depletion of mangroves for aquaculture; overfishing; increased demands for timber/biofuels; hunting/ poaching; green revolution/agrochemicals and run-off.

Source: ESCAP 2000:345.

Table 10.5 National Priorities on the environmental issues in South Asia

Land degradation and desertification	**Bangladesh** Priority	**India** Priority	**Nepal** Priority	**Pakistan** Priority
Water erosion	High	High	High	High
Water logging	Medium	Medium	Low	Low
Salinization	High	Medium	Low	Medium
Loss of Biodiversity	High	High	High	High
Deforestation	High	Low	High	High
Water scarcity	Medium	Low	Low	High
Water Pollution	High	Medium	High	Medium
Need for Water supply & sanitation	High	High	High	Medium
Solid & liquid waste Management in urban centers	High	High	High	High
Degradation of Air quality				
Vehicular emission in urban centers	High	High	High	High
Industrial emission	Medium	Low	N/A	Medium
Domestic cooking	Medium	High	High	High
Environmental Health issues	High	High	High	High
Depletion and degradation of Coastal & Marine Environment	High	Medium	N/A	Medium
Natural disasters				
Droughts	Medium	Low	Medium	Medium
Floods & Land slides	High	High	High	Low
Earthquakes	Low	Medium	High	Low
Sea level rise	Medium	Medium	N/A	High

Source: ESCAP 2000; UNEP 2001; SAARC 2010

respiratory symptoms are high in several metropolitan centers in the region (SAARC 2010:2 website). The Asian countries are aware on the issues and have kept on the high priority to address them, by principle and practice.

In Summary, each of the four countries has given high priority for the overall conservation of natural resources, including wetlands and introduced or being prepared the strong policies and programs to stop further degradation of nature. Among them, India has the established system of conservation mechanism; however, in terms of policy implementation and conservation, Nepal has shown the exemplary cases of natural resource management with the application of the public participation machineries. Bangladesh and Pakistan performance in conservation is relatively weak, even having very strong involvement of international organization to improve their situation.

In addressing the conservation problems, there have been some efforts in the region, coordinated by various organizations such as The South Asian Association for Regional Cooperation (SAARC), Bay of Bengal Initiative for Multi Sectoral Technical and Economic Cooperation (BIMSTEC- Sub Regional mechanism for selected South-South East Asian Member Countries Member Countries: Bangladesh, India, Myanmar, Sri Lanka, Thailand, Bhutan, Nepal), Asian Disaster Reduction and Response Network (ADRRN), IUCN regional office, UN Agency Regional Office, and several other international organization. Among them SAARC's initiatives are very important to address the conservation problems which hold the regional issues particularly, water resources and climate change.

Q: *We really do want to thank you for taking the time to share your thoughts and experiences with us. This has been an absolute pleasure for me talking to you and I am sure that our many readers will enjoy your insights into South Asia.*

A: It is my pleasure to share knowledge and expertise As I noted, earlier, my family, communities, and various societies (wherever I have been), including the nature and culture, traditions combinedly nurtured me, without any expectations. My intention, of life is to give or contribute to the society in fullest whatever I have. I would be more than happy, if readers find this information useful. I am open to engage in any kind of collaborative research, teaching, or any other tasks which can contribute to overcome or minimize the devastating impact of climate change.

I would like to clearly state that, most of the information, I have noted in this interview are based on web-search as well as taken from my published and forth coming books manuscripts:

- Bhandari, Medani P. (2018). *Green Web-II: Standards and Perspectives from the IUCN, case studies from India, Nepal, Bangladesh and Pakistan, River Publishers, Denmark/the Netherlands ISBN: 978-87-70220-12-5 (Hardback) 978-87-70220-11-8 (eBook).* http://www.riverpublishers.com/book_details.php?book_id=568
- State of Environment in South Asia-A comparative study of Bangladesh, India, Nepal and Pakistan, with Reference of the conservation intervention
- India: Oldest Civilizations, Oldest Conservation History- and the Mutuality with the International Organizations

- Bangladesh: The Country of over Population and UN's Role to Address the Vulnerable Environment
- Environmentalism in Nepal- the Pioneer for Environment conservation efforts in The Country of Geological Variation and Conflict.
- The IUCN as Organization of Knowledge- the Roles of Policy formation in the Country of Unrest - Pakistan

Most of the pictures, graphs, tables are taken from the websites. I have tried to provide proper sources, citations, and links of the original sources. However, if I missed to note any source, I apologize in advance to the all concern authors, journalists, government agencies and any other stakeholders whom I have cited in this note.

References-Including Websites: (Source Websites are Listed in Chronological Order as Appears in the Above Tests)

[1] A distance few of type of landscape of my play area of childhood Photo source: Facebook, Khusi Rai; https://www.facebook.com/khushirai.purbeli/photos/
[2] http://documents.worldbank.org/curated/en/570441468763468377/pdf/269751Faith0in0Conservation010paper.pdf
[3] http://documents.worldbank.org/curated/en/570441468763468377/pdf/269751Faith0in0Conservation010paper.pdf
[4] https://en.wikipedia.org/wiki/Ocimum_tenuiflorum
[5] https://en.wikipedia.org/wiki/Desmostachya_bipinnata
[6] https://en.wikipedia.org/wiki/Pterocarpus_santalinus
[7] https://www.gemsratna.com/benefit-of-rudraksha
[8] https://en.wikipedia.org/wiki/Rudraksha
[9] https://en.wikipedia.org/wiki/Buddha
[10] http://hindi.webdunia.com/astrology-articles/importance-of-five-113122600061_1.html
[11] https://en.wikipedia.org/wiki/Saraca_asoca
[12] https://www.britannica.com/plant/banyan
[13] https://www.facebook.com/khushirai.purbeli/photos/a.468013640066789.1073741828.467993226735497/831331163735033/?type=3&theater
[14] https://commons.wikimedia.org/wiki/File:Bar_pipal_by_Mahalaxmi.JPG

[15] VijayalaxmiKinhal https://greenliving.lovetoknow.com/Top_30_Environmental_Concerns
[16] https://www.conserve-energy-future.com/current-environmental-issues.php
[17] ISSUE NO. 29: United Nations Environment Program (2018-01-08); https://wedocs.unep.org/bitstream/handle/20.500.11822/22416/Perspective_No_29_web.pdf?sequence=1&isAllowed=y
[18] Sewage and waste management problem (http://www.yourarticlelibrary.com/environment/10-major-environmental-challenges-faced-by-india/9862).
[19] https://www.edf.org/climate/how-climate-change-plunders-planet
[20] Bangladesh losing 1% of its GDP every year due to air pollution: Md Amjad HossainPublished at 10:37 PM December 10, 2017Last updated at 11:33 PM December 10, 2017-http://www.dhakatribune.com/bangladesh/dhaka/2017/12/10/wb-bangladesh-losing-1-gdp-every-year-due-air-pollution/
[21] www.grida.no/resources/5648
[22] Dhaka Tribune,Tribune Desk Published at 11:15 AM February 23, 2018-http://www.dhakatribune.com/bangladesh/environment/2018/02/23/poor-air-quality-strangles-life-dhaka-residents/
[23] https://thehimalayantimes.com/nepal/nepals-kathmandu-ranks-5th-in-pollution-index-2017/
[24] https://thehimalayantimes.com/opinion/kathmandu-valley-dust-bowl-pollution/
[25] http://kathmandupost.ekantipur.com/news/2016-06-05/world-environment-day-campaigners-symbolically-lay-dead-demanding-clean-air.html
[26] http://kathmandupost.ekantipur.com/news/2018-01-25/nepals-air-quality-is-worst-in-the-world-epi-report.html
[27] Republica (http://www.myrepublica.com/news/35000/?categoryId=81) January 24, 2018 17:13 PM
[28] WWF-https://www.worldwildlife.org/stories/after-devastating-earthquake-nepal-aims-to-reduce-the-risk-of-disaster-through-green-rebuilding
[29] [(Environment Statistics Nepal (2015) http://cbs.gov.np/image/data/2016/Compendium%20of%20Environment%20Statistics%20Nepal%202015.pdf]
[30] Nepal- How the people of Nepal live with climate change and what communication can do (BBCMEDIA http://dataportal.bbcmediaaction.org/site/assets/uploads/2016/07/Nepal-Report.pdf].

[31] (BBCMEDIA http://dataportal.bbcmediaaction.org/site/assets/uploads/2016/07/Nepal-Report.pdf].
[32] Duniyanews.tv-A National Communications Services Company; Last Updated On 18 October,2017 11:35 am http://dunyanews.tv/en/Pakistan/410284-
[33] Duniyanews.tv-A National Communications Services Company; Last Updated On 18 October,2017 11:35 am http://dunyanews.tv/en/Pakistan/410284-
[34] Amna Chaudhry, Herald, Updated Sep 04, 2017 09:00pm https://herald.dawn.com/news/1153824/keeping-trash-alive
[35] BrayshnaKundi https://asiafoundation.org/2017/11/01/pakistans-water-crisis-national-water-policy-needed/
[36] GiorgiaGuglielmiAug. 23, 2017, 4:15 PM: Science: 2018 American Association for the Advancement of Science. All rights Reserved. AAAS is a partner of HINARI, AGORA, OARE, CHORUS, CLOCKSS, CrossRef and OUNTER. http://www.sciencemag.org/news/2017/08/arsenic-drinking-water-threatens-60-million-pakistan
[37] http://www.sciencemag.org/news/2017/08/arsenic-drinking-water-threatens-60-million-pakistan
[38] Posted in: Asia/PacificClimateEarthHealth- doi:10.1126/science.aap7590
[39] August 13, 2017 5:10 AM; By Ayaz Gul- https://www.voanews.com/a/one-billion-trees-planted-in-pakistan-nw-province/3983609.html
[40] Share this article to let the world know about this amazing environment initiative by Pakistan" Source: http://pakiholic.com/billion-tree-tsunami-huge-contribution-pakistan-stop-climate-change/.
[41] http://pakiholic.com/billion-tree-tsunami-huge-contribution-pakistan-stop-climate-change/
[42] (Chandra, Mahesh (2015) "Environmental Concerns in India: Problems and Solutions," Journal of International Business and Law: Vol. 15: Iss. 1, Article 1. Available at: http://scholarlycommons.law.hofstra.edu/jibl/vol15/iss1/1)
[43] VOX: By Umair Irfan Updated Nov 25, 2017, 4:32pm EST https://www.vox.com/energy-and-environment/2017/11/22/16666808/india-air-pollution-new-delhi
[44] CNN: By Huizhong Wu, CNN: Updated 3:32 AM ET, Mon January 15, 2018 https://www.cnn.com/2018/01/15/health/india-air-pollution-study-intl/index.html
[45] Vehicles drive through smog in New Delhi. (Reuters: Saumya Khandelwal)

[46] ABC News-Correspondents Report By South Asia correspondent Siobhan Heanue Posted 23 Feb 2018, 5:52pm http://www.abc.net.au/news/2018-02-24/air-pollution-in-india-risks-becoming-a-humanitarian-catastrophe/9469922
[47] Business Standard- Delhi Last Updated at July 20, 2016 00:21 IST http://www.business-standard.com/article/government-press-release/government-has-taken-series-of-steps-to-address-pollution-related-issues-116071901137_1.html
[48] ADB- https://www.adb.org/sites/default/files/linked-documents/cps-ind-2013-2017-ena.pdf
[49] (Chris Baynes Monday 3 July 2017 09:45 BST) http://www.independent.co.uk/news/world/asia/india-plant-66-million-trees-12-hours-environment-campaign-madhya-pradesh-global-warming-climate-a7820416.html
[50] Skymet Weather Services Pvt. Ltd. 2018-9 April 2015 03:09 PM https://www.skymetweather.com/content/weather-news-and-analysis/6-important-eco-friendly-steps-taken-by-narendra-modi-government/(Main Image Credits: narendramodiblog.wordpress.com) (Featured Image Credits: indiatimes.com) all texts and pictures are taken from Skymet Weather Services Pvt. Ltd. 2018, https://www.skymetweather.com/content/weather-news-and-analysis/6-important-eco-friendly-steps-taken-by-narendra-modi-government/ on 3/18/2018

Index

F

V

About the Author

Prof. Medani P. Bhandari completed his M.A. in Anthropology (Tribhuvan University, Nepal), M.Sc. Environmental System Monitoring and Analysis (ITC-The University of Twente, the Netherlands), M.A. Sustainable International Development (Brandeis University, Massachusetts, USA), M.A. and Ph.D. in Sociology (Syracuse University, NY, USA). He is dedicated to conservation of nature and natural resources and social empowerment through research and action project. *His purpose of life is to give or contribute to the society fullest through whatever he has, earned, learned or experienced.* He has worked with various organizations as consultant- United Nations Environment Program (UNEP)/Adelaide University, the United Nations Development Program (UNDP), the IUCN, the World Wildlife Fund (WWF), the World Resource Institute (WRI), Winrock International, the Japan Environment Education Forum, and the Pajaro Jai Foundation (PJF), along with others. During 2015-17, he served as a Professor of Natural Resources and Environment at the Arabian Gulf University, Bahrain. Prof. Bhandari has spent most of his career focusing on the Sociological Theories; Environmental Sustainability; Social Inclusion, Climate Change Mitigation and Adaptation; Environmental Health Hazard; Environmental Management; Social Innovation; Developing along the way expertise in Global and International Environmental Politics, Environmental Institutions and Natural Resources Governance; Climate Change Policy and Implementation, Environmental Justice, Sustainable Development; Theory of Natural Resources Governance; Impact Evaluation of Rural Livelihood; International Organizations; Public/Social Policy; The Non-Profit Sector; Low Carbon Mechanism; Good Governance; Climate Adaptation; REDD Plus; Carbon Financing; Green Economy and Renewable Energy; Nature, Culture and Power. Prof. Bhandari's major teaching and research specialties include: Sociological Theories and Practices; Environmental Health; Social and Environmental research methods; Social and Environmental Innovation; Social and Environmental policies; Climate Change Mitigation and Adaptation; International Environmental Governance; Green Economy; Sustainability and assessment of the

Economic, Social and Environmental impacts on society and nature. In brief, Prof. Bhandari has sound theoretical and practical knowledge in social science and environment science. His field experience spans across Asia, Africa, the North America, Western Europe, Australia, Japan and the Middle East. Prof. Bhandari has published 4 books on science domain, four volumes of poetry jointly with Prajita Bhandari and over 60 scholarly papers in international scientific journals and many monographs, and views papers as a column writer. Prof. Bhandari serves as an editor/advisor/member of editorial boards for more than 20 international scientific journals. He is active member of many academic organizations/association as well as in the advisory boards of NGOs and CBOs. Currently, he is serving as a Professor of Inter-Disciplinary Department – Natural Resource & Environment/Sustainability Studies, at the Akamai University, USA and Professor of the Department of Finance and Entrepreneurship, Sumy State University (SSU) Ukraine; and International Program Coordinator, Atlantic State Legal Foundation, NY, USA, Executive Director -Human Survival Foundation, UK, Treasurer and General Secretory – Equality Foundation, USA (remotely).